Advances in Astrobiology and Biogeophysics

This series aims to report new developments in research and teaching in the interdisciplinary fields of astrobiology and biogeophysics. This encompasses all aspects of research into the origins of life - from the creation of matter to the emergence of complex life forms - and the study of both structure and evolution of planetary ecosystems under a given set of astro- and geophysical parameters. The methods considered can be of theoretical, computational, experimental and observational nature. Preference will be given to proposals where the manuscript puts particular emphasis on the overall readability in view of the broad spectrum of scientific backgrounds involved in astrobiology and biogeophysics. The type of material considered for publication includes: - Topical monographs - Lectures on a new field, or presenting a new angle on a classical field - Suitably edited research reports - Compilations of selected papers from meetings that are devoted to specific topics.

More information about this series at http://www.springer.com/series/5118

Anna Neubeck • Sean McMahon
Editors

Prebiotic Chemistry and the Origin of Life

Springer

Editors
Anna Neubeck
Department of Earth Sciences
Uppsala University
Uppsala, Sweden

Sean McMahon
UK Centre for Astrobiology
University of Edinburgh
Edinburgh, United Kingdom

ISSN 1610-8957 ISSN 1613-1851 (electronic)
Advances in Astrobiology and Biogeophysics
ISBN 978-3-030-81038-2 ISBN 978-3-030-81039-9 (eBook)
https://doi.org/10.1007/978-3-030-81039-9

This Springer imprint is published by the registered company Springer Nature Switzerland AG.
The registered company address is: Gewerbestrasse 11, 6330 Cham, Switzerland

Preface

This is a book about the beginnings of life on Earth from a geochemical and biochemical standpoint. It has been written by specialists working on the frontiers of several disciplines, and it is intended to provide a readable overview both for trained scientists and for advanced students in geobiology, astrobiology, and related fields. It is by no means a complete or comprehensive introduction to the field, but rather a work-in-progress report from some of its leading protagonists. We have tried to sidestep some well-worn arguments in favour of fresher perspectives and original insights. Nevertheless, the eleven peer-reviewed chapters assembled here outline many of the central questions that drive origins-of-life research today, suggest what the answers may be, and show how future work may settle them.

According to accepted common scientific understanding, life on Earth emerged through the self-organisation of lifeless matter far from thermodynamic equilibrium. Self-catalysing networks of chemical reactions gave rise to protocells: encapsulated, homeostatic units capable of reproducing themselves according to an internal genetic code. How this event or series of events unfolded is deeply unclear; the first half-billion years or so of life's history are missing from the record, and carbonaceous residues in the oldest known rocks have been heavily altered by heat, pressure, and the passage of time, destroying biochemical information. But if the origin of life cannot be observed, perhaps it can be reconstructed through painstaking, interdisciplinary scientific detective work. Clues are found in the chemistry of the elements, in the composition of meteoritic and primordial materials, in the biochemical and genetic makeup of viruses and cellular organisms alive today, in laboratory and computer simulations of prebiotic chemical reaction pathways, and in geochemical evidence from the early Earth. Cellular remains, when they finally appear in the rocks, tell us that life had already become surprisingly diverse, complex, and tolerant of environmental extremes when the fossil record began to be written. Taken together, these clues are beginning to tell a coherent story.

It is a story that begins in space. In **Chap.** 1 of this volume, **Marco Fioroni** describes the synthesis of transition metals in supernovae and their chemical transformations in space and on the young, habitable Earth. He emphasises the importance of transition metals and minerals for organometallic chemistry on the early Earth and also for the origin of life: today, almost 40% of all enzymatic reactions involve metalloenzymes. Several chemical pathways to form simple and complex building blocks and catalysts for life are thoroughly discussed, and the possible contribution of extraterrestrial organometallic/metallorganic compounds is emphasised.

In **Chap.** 2, **Robert M. Hazen and Shaunna M. Morrison** enlarge upon the mineralogical complement of the early Earth and its significance for prebiotic chemistry. They show that many biochemically central elements were supplied not in their rare and exotic mineral forms, but from the more common rock-forming minerals in which they occur as minor and trace elements. Thus, boron, molybdenum, phosphorus, and other elements would have been widely available at reactive surface sites of minerals such as olivine, pyroxene, feldspar, and magnetite, even though borate, molybdate, and phosphate minerals (and so on) were rare. This provides an elegant solution to a long-standing problem in origins-of-life chemistry.

Chapter 3, contributed by **Eva Stueeken and Nicholas Gardiner**, reconstructs the Hadean Earth as a complex, geologically dynamic world. The primordial crust, volcanic and hydrothermal systems, oceans, ice, the atmosphere, and the interfaces and transport pathways between these linked systems may all have played important roles in the operation of prebiotic chemical reaction networks. In this perspective, the major steps leading to the origin of life need not all have occurred in any single environmental setting or "crucible". Rather, we can consider the early Earth as a global chemical reactor for the origin of life.

Chiral molecules—those that occur in left- and right-handed forms—are essential in the chemistry of life. In **Chap.** 4, **Axel Brandenburg** asks why biology prefers left-handed amino acids and right-handed carbohydrates, and whether this homochirality was a prerequisite for life or a consequence of it. We learn about biological and abiotic chiral dynamics and what drives the transition from racemic mixtures to homochirality. Brandenburg takes us through the fascinating story of the discovery of chiral molecules and how they might be used to find life on Mars. Homochirality may have emerged as a consequence of autocatalysis in the hypothetical RNA world, and in **Chap.** 5, **Frank Trixler** discusses the importance of mineral surfaces for concentration, selection, homochirality, and the synthesis of nucleotides in such a world. This chapter confronts the complexity of prebiotic nucleic acid synthesis and the multiple paradoxes that inevitably arise from this given the central role of nucleic acids in biology. Another seemingly indispensable feature of life is encapsulation: enclosure within semi-permeable boundaries is a prerequisite for homeostasis. In **Chap.** 6, **Augustin Lopez, Carolina Chieffo, and Michele Fiore** discuss the origins of these boundaries from a chemical perspective, including the synthesis of amphiphilic molecules and their self-assembly into protocellular boundaries on the early Earth. In **Chap.** 7, **Oliver Trapp** discusses the formation and self-modification of organic catalysts and their importance in the origin of chirality, nucleosides, and Darwinian life. Informed by the latest experimental results, he suggests several routes towards nucleic acid-based evolutionary systems and concludes that DNA and RNA may have arisen simultaneously on the early Earth.

The boundary between life and non-life is a debated question closely linked to our understanding of life's chemical and evolutionary origins. Ever since the discovery of bacteriophages in the 1910s, viruses have been central to these debates. **Donald Pan** considers multiple aspects on the role of viruses in the origin and definition of life in **Chap.** 8. Viruses may have played a crucial role in the onset of life and can serve as useful models of prebiotic replicators, although it remains unclear whether they predate the Last Universal Common Ancestor (LUCA) of cellular life. In **Chap.** 9**, Anthony Poole** examines efforts to reconstruct LUCA using the imperfect record of evolutionary history preserved in the genetic sequences of modern organisms. Time has blurred LUCA's traits, and the interpretation of features shared by its descendants is multifaceted. Horizontal gene transfer, gene displacement, and loss all obstruct the prospect of an unambiguous reconstruction of LUCA. Poole describes how these and other constraints limit the resolution at which LUCA can be reconstructed with commonly used methods and finally suggests a feasible procedure to reconstruct LUCA by focusing on general traits.

The first stages of biological evolution are not preserved in the rock record. Nevertheless, the oldest known fossils provide important insights into the nature of early life and its environment. In **Chap.** 10**, Barbara Cavalazzi and her colleagues** describe the cellular remains, fossil bio-sedimentary structures, and other chemical and mineralogical evidence of life present in some of Earth's oldest rocks, showing how these important but controversial materials can be critically interrogated and understood. The fossil record also contains key evidence for the timing of the origin of the eukaryotes. The nucleated cell and its organelles transformed the early biosphere and eventually permitted the emergence of complex, macroscopic, multicellular organisms like ourselves. Thus, in **Chap.** 11, **Heda Agic** concludes our volume by considering this momentous event (or a series of events) from a palaeontological perspective, presenting some recently described, remarkably well-preserved microfossils that record the early evolution of eukaryotes.

This book offers a partial snapshot of origins-of-life research in the early twenty-first century. Many of its chapters were written under strain in the midst of the global COVID-19 pandemic (another testament to the power of viruses), which caused some inevitable delays; we thank our contributors both for their perseverance and for their patience. We are also hugely grateful to the peer reviewers who provided expert critical feedback on the initial drafts of each chapter, and to Ramon Khanna and Christina Fehling at Springer Nature for their support and guidance. Despite all these efforts, we earnestly hope that this volume will *not* stand the test of time: progress in this field is not merely of academic interest but of fundamental importance to humankind. New data will come from many sources: increasingly sophisticated laboratory experiments, new computational approaches to prebiotic reaction pathways, and new discoveries in the rocks of the Earth, Mars, and other planetary bodies. But it will take many more years, much labour, and the skills and ideas of diverse people around the world to understand, finally, where we came from. We hope this book will inspire its readers to take part in this great adventure.

Uppsala, Sweden | Anna Neubeck
Edinburgh, UK | Sean McMahon

Contents

Chapter 1
Transition Metal Organometallic/Metallorganic Chemistry: Its Role in Prebiotic Chemistry and Life's Origin

Marco Fioroni

Abstract Without transition metals (TM) Life would not exist, at least not in the form we know. In fact, first row TM share an important ubiquitous presence in all Phyla where 30% of all proteins are ligated to a TM and 40% of enzymatic reactions are processed by a metallo-enzyme. Consequently it can be assumed that TM and their organometallic/metallorganic compounds played an important role in the prebiotic world as well as in the chemistry of Life's origin. In this chapter, I address the involvement of TM in prebiotic chemistry and the origin of Life, including the origin of both metabolism and genetics. I discuss both exogeneous (astrochemical) and endogenous synthesis and self-organization of organic building blocks, emphasizing the importance of TM mineral catalysts and inorganic-organic interactions. I highlight the presence of organic-inorganic conglomerates in meteorites, and argue that while the role of TM in endogenous synthesis is recognized and attracting much scientific interest, "organometallic-metallorganic astrochemistry" in exogenous synthesis requires further elaboration to understand its significance for prebiotic and origin-of-life chemistry.

1.1 Transition Metals: Elegance in Chemistry

The IUPAC (International Union of Pure and Applied Chemistry) defines a transition metal (TM) as:

> An element whose atom has an incomplete d sub-shell, or which can give rise to cations with an incomplete d sub-shell (Hartshorn et al. 2015).

M. Fioroni (✉)
Smith Chemistry Building, The University of Memphis, Memphis, TN, USA

Department of Earth Sciences, Palaeobiology Uppsala University Geocentrum, Uppsala, Sweden
e-mail: mfioroni@memphis.edu

A. Neubeck, S. McMahon (eds.), *Prebiotic Chemistry and the Origin of Life*, Advances in Astrobiology and Biogeophysics,
https://doi.org/10.1007/978-3-030-81039-9_1

In such a short, simple sentence there is an opening towards a complex, elegant and unique chemical world!

In fact it can be well claimed that the importance of transition metals in the chemistry field is such that Life, industrial chemistry and technology; at least as we know them; would not exist.

In the history of Science TM have contributed to the understanding of the periodicity and relationship between the elements (Constable 2019) while in technology (Komskii 2014), medicine (Lippard and Berg 1994) and industrial chemistry (Beller and Bolm 2004; Elschenbroich 2006) their chemical-physical properties together with their catalytic abilities made them essential elements in the modern society and economy.

However when dealing with the prebiotic chemistry and Life's origin TM have, till now, attracted an oscillating interest due to the close relationship to the astrochemistry research field where TM have not been considered much yet.

In fact because of the increasing consensus between the researchers on the partial extraterrestrial origin of the Earth Life's building blocks (exogenous delivery), to understand how Life's building blocks have been synthesized means to unify astrochemistry and prebiotic chemistry within the special class of reactions resulting in the production of Life's chemical constituents *on Earth and in Space*.

As a consequence the "trio" astrochemistry-prebiotic chemistry-Life's origin, based on the actual knowledge piled up during the last decade, is increasingly entangled; though the "Life's origin" terms are, mainly, referred to the processes, evolution and flow of information organizing the chemical building blocks in a living system (Adami 2015; Seoane and Solé 2018).

Focusing on the exogenous delivery, organic and prebiotic molecules are proposed to be synthesized on extraterrestrial bodies from the interstellar medium (ISM), to circumstellar disks and planetary systems, as a consequence entering in the "classical" astrochemistry field.

However within the "classical" astrochemistry, TM have not played a consistent role though since the '80s there is an "underground" interest in their involvement in the astrochemistry reaction networks. *De facto* deleting TM from the modern astrochemistry knowledge would not make a huge difference, while in Life's chemistry, chemical industry and technology the impact would be enormous.

In synthesis "Earth" based chemistry recognizes the important role of TM while in more "celestial" matters TM seem not to be considered much.

When shifting the attention to the "endogenous delivery" hypothesis, stating the basic chemical constituents of Life's origin have been synthesized on Earth, TM do have some importance being presumably involved in ancient bio-synthetic pathways.

Hopefully, in a kind of "backfire" the prebiotic and Life's origin chemistry can help to rekindle interests toward TM in the astrochemistry field.

In the chapter development there will be a focus on a subset of the vast TM chemistry important to prebiotic and Life's origin chemistry. Such subset includes the CHNOPS elements as main constituents of the living matter and by definition the

TM-C(HNOPS) bonds will determine the organometallic (TM-C) and metallorganic (TM-X-C, X=H,N,O,P,S) chemistry of the astro-prebiotic-Life's origin chemistry.

Furthermore because "we are stardust", the link between exogenous delivery and endogenous synthesis applied only to TM related chemistry, will be better appreciated through knowledge of TM formation by stellar nucleo-synthesis (Sect. 1.2) and the condensation and formation of TM compounds after stellar explosions (Sect. 1.2.1).

Further paragraphs on TM-physical-chemistry Sect. 1.2.2, organometallic compounds Sect. 1.3, astro-organometallics Sect. 1.4 will be specially devoted to the description by molecular orbitals theory (MO) and some basic concepts to better understand TM related chemistry.

Finally a section referred to the TM organometallics and metallorganic chemistry applied to prebiotic and Life's origin chemistry classified by "metabolism first" or "genetic first" will follow Sect. 1.5.

1.2 TM Astro-Genesis

The Periodic Table starts with the Big Bang (^{1}H, ^{3}He, and ^{4}He) (Galli and Palla 2013). All the elements besides the ones produced during the Big Bang, apart from ^{6}Li, ^{9}Be, ^{10}B and ^{11}B continuously forming by *spallation* processes (Ramaty et al. 1997), are born during the stellar evolution and ejected in space during the stars life span or death.

The star type, i.e. giant, main sequence or dwarf, to mention only a few, and its correlated evolution is function of some initial parameters such as mass and metallicity (in astrophysics all elements but H and He are classified as "metals") well represented by the Hertzsprung-Russell temperature-luminosity diagram (Rose 1998).

Life of a star starts with the H fusion phase where He is produced by a proton-proton (pp) chain and/or by a CNO-cycle. When the H core is exhausted and enriched in He, H fusion continues in a shell enveloping the He core and the following gravitational contraction increases the core temperatures till igniting He [*Asymptotic Giant Branch* (AGB) stage]. Depending on the mass of the star this process can continue till the synthesis of ^{56}Fe in an *onion* layered structure (Limongi and Chieffi 2010), where a series of shells made of ^{4}He, ^{12}C, ^{20}Ne, ^{16}O and ^{28}Si are concentrically organized around a central ^{28}Si/^{56}Fe core (see Fig. 1.1).

The AGB stars are probably the main sources of C and N present in the actual Universe as well as of some of the heavier nuclides (Habing and Olofsson 2004; Herwig 2005). For example one AGB subclass *Thermal Pulsating Asymptotic Giant Branch* (TP-AGB) is characterized by a strong mix between the deep layers and the surface experiencing mass loss via strong stellar winds, thereby enriching the interstellar medium with the products of nucleosynthesis (Jordi and Iliadis 2011).

Elements heavier than ^{56}Fe are mainly produced in explosive nucleosynthesis classified as: r-process (rapid neutron capture) (Seeger et al. 1965); rp-process

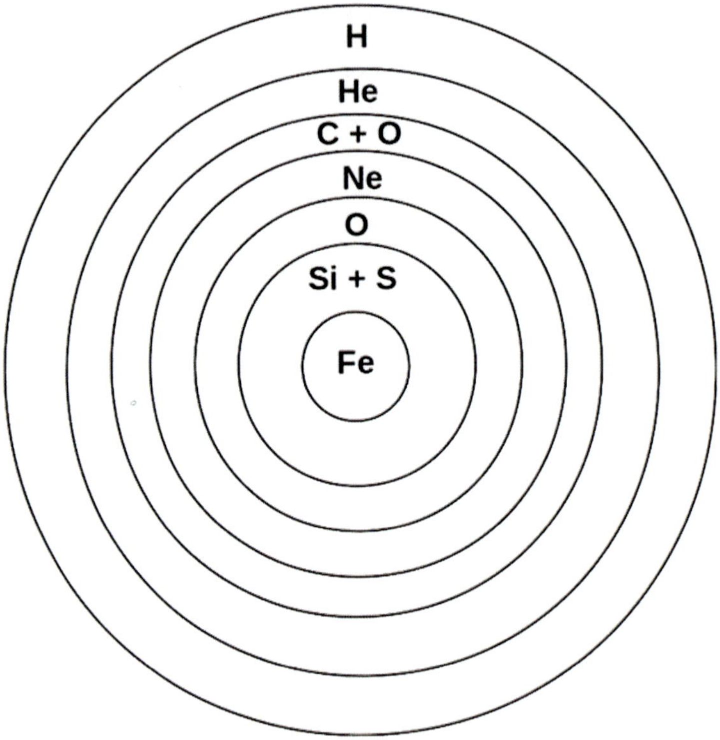

Fig. 1.1 The "onion like" structure of a massive star just prior to its death as "supernova" (type II, Ib, Ic). See ref. Nomoto et al. (2013)

(rapid proton capture) (Jordi and Iliadis 2011) and s-process (slow neutron capture), responsible for the nucleosynthesis in the stellar core (Arlandini et al. 1999; Busso et al. 1999) and the latter responsible for the synthesis of about half of the elements heavier than ^{56}Fe.

In particular, explosive nucleosynthesis, characterized by p-, ν-, νp-, α- and r-processes referring to core-collapse supernovae, (type II, Ib, Ic) are of outstanding importance being one the most prolific sources of Galactic elements (Nomoto et al. 2013). In general the complex phenomenon of explosive nucleosynthesis is predicted to be responsible of the majority of nuclides with $A > 12$.

TM are detected within supernova remnants with an overwhelming presence of Fe^{n+}, Ni^{n+} and Co^{n+} or in star atmospheres and envelopes forming molecules such as FeO (Furuya 2003), FeCN (Zack et al. 2011) and TiO/TiO_2 (Kamínski 2013).

The amounts of first row TM produced in a supernova explosion are "astronomical". A total of $\sim$0.07 $M_\odot$ ($M_\odot$ = solar masses) of ^{56}Co in the ejecta (Leising and Share 1990) of the SN 1987A supernova were estimated, while an estimate for the ^{56}Ni masses deduced from 17 type Ia supernovae sets a range of 0.1–1.0 $M_\odot$ (Stritzinger et al. 2006).

Fractional TM abundances such as Fe, Co, Ni are about $\sim 1\text{–}10^{-2/-3}$ as compared to Si (Cowley 1995). Higher atomic weight TM can be set with Si fractional abundances comprised between the Pb peak and the Fe peak at $10^{-5} < \text{TM} < 1$ (Cowley 1995).

It should be mentioned the relative TM abundances have been deduced by observational data as well as calculated on evolutionary models of stars/galaxies as refractory elements are not easy to be detected (Lodders 2003; Nomoto et al. 2013; Wood et al. 2019). In fact as reported in the next Sect. 1.2.1, all the heavy and refractory elements are depleted (Kuljeet and Piyush 2018; Thiabaud et al. 2014) and condensed into dust grains, resulting in an interesting mix of TM to develop a complex (organic) chemistry.

1.2.1 TM Condensation

Once dispersed in space, elements undergo a series of physical-chemical processes producing the "stardust".

The TM abundances, physico-chemical state, evolution in time, location and distribution within the variety of astrophysical objects is still a work in progress and researchers are using a large variety of observational and theoretical tools (Ascasibar et al. 2015; Asplund 2005; Lodders 2003; Nomoto et al. 2013).

Low-mass AGB stars and *novae* are predicted to be the most prolific sources of dust in the Galaxy being characterized by dust forming episodes in the ejecta (Gehrz et al. 1998) posing the best conditions for gas/dust chemistry (Tielens 2013). Starting near the stellar photosphere, chemical species and dust condensates are produced in concentric shells characterized by a non-thermodynamic equilibrium due to steep temperature-density-radiation gradients (Schöier et al. 2011). As a consequence, circumstellar envelopes and disks can be well defined as *chemical factories* (Jones 2016; Li et al. 2020) also producing biologically relevant molecules (Bekaert et al. 2018). For example, in the C-rich shell of the AGB star IRC+10216 and VY Canis Majoris, over 70 different chemical compounds have been found (Tenenbaum et al. 2010a,b).

Most importantly, refractory-element-bearing species such as TM are predicted to be contained in mineral grains (Henning and Semenov 2013), as supported by condensation models (Lodders and Fegley 1999; Wood et al. 2019).

In fact, the difficulty in the detection of TM in their elemental and molecular forms (gas-phase) mainly derives from their inherent physical-chemical properties behaving as refractory elements easy to deplete and condense into dust grains (Schwander et al. 2014; Thiabaud 2014). For example, even if depleted the detected amount of Fe is 100 times less than what is expected justified by further condensation in silicate dust grains (Jenkins et al. 1986). As a consequence Fe together with its FeS derivative has been proposed to be included in interstellar silicates based on lithophile elements such as O, Si, Mg, Ca and Al (Savin et al.

2012) or included in interstellar silicates in the form of FeS (Köhler 2014; Wood et al. 2019) though by examining colder regions some evidence has been collected on the presence of iron in the ISM (Inter-Stellar Medium) in its neutral or ionic forms (Snow et al. 2002).

However refractory elements in circumstellar environments are not all contained in dust grains. Si containing molecules have been found in circumstellar shells together with molecules containing metals (Savin et al. 2012). In C-rich envelopes no-TM halides or cyanides (Pulliam et al. 2010) have been found, while in O-rich shells, oxides and hydroxides such as AlO and AlOH dominate with Al proposed to be condensed in Al_2O_3 (Tenenbaum and Ziurys 2010) where the presence of AlO and AlOH indicates that photospheric shocks are likely to disrupt grains.

Furthermore in C-rich shells, silicon is primarily present as SiC, when in O-rich objects in oxide condensates while P is probably present in the form of *schreibersite*, $(Fe,Ni)_3P$ and Mg is contained in silicon and aluminum oxides (O-rich shells) and as MgS (C-rich shells).

The detection of silicate dust grains containing O, Si, Fe and Mg, as well as some Ca and Al, provides an important clue on dust chemical composition and evolution (Savin et al. 2012). For example to unveil the cooling history of the Solar Nebula, new kinetic condensation models have been applied to simulate the condensation of refractory metals based on the elemental analysis of Refractory Metal Nuggets (RMNs; i.e. submicrometer highly siderophile element rich metal alloys) in primitive meteorites (Mingen 2020).

Since all the heavy and refractory elements are condensed into dust grains (Kuljeet and Piyush 2018; Thiabaud et al. 2014), it is interesting to understand how TM can contribute to the development of a complex (organic) chemistry working as catalysts. In fact TM due to their bonding abilities with a vast set of organic and small inorganic molecules such as water or ammonia, open the possibility to be fixed not only in "dusty" inorganic particles condensates but also in organometallic or metallorganic compounds.

In the next Sect. 1.2.2 the physical-chemistry of TM is briefly reviewed.

1.2.2 Physical-Chemistry of TM

The TM physical-chemistry in the pure state or as part of chemical compounds is out of scope of the present study and there are excellent reviews and books dealing with it (Albright et al. 2013; Miessler et al. 2014; Vogiatzis et al. 2019). However it follows a short presentation on why TM have such a rich and flexible chemistry.

Basically the physical-chemical behaviour of TM is determined by their d shell electrons (d orbitals). In Fig. 1.2 a pictorial representation of the five d orbitals is shown.

In fact some TM properties like: (a) wide variety in the oxidation number, due to the relatively low energy gap between the different oxidation states; (b) magnetic properties such as paramagnetism due to the unpaired electrons; (c) color,

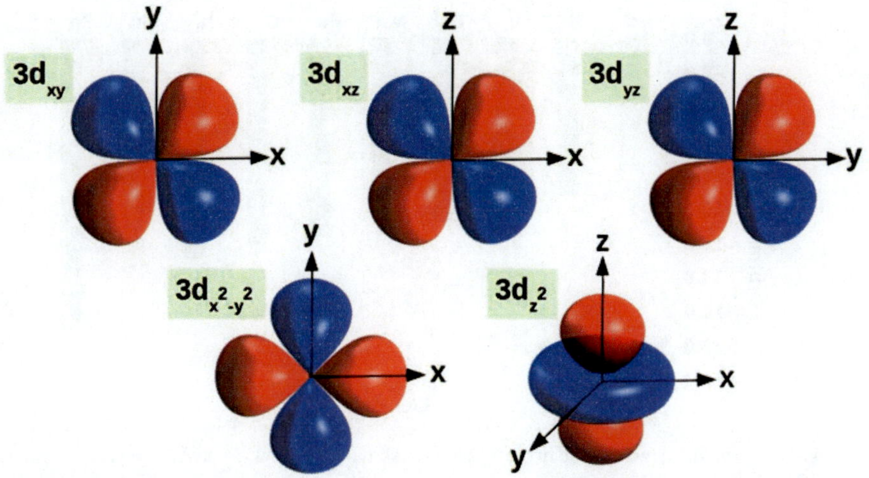

Fig. 1.2 Pictorial representation of the five d orbitals of a first row TM. Rendered by Avogadro Hanwell et al. (2012)

Table 1.1 Periodic table of the first row transition metals

Element	Sc	Ti	V	Cr	Mn	Fe	Co	Ni	Cu	Zn
Atomic Number	21	22	23	24	25	26	27	28	29	30
Electron configuration	$3d^1 4s^2$	$3d^2 4s^2$	$3d^3 4s^2$	$3d^5 4s^1$	$3d^5 4s^2$	$3d^6 4s^2$	$3d^7 4s^2$	$3d^8 4s^2$	$3d^{10} 4s^1$	$3d^{10} 4s^2$

due to the *d-d* electronic transitions and most importantly for our analysis (d) catalytic abilities, due to the low energy difference between the oxidation states and involvement of the d orbitals (see Sect. 1.3).

The general electronic configuration of the d-block elements of which TM are part is: [Inert gas] (n-1)d^{1-10} ns^{0-2}. In Table 1.1 the electronic configuration of only the first row TM is reported.

The particular focus on first row TM is because in addition to Mo, Cd (both second row TM), W (third row TM) and some Lanthanides (f block) (Daumann 2019) many are essential to living organisms (Nielsen 1999).

A fast look at Table 1.1 tells us that Zn is not included in the TM family. In fact by definition a TM has a partially filled *d* shell while Zn has also in the oxidation state 2^+ a completely filled d shell. Furthermore with the exception of Cr and Cu where one s electron moves into the d shell once an electron is added during the *aufbau* process, the filling of the d orbitals follows the *Madelung* rule: [Inert gas] ns^2 (n-1) d^{1-10} (not valid for the TM ions). It should be underlined that the n*s* and n*p* orbitals contribute very little to TM chemical properties moving in a row from left to the right.

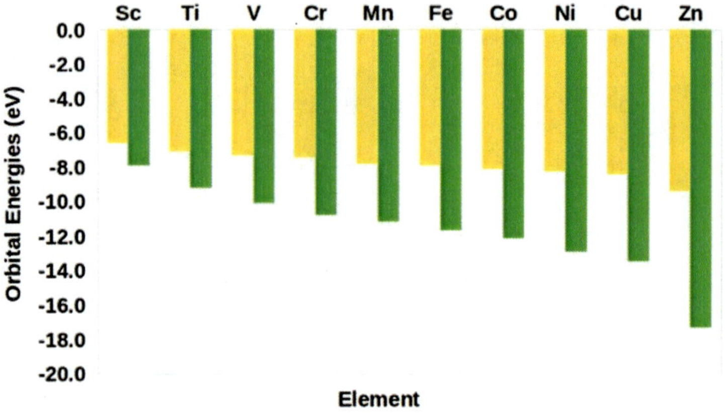

Fig. 1.3 Orbital energies of the first row TM. 4s orbitals = yellow; 3d orbitals = green. Data from ref. Mann et al. (2000)

This phenomenon is particularly important considering TM ions with charges ≥1+. In fact TM ions with 1+/2+ or greater charges have no *s* electrons but only *d* electrons (Mann et al. 2000; Rich 1965) (see Fig. 1.3).

Furthermore and in particular in case of first row TM spin-forbidden reactions are easy to be accomplished. In fact the classical reaction scheme is based on a single determined Potential Energy Surface (PES) characterized by one spin. The jumping to a different PES (inter-system crossing) with a different spin is generally forbidden. However when TM are involved the inter-system crossing is easily allowed due to the strong spin-orbit coupling opening a new realm of reaction channels otherwise impossible to be reached (Miessler et al. 2014). Inter-system crossing is quite frequent for the first-row TM due to the large exchange constants between the 3d-orbitals, resulting in a large spin-orbit coupling.

To sum up, the ability to easily shuffle electrons between the d orbitals as well as their ability to mix with other donor/acceptor electron molecules (see Sect. 1.3) makes TM in both neutral and ionic forms "catalytic machines".

1.3 Organometallic and Metal-Organic Compounds

The IUPAC definition of organometallic compounds (OC):

> Classically compounds having bonds between one or more metal atoms and one or more carbon atoms of an organyl group. In addition to the traditional metals and semimetals, elements such as boron, silicon, arsenic, selenium and litium are considered to form organometallic compounds (Moss et al. 1995).

In "common jargon" a metal-organic compound differs from an organometallic one because of the metal center bond to an organic molecule not by a C atom but by a hetero-atom. For example TM-alkoxides, TM-phosphinine, TM-alkylamides and

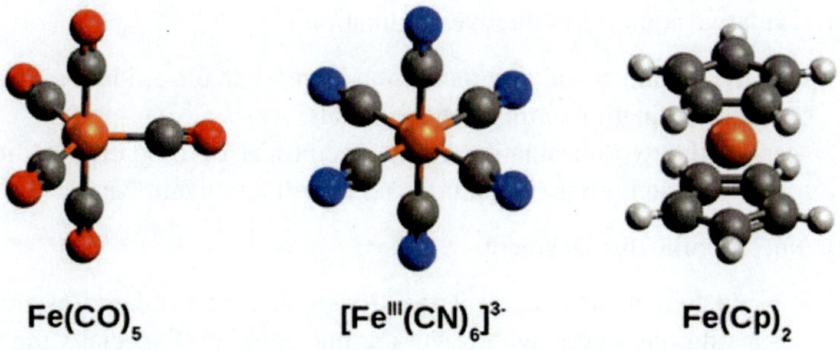

Fig. 1.4 "Classical" organometallic compounds: iron-pentacarbonyl; [iron(III) hexacyanide]$^{3-}$; ferrocene. Rendered by Avogadro (Hanwell et al. 2012)

TM-NHC (NHC=N-heterocyclic carbene) are all metal-organic compounds where the organic molecule or ligand is bond to an O, P and N atom, respectively.

In the following explanation, if not directly cited, the OC term is used indifferently for the metal-organic compounds too.

In an OC the TM center is "surrounded" by ligand molecules (L) as in classical carbonyl compounds like $Fe(CO)_5$, $[FeIII(CN)_6]^{3-}$ or in a sandwich compound like ferrocene (see Fig. 1.4).

The importance of OC is due to their catalytic abilities. Especially in industrial processes (homogeneous phase) like hydrogenation, hydrosilylation, hydroformylation, hydrocyanation, olefin metathesis, alkene polymerization, alkene oligomerization, carbonylation and hydrocarboxylation would be impossible or extremely difficult to be performed if not for the OC (Cornils et al. 2017).

OC are not only used and studied in homogeneous catalysis but in the heterogeneous phase as well, where intermediates on metallic surfaces, metallic oxide and MOF (Metal-Organic Frameworks) are proposed to be similar to the homogeneous phase OC (Li and Liu 2014), one example is the Fischer-Tropsch process (van de Loosdrecht et al. 2013).

OC are also used in fine chemical or laboratory synthesis especially in cross-coupling reactions forming carbon-carbon bonds between complex substrates i.e. Suzuki-Miyaura coupling, Buchwald-Hartwig and Sonogashira coupling to cite only a few (de Meijere et al. 2013).

The chemistry of OC compounds is characterized by a series of important reactions such as gain or loose of one or several ligands, undergoing molecular rearrangement or the formation and breaking of bonds like TM-C, TM-X and TM-TM (Wilkins 2002). Here follows a list of the most typical reactions based on the scheme of Miessler et al. (2014). Details of the kinetics and reaction mechanisms can be studied in several books (Miessler et al. 2014; Wilkins 2002).

1. Reactions involving gain or loss of ligands

   (a) Ligand dissociation and substitution

      – one or more of the ligands is lost and substituted by a different one;

(b) Oxidative addition/Reductive elimination

- in the oxidative addition there is an increase in the oxidation state (formal) and coordination number of the metal
- the reductive elimination is the reciprocal of oxidative addition i.e. a decrease in both the oxidation and coordination number.

(c) Nucleophilic displacement

- some ligand displacement reactions can be considered as nucleophilic substituions where ligands work as nucleophiles. Especially the negatively charged OC can be considered nucleophiles.

2. Reactions involving modifications of ligands

(a) Insertion

- insertion reactions are defined as formal insertions of one ligand (generally unsaturated) into another metal-ligand bond on the same complex;

(b) Hydride elimination

- transfer of a hydrogen atom from a ligand to the metal center;

(c) Abstraction

- defined as elimination reactions in which the coordination number of the metal does not change;

In a catalytic cycle where OC function as a catalyst many of the aforementioned single reactions are involved and each single reaction can be characterized by many reactions steps (Wilkins 2002).

Such an ensemble of reactions is possible due to the extreme flexibility of TM i.e. oxidation state, spin state, ligand geometry and the ability to accept/donate electrons to/from $\sigma - \pi$ donors/acceptors ligand molecules (Albright et al. 2013).

To illustrate the rational behind single reactions based on a Molecular Orbital (MO) approach would be out of the scope of the present work and there exist excellent reviews (Albright 1982) as well as textbooks (Albright et al. 2013) on the argument.

However a short introduction to the MO theory of TM complexes on a "classical" octahedral complex made by six σ donor ligands (L) can give a hint on the complexity and elegance of TM compounds.

There are several steps needed to build an MO diagram such as is reported in Fig. 1.5 like: identify the symmetry group from the geometry, fix a reference axial system, determine the fragments, energy levels and symmetry of the fragment orbitals, combine fragment orbitals with same symmetry, electron insertion, check for MO mixing (Albright et al. 2013).

The amount of qualitative information to be obtained from such MO diagrams is quite detailed and gives a first impression on what properties the OC compounds examined has.

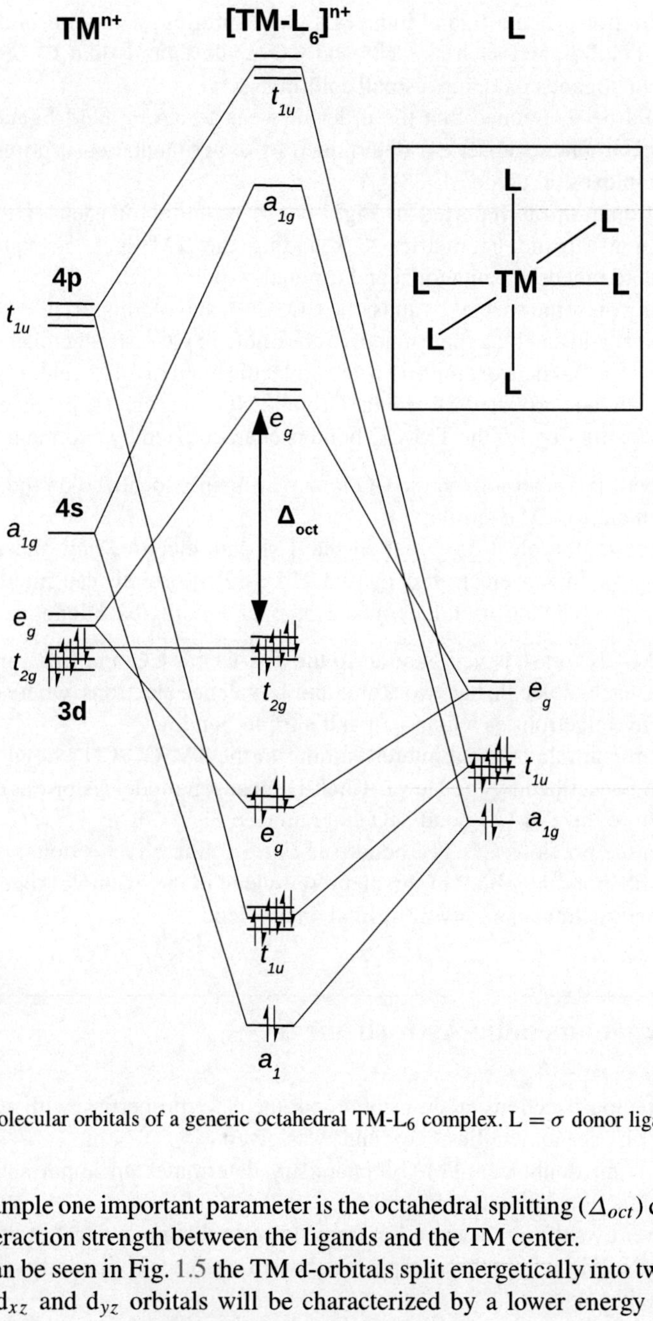

Fig. 1.5 Molecular orbitals of a generic octahedral TM-L_6 complex. L = σ donor ligand

For example one important parameter is the octahedral splitting (Δ_{oct}) depending on the interaction strength between the ligands and the TM center.

As it can be seen in Fig. 1.5 the TM d-orbitals split energetically into two groups. The d_{xy}, d_{xz} and d_{yz} orbitals will be characterized by a lower energy compared to the d_{z^2} and $d_{x^2-y^2}$. This is due to the fact that the d_{xy}, d_{xz} and d_{yz} orbitals are subject to less repulsion compared to the d_{z^2} and $d_{x^2-y^2}$.

By definition when a ligand induces a large splitting of the d-orbitals (Δ) it is a "strong-field" ligand such as CN^- and CO while ligands like I^- or SCN^- are "weak-field" ligands inducing a small splitting.

It should be mentioned that the order of weak to strong field ligands is based on the "spectrochemical series" determined by experimental absorption spectra of cobalt complexes (Tsuchida 1938).

A MO diagram as reported in Fig. 1.5 can be built for each of the possible "fundamental" ligand geometries surrounding the TM such as square planar, pyramidal, tetrahedral, pentagonal and trigonal bi-pyramidal.

Focusing on some special ligands, the CO, CN^- and olefins (R_1-C=C-R_2) are of main interest and can be a representative ensemble in OC-astrochemistry.

In Fig. 1.6 a classical example of a CO molecule bond to a TM center is illustrated together with the correlated Molecular Orbital (MO) diagram (Albright et al. 2013).

By analyzing Fig. 1.6 the TM-CO bond is characterized by two main parts:

(a) an electron donation from the CO lone pair mainly localized on the CO carbon into an empty TM d-orbital;
(b) because of the electron donation the TM gets electron rich and in order to release the higher electron density a TM d-filled orbitals can interact with an unoccupied CO π^* orbital, defined as π-backbonding/backdonation.

The TM-CN^- case is very similar to the TM-CO as CO and CN^- are "isoelectronic" i.e. each molecule has two atoms and 10 valence electrons, where each single atom has five electrons as a lone-pair and a triple-bond.

Another example of great interest similar to the TM-CO/CN^- bond is the TM-olefin bond described by the Dewar-Chatt-Duncanson model (Albright et al. 2013) and similar to the TM-CO bond/MO diagram (see Fig. 1.7).

Like in the previous CO-TM bond (see Fig. 1.6) there is a σ donation from the C=C double bond π orbital of the olefin (ethylene in the example) together with a π-backbonding into an empty π^* orbital on ethylene.

1.4 Organometallic-Astrochemistry

In the previous sections a description of the TM properties with a particular emphasis on organometallic compounds was given.

There is no doubt that TM/OC chemistry determines an important *corpus* of chemical reactions in modern chemistry.

However by shifting the attention from "terrestrial" to more "celestial" problems the interest in TM within the astrophysicists/astrochemists community still occupies a light-weight place.

Citing Irikura et al. (1990):

> Transition metals traditionally have been ignored in chemical models of interstellar clouds. Thermochemical considerations indicate that the formation of several important species can be catalyzed by transition metal ions...

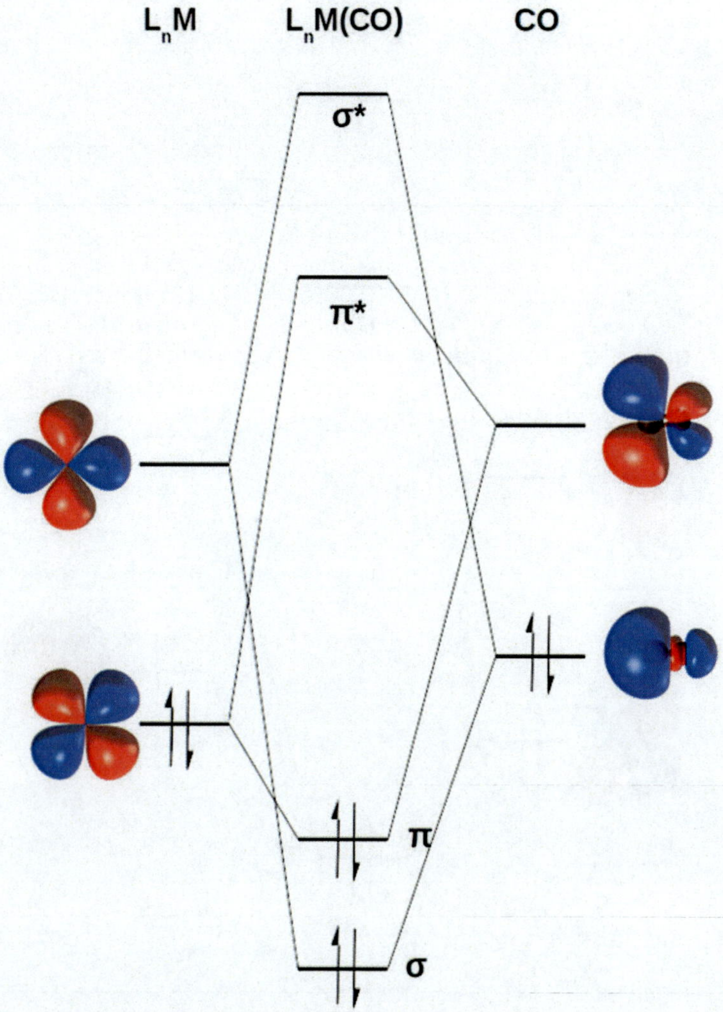

Fig. 1.6 MO diagram of the CO σ bond and π-backbonding with a TM center. Rendered by Avogadro Hanwell et al. (2012)

In fact seminal works considering the importance of TM in space date back to the 1980–1990s decades of the last century including the analysis of the formation of FeH^+ by radiative association in diffuse interstellar clouds (Irikura et al. 1990), the gas phase chemistry in interstellar clouds where Mg and Fe are included (Prasad et al. 1980), the PAH (Polycyclic Aromatic Hydrocarbons) and fullerene ions reactions in interstellar and circumstellar chemistry (Bohme 1992), the role of organometallic molecules in the interstellar and circumstellar chemistry (Klotz et al. 1994), the first analysis if organometallic reactions are efficient processes to be included into astrochemical reaction pathways (Serra et al. 1992), theoretical

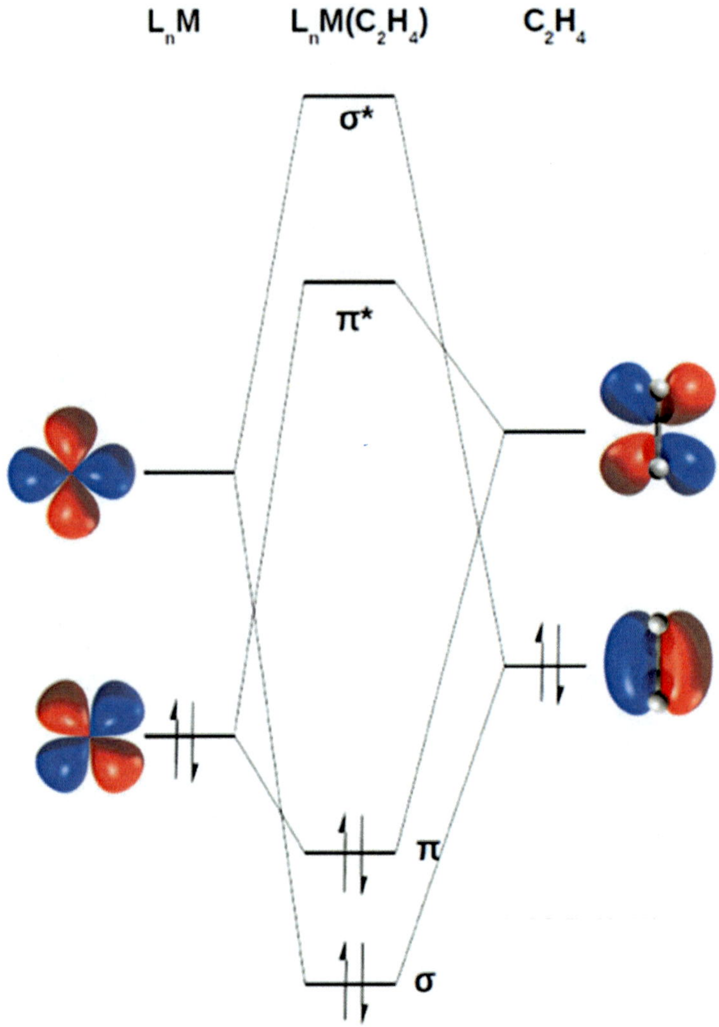

Fig. 1.7 MO diagram of an olefin double bond (C_2H_2) involved in a σ bond and π-backbonding with a TM center. Rendered by Avogadro Hanwell et al. (2012)

analysis of the coordination of Fe-PAH (Chaudret et al. 1991), experimental investigations on gas-phase reactions of $[Fe(CO)_n]^+$ with H and N (Mestdagh et al. 1992), the possible contribution of organometallic molecules in the solar system ices (Klotz et al. 1996), experimental investigations on PAH-$[TM]^+$ adducts in the gas phase (Boissel 1994) and photochemical reactions of the Fe oxide series (Fe, FeO, Fe_3O_4, Fe_2O_3) (Duley 1980).

In the last decade there happened a slow growth of interest in TM and OC compounds as to explain iron depletion by iron-carbonyl species (Halfen and Ziurys

2007) or Fe-PAH (Szczepanski et al. 2006), Si-PAH π complexes (Joalland 2009), the 21 μm line by Ti-fullerene adducts (Kimura et al. 2005) and FeO (Li et al. 2013), the refractory components chemical evolution and condensation sequence in stars and disks (Schwander et al. 2014; Thiabaud et al. 2014), H_2 formation on Fe^+ centers grafted on a silica surface (Fioroni and DeYonker 2016) and a consistent ensemble of laboratory based millimeter–submillimeter spectra of TM (Cr, Ni, Co) nitrides and carbides (Brewster and Ziurys 2001; Sheridan et al. 2002, 2003) whose detection would give further details on the AGB nucleosynthesis, dust-grain composition and gas-phase chemistry in circumstellar material.

Obviously once TM are introduced the correlated TM "organometallic chemistry" is implicitly proposed as some of the aforementioned compounds are organometallic in nature though in some pioneer studies the term "organometallic" or "coordination" was mentioned within the main title (Boissel 1994; Klotz et al. 1996, 1995, 1994; Marty et al. 1996; Serra et al. 1992).

More recently TM have been proposed to be players in the net of reactions involving the formation of organic/organometallic compounds in space by co-determining the homogeneous/heterogeneous chemistry represented by the gas/gas and gas/dust grain interactions function of the astrophysical object (Fioroni 2014, 2016).

Several classes of organometallic compounds based on the most abundant cosmic TM as Fe, Co and Ni and ligands such as CO, CN^-, H_2O and PAH have been analyzed in their IR spectra to characterize the main features for a possible observational survey (Fioroni 2016).

Another class of OC analyzed are Fe-pseudocarbynes (Tarakeshwar et al. 2019). In fact, the presence of such compounds in circumstellar and interstellar environments can partially explain Fe depletion and they can act as catalysts to explain the composition gap between molecules containing nine or fewer carbon atoms and complex molecules.

Regarding $Fe(CO)_n$ complexes, the interaction of Fe and CO at cryogenic conditions results in the formation of FeCO complex which in presence of an excess of CO ends up in the formation of CO_2 (Methikkalam et al. 2020).

The aforementioned list of literature shows a clear interest towards TM and their OC chemical derivatives but a main problem remains: no organometallic species could be observed within the inter-stellar medium with the exception of FeCN found in the dust envelope of the carbon star IRC+10216/CW *Leonis* (Zack et al. 2011).

TM containing molecules not organometallic are FeO, observed in the giant molecular cloud Sagittarius B2 (Furuya 2003) and TiO/TiO_2, observed in the red supergiant VY Canis Majoris (Kamínski 2013).

This is quite surprising since the ability of TM to bind to organic/inorganic molecules is a well established knowledge and this ability enables TM to produce stable chemical complexes with ligands such as the most abundant molecule in space i.e. CO; the organometallic "ligand prototype"; as well as other molecules/ligands like H_2O, CO or HCN/CN to mention few.

However main limitations exist in the *chemical analysis* of extraterrestrial bodies when such analysis is mainly (or only) based on the study of emission/absorption (UV/VIS/IR) and roto/vibrational transition spectra.

In fact, of the ≈204 molecules detected in space (interstellar and circumstellar medium), made of 16 different elements and ranging from a size of 2–70 atoms all have been detected across the electromagnetic spectrum i.e. MW to UV (McGuire 2018) in the gas-phase (Kwok 2016).

As a consequence there is an observational bias towards molecules that are easily detected by radio telescopes and thousands of observed molecular lines remain unassigned, due to the lack of laboratory data for comparison (Smith 2019).

In fact using only UV/VIS/IR/MW spectroscopical methods makes it extremely difficult to resolve and to obtain the complete data composition on a complex system especially if in the solid phase.

It is not by chance that the only discovery of organo-metallic molecule FeCN is based on the analysis of the 2 and 3 mm wavelength with the FeCN present in gas-phase (Zack et al. 2011).

It is important to note that the big step toward the assignment of detected lines as well as the modeling of possible complex inorganic/organic mixtures is based on the huge set of laboratory as well as theoretical/computational data that has been produced within the last 10 years (Ruf et al. 2018).

Comparison between the laboratory/theoretical data and the observational one made possible to partially elucidate the complex chemistry found in circumstellar/interstellar environments (Herbst et al. 2020).

However and most importantly the "breakthrough" in the understanding of the complex mixtures and interactions between the inorganic and the organic material is based on the in situ analysis of meteorites (Kebukawa et al. 2019; Schmitt-Kopplina et al. 2010; Smith et al. 2019).

The in situ analysis of meteorites is the link between OC astro-/pre-/Life' s origin chemistry.

In the next Sect. 1.5 the known OC compounds involved in the prebiotic and Life's origin are listed, and future potentials of the topic will be developed.

1.5 OC in Pre-biotic and Life's Origin

In 2010 Morowitz et al. published a paper with the title: *Ligand Field Theory and the Origin of Life as an Emergent Feature of the Periodic Table of Elements* (Morowitz et al. 2010). In summary the hypothesis is that by TM catalysis it is possible to build up the monomers that can be used to make more efficient and large protein catalyst. The title is a direct "hit" clearly pointing to the main contribution given by the TM and related OC to prebiotic reactions.

It should be underlined that not so rarely organometallic compounds characterized by a TM-C bond are confused with "metal organic compounds", previously defined. For example the Fe-CN is an organometallic compound while Fe-NC is data on the organic-mineral phase was obtained. The organic matter in mixed phase with partially hydrated amorphous silicate phase and iron under different phases with $Fe_n(OH)_n$ is the most interesting.

Furthermore the $Fe_n(OH)_n$ phase is in association with metal-containing organic moieties underlining the specific interactions between the mineral and the organic phase.

Other studies analyzed the Renazzo (CR2), Murchison (CM2) and Orgueil (CI) meteorites where interactions between aromatics and carboxylic functional groups with phyllosilicates have been proposed (Le Guillou et al. 2014) probably due to organic matter by clay-mediated reactions (Pearson et al. 2002), while in the Tagish lake meteorite (C2-ung chondrite, ungrouped carbonaceous meteorite) associations of aliphatic CH_n and OH in phyllosilicates have been found (Yesiltas and Kebukawa 2016).

1.5.1.2 First Organometallics in Meteorites

Probably among the most "pure" organometallic compounds found on extraterrestrial bodies, the iron-carbonyl-cyanide series is the most representative one as found in a study to determine the content of CN^- and releasable cyanide (total cyanide) (Smith et al. 2019).

In an ensemble of primitive meteorites (carbonaceous chondrite) and in particular in the Lewis Cliff 85311 meteorite, the releasable cyanide is primarily in the form of $[Fe^{II}(CN)_5(CO)]^{3-}$ and $[Fe^{II}(CN)_4(CO)_2]^{2-}$ as determined by liquid chromatography-high resolution orbitrap mass spectrometry.

Most importantly both $[Fe^{II}(CN)_5(CO)]^{3-}$ and $[Fe^{II}(CN)_4(CO)_2]^{2-}$ have a double characteristics:

(a) they are "classical" organometallic compounds with a ligand sphere made of CO and CN^-;
(b) both OC have similarities to the active sites of bacteria hydrogenases: $[Fe^{II}(CN)_5(CO)]^{3-}$ to the [NiFe]-hydrogenase from *Desulfovibrio gigas*, $[Fe^{II}(CN)_4(CO)_2]^{2-}$ to the [FeFe]-hydrogenase from *Clostridium pasteurianum*.

It is conceivable that on early Earth H_2 was one of the earliest energy sources, and this hypothesis is fortified by the ubiquity of the hydrogen metabolism in both *Archea* and bacteria (Peters et al. 2015). In both [NiFe]- and [FeFe]-hydrogenases though phylogenetically unrelated, Fe atoms of the active sites are coordinated to the unusual CO and CN^- ligands not found in other metalloenzymes and are assumed to be essential for enzyme activity (Casalot and Rousset 2001; Volbeda et al. 1996).

Furthermore it is reasonable that iron-cyanocarbonyl complexes worked as building blocks *via* a dimerization to form the [FeFe]-hydrogenases active site and by association/dimerization with nickel-containing species for [NiFe]-hydrogenases.

The importance of cyanides is reported in a series of studies where RNA, proteins and lipids are proposed to be synthesized in a "one pot" chemistry by precursors in a cyanosulfidic protometabolism (Patel et al. 2015; Ritson et al. 2018; Sutherland 2016).

It is conceived that during the Late-Heavy Bombardment iron-rich meteorites provided FeS, (FeNi)S and $(Fe,Ni)_3P$ and by their high energy impacts large quantities of atmospheric HCN through shock waves and electric discharge plasma would have produced radical CN. The following "CN rain" would have corroded meteoritic metal inclusions providing cyanometallates, phosphate and HS^--H_2S (Patel et al. 2015; Ritson et al. 2018; Sutherland 2016).

For example by UV irradiation on an aqueous mixture of glycolonitrile, sodium phosphate, and hydrogen sulfide (reductant) the addition of CuCN generates by a redox cycling based on Cu(I)-Cu(II) the aldehyde precursors for the building blocks of RNA and proteins (Patel et al. 2015; Ritson and Sutherland 2013).

Such a scheme coupled with the previous *scenario* of meteorites rich in ferrocyanides would be the perfect situation for a cyanosulfidic prebiotic chemistry.

1.5.1.3 First Metallorganics in Meteorites

In another study conducted on three representative meteorites, the achondrite Northwest Africa 7325 [NWA 7325, ungrouped], ordinary chondrites Novato and Chelyabinsk and carbonaceous chondrite Murchison [CM2], first presence of (CHOMg) based metallorganic compounds was reported (Ruf et al. 2017). The same (CHOMg) series was found in another set of 61 meteorites of various petrological classes.

Specifically within the generic (CHOMg) chemical class dihydroxymagnesium carboxylates $[(OH)_2MgO_2C\text{-}R]^-$ where R is the organic residue, were found in the meteoric soluble organic matter.

Organomagnesium compounds are known to have high thermostability and as a consequence such compounds have the potential to stabilize organic molecules found in the mixed mineral-organic matrix as found in carbonaceous meteorites (Pizzarello and Shock 2010; Schmitt-Kopplina et al. 2010) on a geological time scale.

In fact metal ions can play a double role to catalyze reactions or to stabilize organic molecules, as the life-organic molecules are often thermolabile and due to the high T phases that celestial bodies experience (Sandford et al. 2020) such stabilization is necessary.

Furthermore these stable metallorganic chemical systems might have been important intermedia on which the selection of the future life-organic molecules acted.

For example a fatty acid concentration-fractionation by the stabilization of their correlated organomagnesium compounds can be proposed, as an important process in the compartmentalization-vesicle formation in membranes for late cellular formation.

1.5.2 Endogeneous Synthesis

While in the exogeneous delivery of prebiotic compounds the TM role is still quite restricted, in the prebiotic endogeneous synthesis as well as in the set up of the first ancient metabolism, TM have already attracted some attention.

In fact since the appearance of the "iron-sulfur world hypothesis", originally suggested by Wächtershäuser proposing that early life may have formed on the surface of iron-nickel sulfide minerals (Huber and Wächtershäuser 1997; Wächtershäuser 1990), TM came into the scene. One of the great challenges of the twenty-first century science is how life transitioned from geochemistry into biochemistry (Pross 2012).

To resolve such a problem modern research is following a general scheme (Muchowska et al. 2020):

1. what occurred first, genetics and metabolism? who gave rise to the other?
2. no "singularity" between prebiotic chemistry and biochemistry;
3. does prebiotic chemistry start as kinetically coupled reactions or are prebiotic synthesis scenarios based on sequential chance events?

Of the three points point 1 seems to separate and polarize the prebiotic-life origin research community.

In the "genetics first" hypothesis the prebiotic synthesis of the first genetic molecule, generally RNA, undergoing to self-replication into a compartment subjected to Darwinian selection is proposed (RNA world) (Joyce and Szostak 2018; Robertson and Joycek 2012).

In the "metabolism first" hypothesis a non-enzymatic version of the metabolism is suggested catalyzed by minerals and metal ions (Muchowska et al. 2020; Peretó 2012; Wächtershäuser 1990).

The systematic presentation of the TM contribution to the prebiotic as well as Life' s origin chemistry will follow the two guiding principles "genetics first" and "metabolism first" approaches.

The metabolism and genetic approaches should give a better picture of the single TM contribution to the prebiotic-Life's origin chemistry by a consecutive logic between the two levels.

It should be noticed that in the literature the term "metal" is used for alkali-metals (Na, K), alkali-earth metals (Mg, Ca), TM, post-TM (Zn, Al) and lanthanides (La-Lu) and the terms "organometallic" or "metallorganic" are not commonly used and are often "hidden" in the text (Beck 2011; Brasier et al. 2011; Cody et al. 2000; Peters et al. 2010).

In the literature TM are often referred to as "cations", this does not exclude organometallic or metallorganic intermedia and reactants. Sometimes this is specifically analyzed like in the work of Beck (2011) where the peptide formation within the coordination sphere of metal ions and organometallic complexes is linked to prebiotic chemistry.

Before analyzing the TM contribution to prebiotic chemistry, it will be helpful to shortly analyze the minerals distributions containing TM on land as well as in the sea during the Hadean Eon.

1.5.2.1 TM Distribution in the Hadean Eon

The late Hadean or early Archean are the Eons where Life probably originated (Pearce et al. 2018). The evolution and availability of mineral phases during the Hadean were well analyzed by Hazen (2013), Morrison et al. (2018) in relationship to prebiotic chemistry.

An early list of 420 mineral species were proposed (Hazen 2013) to be completed by a successive analysis of the first row TM i.e. Ti, V, Cr, Mn, Fe, Co, Ni, Cu and Zn as well as As, Se, Mo, Ag, Sb, Au, Pb, Bi and the Pt group elements associated with sulphur deposit (Hazen 2019; Morrison et al. 2018).

Part of the mineral classification as defined by Hazen is Hazen (2019), Morrison et al. (2018):

(I) biominerals: based on carbon, nowadays mostly derived from biological material (Hazen et al. 2013; Perry et al. 2007). In the Hadean when carbon-rich meteorites (chondrites) and comets impacting on Earth enriched the surface with organic molecules (exogeneous delivery) and (sub)surface organic synthesis occurred in carbon-rich aqueous fluids (endogeneous synthesis), numerous organic crystals may have existed (Morrison et al. 2018).

(II) impact mineralization: the impact of meteorites or comets on the Earth's rocky ground produces extremely high temperature and pressure conditions rendering local minerals "metamorphic". For example typical high-pressure shock minerals, such as *lonsdaleite* an allotrope of carbon, *coesite* and *stishovite*, different crystalline forms of SiO_2, are important marker minerals for impact events. Furthermore large impacts can excavate and scatter rocks from the deep, thus mixing the deep metamorphic minerals with the surface material.

Most importantly the impacts start hydrothermal systems in circumferential fracture zones with a lifetime of 10^5 years (Pirajno 2009).

(III) prebiotic redox gradients: planets potentially have physico-chemical processes creating redox gradients at various scales, shifting electrons between reduced-oxidized sites. For example such a process makes the formation of minerals with relatively oxidized TM by abiological redox processes i.e. photo-oxidation possible.

Based on the previous classification, rare elements presumably important for prebiotic and Life's origin chemistry like B (ribose stabilization Ricardo et al. 2004 and regioselective phosphorylation of ribonucleosides Kim et al. 2016), Mo (carbohydrate rearrangement Ziegler et al. 2018) and P (phosphorylation by *schreibersite* corrosion Pasek et al. 2007), were relatively "abundant".

Minerals containing the previous list of elements as essentials such as *colemanite*, $Ca_2B_6O_{11}$ $5H_2O$, *powellite*, $CaMoO_4$ or *lüneburgite*, $Mg_3B_2(PO_4)_2(OH)_6$ $5(H_2O)$ are quite rare though, challenging the idea of their contribution to prebiotic chemistry. However there is an "ubiquity of trace and minor elements in common rock-forming minerals" (Morrison et al. 2018). For example B has a concentration of ≈10 ppm on the Earth crust while in the feldspar group of aluminosilicates B can reach 5000 ppm (Morrison et al. 2018).

Application of the same concept to other important rare elements and rare TM involved in the prebiotic or Life's origin chemistry, suggests sufficient amounts of these elements to start the sequence of chemical reactions to give birth to Life.

Regarding the Sea in the Hadean or early Archean Eons, due to the reducing and acidic pH (H_2 presence and high CO_2, though there is some evidence suggesting that such conditions did not exist Krissansen-Totton et al. 2018), surface waters were rich in Fe and Mn (Arndt and Nisbet 2012) with $Fe^{2+} = 10$ mM (Halevy and Bachan 2017) and $Mn^{2+} = 10^{-6}$ mM (Saito et al. 2003).

Furthermore due to the probable important role of hydrothermal vents (Shock and Canovas 2010) in the synthesis of prebiotic molecules as well as the location of Life's origin, the high local concentrations of TM, in particular Fe, Mn and Ni should have increased the adsorption of nucleotides on clays (Hao et al. 2019). Such a process would have enhanced the local concentrations of organics and protected the same from UV damage and heat (Hao et al. 2019).

In the next Paragraphs a list of TM contribution in prebiotic chemistry categorised in the "Metabolism first" and "Genetics first" group is reported.

1.5.2.2 Metabolism First

One-third of all known enzymes are metallo-enzymes, defined as proteins containing metal ions and-or metallo-organic cofactors (Bartnikas and Gitlin 2001). However metallo-proteins are not only working as enzymes but are also involved in metal storage (ferritin for iron), in electron transfer (cytochromes) and as transport proteins (transferrin for iron).

Focusing only on the enzymatic properties the question is: why enzymes involved in many of the metabolic pathways in "modern" organisms are based on transition metals?

In general acceleration rates due to enzymes are measured against the same not-catalyzed reactions. However a strong rate acceleration is obtained by the use of metals ions, minerals and small models of the active site based on an organometallic moiety (Brazzolotto et al. 2016; Campitelli and 2020; Junhyeok et al. 2017; Schilter et al. 2016; Stockbridge et al. 2010).

It has been suggested that the catalytic effect of "proto-enzymes" were mainly enthalpic in nature, a fundamental characteristic to maintain a high reaction rate once the environment became cooler (Stockbridge et al. 2010).

Many similarities between free metal ions, minerals, and enzymes are known and it is reasonable to assume that Life in its ancient and modern forms have conserved such prebiotic metallocomplexes in modern enzymes (Belmonte and Mansy 2016).

Summing up:

> Enzymes do not perform feats of magic; they just accelerate and add specificity to reactions that tend to occur anyway. It suggests that the basic underlying chemistry of the (biochemical) pathway is older than the enzymes that catalyze it (Sousa and Martin 2014).

To systematize the TM contribution to prebiotic and Life's origin chemistry, the method of Smith and Morowitz (2016, 2004) placed into a systematic view by Muchowska et al. (2020) will be followed.

Living organism base their biochemistry on five compounds defined as "universal metabolic precursors" (Muchowska et al. 2020):

(I) acetate-acetyl: biosynthetic precursor to lipids and terpenoids;
(II) pyruvate: sugars and (some) aminoacids precursor,
(III) oxaloacetate: precursor to various amino acids and pyrimidines,
(IV) succinate: precursor to various cofactors,
(V) α-ketoglutarate: precursor to various amino acids.

Acetyl-CoA and Pyruvate Pathway

The acetyl-CoA pathway is the one which is most dependent on transition metals and is part of the six autotrophic CO_2 fixation biochemical pathways (Muchowska et al. 2020). The active sites of the metallo-enzymes involved in the acetyl-CoA pathway have TM centers based on Fe, Ni, Mo and W (Darnault et al. 2003; Ragsdale 2011).

Many experiments have been conducted under various conditions to understand the link of the acetyl-CoA biochemical pathway to pre-biotic C_1 chemistry. Some experimental conditions considered neutral metallic Fe as a catalyst under a pressure of 1–35 bars of CO_2 as carbon source in water (Varma et al. 2018) and Ni_3Fe, Fe_3O_4 or Fe_3S_4 catalysts under a pressure of 15–25 bars in CO_2 with H_2 as reducing agent in water (Preiner et al. 2020).

Near to the "iron-(Ni)-sulphur world", experiments were conducted to model the acetylcoenzyme A pathway under hydrothermal conditions, where a mud of co-precipitated NiS and FeS converted CO and CH_3SH into the thioester CH_3-CO-SCH_3 readily hydrolyzing to CH_3COOH (Huber and Wächtershäuser 1997). Furthermore when the FeS-NiS was doped with Se, CH_3COOH was directly formed from CH_3SH and H_2S (Huber and Wächtershäuser 1997).

Acetate and pyruvate were synthesised by heating a gold tube containing neat formic acid, nonyl thiol and FeS at 250 °C under $500 \leq P \leq 2000$ bar (Cody et al. 2000).

In summary, there is a quite considerable ensemble of reducing conditions and catalysts able to reduce CO_2 to acetate and pyruvate giving strength to the hypothesis that such a pathway emerged from geochemistry were TM played a fundamental role as catalysts.

Oxaloacetate; TCA, rTCA

The reductive tricarboxylic acid (rTCA) cycle or reductive Krebs cycle together with the tricarboxylic acid (TCA) cycle or Krebs cycle are proposed to be one of the first metabolic pathways dating back to geochemistry. However if the rTCA is ancient it must have been catalyzed by minerals delimiting the transition from the geochemical to biochemical stage. TM sulfides like FeS, CuS under early ocean alcalin hydrothermal systems promote a rTCA cycle suggesting that metal production and metal-sustained primordial metabolism probably occurred in the active hydrothermal processes on the Hadean Earth (Kitadai et al. 2019). Other experiments showed that Fe^0, Cr^{3+} and Zn^{2+} in acidic aqueous solutions are able to promote multiple reactions of the rTCA cycle in consecutive sequence. The reaction network points toward the feasibility of primitive anabolism in an acidic, metal-rich reducing environment (Muchowska et al. 2017).

The "modern" TCA cycle involves several FeS and Fe^{2+} dependent enzymes (McMurry and Begley 2016). Furthermore due to its determinant role in the metabolism, it has been suggested to be the TCA cycle's prebiotic origin (Camprubi et al. 2017; Chevallot-Beroux et al. 2019; Hartman 1975; Keller et al. 2017; Lin et al. 2005).

Regarding the TCA prebiotic origin in a recent work of Muchowska et al. (2019) a nonenzymatic reaction network originating from glyoxylate and pyruvate in Fe^{2+}-rich water at 70 °C was found.

The same authors, based on the aforementioned study together with the work of Springsteen et al. (2018), propose the plausibility that both the (r)TCA and glyoxylate cycles had a probable common precursor that originated as prebiotic chemistry where iron played a major role as a catalyst.

Succinate and α-Ketoglutarate

In the succinate pathway involving the succinyl-CoA hydrolysis seemingly metallo-enzymes are not involved. However CoA, relying on a thioester chemistry, has a possible prebiotic origin (Atomi et al. 2013; Fuchs 2011) of which the (FeNi)S based synthesis of acetyl-thioester from CO and MeSH is the TM correspondent of the acetyl-CoA pathway (Huber and Wächtershäuser 1997).

Regarding the α-ketoacid pathway, the prebiotic synthesis of amino acids from ketoacids are promoted by a series of TM as Fe^{2+}, Fe^{3+}, Ni^{2+}, Cr^{3+}, VO^{2+}, $[VO_3]^-$ (Meisch et al. 1978), by ferredoxin (a non-haem iron–sulphur protein) (Nakajima et al. 1975), precipitated FeS or $Fe(OH)_2$ (Huber and Wächtershäuser 2003), partially electroreduced FeS (FeS-Fe^0) under alkaline conditions (Kitadai et al. 2019), on iron oxyhydroxide mineral (Barge et al. 2019).

Carbohydrate

Carbohydrate prebiotic chemistry has been split from the "five metabolic pillars" having deep roots in many of the pillars previously mentioned. Carbohydrate chemistry is affected by the "non-enzymatic phosphorilation problem". In fact the non-enzymatic phosphorylation in gluconeogenesis emphasizes the need of an efficient mineral catalyst for the phosphorylation by donor molecules such as polyphosphates.

However and notably many of the steps characterizing the glycolysis and pentose phosphate pathway can be non-enzymatic and based on Fe^{2+} (Keller et al. 2014, 2016).

Carbohydrate polymerization reactions have been obtained starting from glycoaldehyde-P and glyceraldehyde-P on a double layer of hydroxides based on Co^{2+} and Mn^{2+} part of the TM family and Mg^{2+}, Zn^{2+} (Pitsch et al. 1995; R et al. 1999).

1.5.2.3 Genetics First

In the "RNA world" hypothesis (Rich 1962) the abiotically synthesized RNA is able to work in a self-sustaining and self-replicating way as substrate for the genetic material (genotype) and as a catalyst (phenotype) (Jack 2009; Joyce 2007).

The RNA synthesis is based on the contemporaneous formation of the sugar skeleton and the basis. The review of Yadav et al. (2020) gives a general overview and state of the art of the complex chemistry involved.

Many synthetic routes of the RNA-DNA bases have been proposed based on HCN, cyanoacetylene, cyanoaldehydes, formamide and urea in presence of various (heterogeneous phase) catalysts like clay, $CaCO_3$ and ferrite, to mention a few (Saladino et al. 2004, 2005; Yadav et al. 2020).

Another set of synthetic routes based on formamide and urea toward single bases have been proposed to work efficiently on divalent cations present in silica, alumine, kaolin, and zeolite (Y type), $Ca(CO)_3$, CaO and TiO_2 (R et al. 2001; Saladino et al. 2003).

Near to the "iron-sulphur hypothesis", experiments conducted heating formamide in presence of different minerals such as *pyrrhotine* ($Fe_{1-x}S$), *pyrite* (FeS_2), *chalcopyrite* ($FeCuS_2$), *bornite* ($FeCu_5S_4$), *tetrahedrite* [(Fe,Cu,Sb)S], and *covellite* (CuS) at 160°C for 48h, adenine, purine, 2(1H)-pyrimidinone, isocytosine, urea, carbodiimide, and oxalic acid in different ratios were synthesized (Saladino et al. 2008).

Other synthetic pathways to pyrimidine and purine RNA building blocks with the help of Fe^{2+} have been reported (Becker et al. 2019, 2018).

Furthermore the synthesis of nucleotides necessary to the further polymerization has been investigated under various prebiotic conditions and precursors (Fernández-García et al. 2018; Mariani et al. 2018; Roberts et al. 2018; Stairs et al. 2017; Yi et al. 2018).

However and most importantly nucleotides polimerization from imidazole-activated nucleotides is strongly enhanced when divalent cations of various metals such as Pb^{2+}, Zn^{2+}, $[UO_2]^{2-}$, Pb^{2+}, Lu^{3+}, catalyze oligomer formation with maximum chain lengths from trimers to pentamers pentamers (Sawai 1976; Sawai et al. 1992, 1989; Sawai and Orgel 1975; Sawai and Yamamoto 1996; Sleeper and Orgel 1979).

Regarding the origin and evolution of the ribosome, specific analysis on the effect of metal ions on the structure-activity has been extensively conducted especially referred to the metal function in the LUCA ribosome (Last Universal Common Ancestor) (Bowman et al. 2020).

For example nowadays bacterial riboswitches cooperatively bind to Ni^{2+} or Co^{2+} ions controlling the expression of heavy metal transporters (Furukawa et al. 2015)

The divalent ion Mg^{2+} is the main cofactor for RNA. However it has been conjectured because ribosomes evolved well before the Great Oxidation Event (GOE) and because Fe^{2+} should have been the most abundant metal ion compared to Mg^{2+}, ribosomes should have interacted preferentially with Fe^{2+}.

This hypothesis has been analyzed in a series of studies.

In the first study (Bray et al. 2018) the structure, function, and cation content of the ribosome (low O_2, high Fe^{2+}, and high Mn^{2+}) was studied. Interestingly it has been shown that Mg^{2+}, Fe^{2+}, and Mn^{2+} have very similar effects on the rRNA fold and Fe^{2+} and Mn^{2+} are able to replace Mg^{2+} leaving the ribosome active during the translation of mRNA and both Fe^{2+} and Mn^{2+} strongly associate with the ribosome. Because the translation system originated and matured when Fe^{2+} and Mn^{2+} were abundant, previous findings suggest that Fe^{2+} and Mn^{2+} played a role in ancient ribosomes.

In another set of studies (Guth-Metzler et al. 2020; Hsiao et al. 2013; Lin et al. 2019) it has been reported that Fe^{2+}, like Mg^{2+}, can form ribosomal microclusters, containing paired metals bridged by a single phosphate group and Fe^{2+} transforms the ribosome into oxidoreductase-ribozyme.

The exchange of Fe^{2+} to Mg^{2+} is suggested to be placed in time before the GOE by a "casting" process i.e. the replacement of a metal ion with another however maintaining the original structure-function properties (Bowman et al. 2020; Bray et al. 2018).

1.6 Final Considerations

During the journey to understand what is the "state of the art" of the TM applied to astrochemistry and the prebiotic/Life's origin chemistry several conclusions can be drawn.

1. During the last decade there has been a growing interest toward the "exogeneous delivery" hypothesis of organic and prebiotic molecules on Earth.

This made the "classical" astrochemistry come closer to prebiotic chemistry, which was for a long time mainly based on the "endogeneous synthesis".

By "classical astrochemistry" one intends that *corpus* of observational data; mostly based on micro-waves (MW) and Infrared (IR) corresponding to the roto-vibrational transitions of molecules (mainly) in the gas-phase; to be compared and analyzed with laboratory and theoretical methods to build sophisticated chemical models of reaction networks from circumstellar disks to the deep interstellar medium (ISM) (McGuire 2018; Puzzarini 2020).

Today more than 200 molecules have been discovered in the ISM and circumstellar environments, composed of up to 13 atoms with the exception of the C60-C70 fullerenes (Cami et al. 2010) which are mainly organic in nature forming part of the Complex Organic Matter (COM) (Herbst and van Dishoeck 2009; McGuire 2018). With still thousands of astronomically observed molecular lines not yet assigned the previous list can increase consistently (McGuire 2018).

From a chemical point of view many of the detected molecules such as H_2CO, HCN, NH_3 and various organic radicals would be the perfect candidates to start a complex organic chemistry and the prebiotic synthesis.

It remains some debate on whether the most simple aminoacid glycine was found in the ISM or not (probably due to the low intensities of the rotational transitions) while it was detected in some meteorites like (Schmitt-Kopplina et al. 2010) and in the coma of the 67P/Churyumov-Gerasimenko comet (Hadraoui 2019).

The "glycine case" is mainly due to the spectroscopic vs. in situ chemical-analytical techniques applied to understand the composition of a sample. For example for both Murchinson and the 67P/Churyumov-Gerasimenko, for the former sophisticated techniques like electrospray ionization (ESI), Fourier-transform ion cyclotron resonance/mass spectrometry (FTICR/MS) or atomic force microscopy (AFM) based IR nanospectroscopy (Kebukawa et al. 2019; Schmitt-Kopplina et al. 2010) were applied while in case of the latter the ROSINA mass spectrometer on board the Rosetta probe was used (Hadraoui 2019).

The over 10^4 organic molecules found on the Murchinson meteorite intimately mixed in a mineral-organic matrix (Kebukawa et al. 2019; Schmitt-Kopplina et al. 2010) are an example of the enormous complexity found in extraterrestrial bodies broadening the "classical astrochemistry" allowing a merge of the astrochemistry field with the prebiotic and Life's origin chemistry.

2. The aforementioned conclusion is particularly well adapted if applied to the first row TM and the related organometallic and metallorganic chemistry.

   Up to now the only gas-phase organometallic compound found in the ISM and circumstellar enviroments is iron cyanide (FeCN) (Zack et al. 2011).

   Other first row TM compounds that are not organometallic or metallorganic in nature are FeO (Furuya 2003), TiO and TiO_2 (Kamínski 2013).

   Furthermore the involvement of TM in the astrochemistry reaction networks was and still is not much considered as a sort of "baseline" research since the 1980s (Bohme 1992; Boissel 1994; Chaudret et al. 1991; Duley 1980; Fioroni

2014, 2016; Fioroni and DeYonker 2016; Irikura et al. 1990; Klotz et al. 1996, 1995, 1994; Marty et al. 1996; Mestdagh et al. 1992; Prasad et al. 1980; Serra et al. 1992).

This is probably due to the physical-chemical character of TM expected to be, mainly, in condensed phases like minerals and, based on the last knowledge, in organometallic or metallorganic compounds as found on meteorites (Ruf et al. 2017; Smith et al. 2019).

However due to the increasing interest in the "exogeneous delivery" hypothesis, the new findings in meteorites where there is a deep mix between the mineral-organic phases, "demands" for a more detailed analysis of the "mineral surface" contribution especially if TM are present due to their known catalytic behaviour.

It is time to follow the same "positive" approach as used in the "endogeneous synthesis" regarding TM based on a "reverse engineering" of the modern metabolisms. Here many of the first row TM, with a predominance of Fe, Ni and Mn, do play an important role in the synthesis of Life's chemical bricks as well as in the appearance of the first metabolism or genetics.

There are still many TM to be analysed for their involvement in the prebiotic chemistry and not only first row TM. For example W is present in many bacteria and in some prokaryotes though only some hyperthermophilic archaea appear to be tungsten-dependent (Kletzin and Adams 1996), Mo is essential for all Life's forms (Schwarz et al. 2009), Cd based enzymes found in marine diatoms (Lane and Morel 2000) and the f-group lanthanides methanotrophic and methylotrophic bacteria (Daumann 2019).

The elucidation of a reaction mechanism on a mineral surface (heterogeneous phase) or for a single organometallic compound (homogeneous phase) is not an easy task.

However there is a vast ensemble of experimental-spectroscopic techniques to study the surface as well as the bulky properties of the TM-mineral surface catalysts during reaction conditions like FTIR (Fourier-Transform Infrared spectroscopy), Raman, DRF (Diffuse Reflectance spectroscopy), photoacustic spectroscopy, NMR, EPR, XANES (X-ray Absorption Near Edge Structure), EXAFS (Extended X-Ray Absorption Fine Structure), XRD (X-ray powder Diffraction), Mössbauer and characterization techniques like XPS (X-ray Photoelectron spectroscopy) and TEM.

The theoretical approach has now a vast arsenal of possible techniques to analyze the reaction potential energy surface like DFT and ab-initio molecular dynamics to mention a few (Berendsen 2007).

Probably the TM contributions to the prebiotic and Life's origin chemistry starting with a journey from "space to Earth" are ready to be unraveled.

References

Adami, C.: Information-theoretic considerations concerning the origin of Life. Orig. Life Evol. Biosph. **45**, 309–317 (2015). https://doi.org/10.1007/s11084-015-9439-0

Albright, T.A.: Structure and reactivity in organometallic chemistry. An applied molecular orbital approach. Tetrahedron **38**(10), 1339–1388 (1982). https://doi.org/10.1016/0040-4020(82)80217-2

Albright, T.A., Burdett, J.K., Whangbo, M.: Orbital Interaction in Chemistry. Wiley VCH, Weinheim (2013). https://doi.org/10.1002/9781118558409

Arlandini, C., Käppeler, E., Wisshak, K., Gallino, R., Lugaro, M., Busso, M., Straniero, O.: Neutron capture in low-mass asymptotic giant branch stars: cross sections and abundance signatures. Astrophys. J. **525**, 886–900 (1999). https://doi.org/10.1086/307938

Arndt, N.T., Nisbet, E.G.: Processes on the young Earth and the habitats of early Life. Annu. Rev. Earth Planet. Sci. **40**(1), 521–549 (2012). https://doi.org/10.1146/annurev-earth-042711-105316

Ascasibar, Y., Gavilán, M., Pinto, N., Casado, J., Rosales-Ortega, F., Díaz, A.I.: Understanding chemical evolution in resolved galaxies—I. The local star fraction–metallicity relation. Mon. Not. R. Astron. Soc. **448**(3), 2126–2134 (2015). https://doi.org/10.1093/mnras/stv098

Asplund, M.: New light on stellar abundance analyses: departures from LTE and homogeneity. Ann. Rev. Astron. Astrophys. **43**(1), 481–530 (2005). https://doi.org/10.1146/annurev.astro.42.053102.134001

Atomi, H., Tomita, H., Ishibashi, T., Yokooji, Y., Imanaka, T.: CoA biosynthesis in archaea. Biochem. Soc. Trans. **41**(1), 427–431 (2013). https://doi.org/10.1042/BST20120311

Barge, L.M., Flores, E., Baum, M.M., VanderVelde, D.G., Russell, M.J.: Redox and pH gradients drive amino acid synthesis in iron oxyhydroxide mineral systems. Proc. Nat. Ac. Sci. **116**(11), 4828–4833 (2019). https://doi.org/10.1073/pnas.1812098116

Bartnikas, T., Gitlin, J.: How to make a metalloprotein. Nat. Struct. Mol. Biol. **8**, 733–734 (2001). https://doi.org/10.1038/nsb0901-733

Beck, W.: Metal complexes of biologically important ligands, CLXXVI. Formation of peptides within the coordination sphere of metal ions and of classical and organometallic complexes and some aspects of prebiotic chemistry. Z. Anorg. Allg. Chem. **637**(12), 1647–1672 (2011). https://doi.org/10.1002/zaac.201100137

Becker, S., Feldmann, J., Wiedemann, S., Okamura, H., Schneider, C., Iwan, K., Crisp, A., Rossa, M., Amatov, T., Carell, T.: Unified prebiotically plausible synthesis of pyrimidine and purine RNA ribonucleotides. Science **366**(6461), 76–82 (2019). https://doi.org/10.1126/science.aax2747

Becker, S., Schneider, C., Okamura, H., Crisp, A., Amatov, T., Dejmek, M., Carell, T.: Wet-dry cycles enable the parallel origin of canonical and non-canonical nucleosides by continuous synthesis. Nat. Commun. **9**, 163 (2018). https://doi.org/10.1038/s41467-017-02639-1

Bekaert, D.V., Derenne, S., Tissandier, L., Marrocchi, Y., Charnoz, S., Anquetil, C., Marty, B.: High-temperature ionization-induced synthesis of biologically relevant molecules in the protosolar nebula. Astrophys. J. **859**(2), 142 (2018). https://doi.org/10.3847/1538-4357/aabe7a

Beller, M., Bolm, C.: Transition Metal for Organic Synthesis. Wiley-VCH, Weinheim (2004)

Belmonte, L., Mansy, S.S.: Metal catalysts and the origin of Life. Elements **12**(6), 413–418 (2016). https://doi.org/10.2113/gselements.12.6.413

Berendsen, H.J.C.: Simulating the Physical World: Hierarchical Modeling from Quantum Mechanics to Fluid Dynamics. Cambridge University Press (2007). https://doi.org/10.1017/CBO9780511815348

Bohme, D.K.: PAH [polycyclic aromatic hydrocarbons] and fullerene ions and ion/molecule reactions in interstellar and circumstellar chemistry. Chem. Rev. **92**(7), 1487–1508 (1992). https://doi.org/10.1021/cr00015a002

Boissel, P.: Organometallic chemistry and the interstellar medium: experimental evidence of coordination between metal cations and polycyclic hydrocarbon in the gas phase. Astron. Astrophys. **285**, L33–L36 (1994)

Boogert, A.A., Gerakines, P.A., Whittet, D.C.: Observations of the icy universe. Ann. Rev. Astr. Astrophys. **53**(1), 541–581 (2015). https://doi.org/10.1146/annurev-astro-082214-122348

Bowman, J.C., Petrov, A.S., Frenkel-Pinter, M., Penev, P.I., Williams, L.D.: Root of the tree: the significance, evolution, and origins of the ribosome. Chem. Rev. **120**(11), 4848–4878 (2020). https://doi.org/10.1021/acs.chemrev.9b00742

Brasier, M.D., Matthewman, R., McMahon, S., Wacey, D.: Pumice as a remarkable substrate for the origin of Life. Astrobiology **11**(7), 725–735 (2011). https://doi.org/10.1089/ast.2010.0546

Bray, M.S., Lenz, T.K., Haynes, J.W., Bowman, J.C., Petrov, A.S., Reddi, A.R., Hud, N.V., Williams, L.D., Glass, J.B.: Multiple prebiotic metals mediate translation. Proc. Nat. Ac. Sci. **115**(48), 12164–12169 (2018). https://doi.org/10.1073/pnas.1803636115

Brazzolotto, D., Gennari, M., Queyriaux, N., Simmons, T.R., Pécaut, J., Demeshko, S., Meyer, F., Orio, M., Artero, V., Duboc, C.: Nickel-centred proton reduction catalysis in a model of [NiFe] hydrogenase. Nat. Chem. **8**, 1054–1060 (2016). https://doi.org/10.1038/nchem.2575

Brewster, M.A., Ziurys, L.M.: The millimeter-wave Spectrum of NiC and CoC. Astrophys. J. **559**, L163–L166 (2001). https://doi.org/10.1086/323515

Burton, A.S., Stern, J.C., Elsila, J.E., Glavin, D.P., Dworkin, J.P.: Understanding prebiotic chemistry through the analysis of extraterrestrial amino acids and nucleobases in meteorites. Chem. Soc. Rev. **41**, 5459–5472 (2012). https://doi.org/10.1039/C2CS35109A

Busso, M., Gallino, G., Wasserburg, J.: Nucleosynthesis in asymptotic giant branch star: relevance for galactic enrichment and solar system formation. Annu. Rev. Astron. Astrophys. **37**, 239–309 (1999). https://doi.org/10.1146/annurev.astro.37.1.239

Cami, J., Bernard-Salas, J., Peeters, E., Malek, S.E.: Detection of C60 and C70 in a young planetary nebula. Science **329**(5996), 1180–1182 (2010). https://doi.org/10.1126/science.1192035

Campitelli, P., Crucianelli, M.: On the capability of oxidovanadium(IV) derivatives to act as all-around catalytic promoters since the prebiotic world. Molecules **25**(13), 3073 (2020). https://doi.org/10.3390/molecules25133073

Camprubi, E., Jordan, S.F., Vasiliadou, R., Lane, N.: Iron catalysis at the origin of Life. IUBMB Life **69**(6), 373–381 (2017). https://doi.org/10.1002/iub.1632

Casalot, L., Rousset, M.: Maturation of the [NiFe] hydrogenases. Trends Microbiol. **9**(5), 228–237 (2001). https://doi.org/10.1016/S0966-842X(01)02009-1

Chaudret, B., LeBeuze, A., Rabaa, H., Saillard, J., Serra, G.: Organometallic chemistry in the interstellar medium. II: theoretical study of the coordination of Iron polycyclic hydrocarbons. New J. Chem. **15**, 791–794 (1991)

Chevallot-Beroux, E., Gorges, J., Moran, J.: Energy conservation via thioesters in a non-enzymatic metabolism-like reaction network. ChemRxiv (2019). https://doi.org/10.26434/chemrxiv.8832425.v1

Chyba, C., Sagan, C.: Endogenous production, exogenous delivery and impact-shock synthesis of organic molecules: an inventory for the origins of life. Nature **355**, 125–132 (1992). https://doi.org/doi.org/10.1038/355125a0

Chyba, C.F., Thomas, P.J., Brookshaw, L., Sagan, C.: Cometary delivery of organic molecules to the early Earth. Science **249**, 366–373 (1990). https://doi.org/10.1126/science.1217291

Ciesla, F.J., Sandford, S.A.: Organic synthesis via irradiation and warming of ice grains in the solar nebula. Science **336**(6080), 452–454 (2012). https://doi.org/10.1126/science.1217291

Cody, G.D., Boctor, N.Z., Filley, T.R., Hazen, R.M., Scott, J.H., Sharma, A., Yoder, H.S.: Primordial carbonylated iron-sulfur compounds and the synthesis of pyruvate. Science **289**(5483), 1337–1340 (2000). https://doi.org/10.1126/science.289.5483.1337

Constable, E.C.: Evolution and understanding of the d-block elements in the periodic table. Dalton Trans. **48**, 9408–9421 (2019). https://doi.org/10.1039/C9DT00765B

Cornils, G., Herrmann, W.A., Beller, M., Paciello, R.: Applied Homogeneous Catalysis with Organometallic Compounds: A Comprehensive Handbook in Four Volumes. Wiley VCH, Weinheim (2017)

Cowley, C.: An Introduction to Cosmochemistry. Cambridge University Press, Cambridge (1995)

Darnault, C., Volbeda, A., Jin Kim, E., Legrand, P., Vernède, X., Lindahl, P.A., Fontecilla-Camps, J.C.: Ni-Zn-[Fe4-S4] and Ni-Ni-[Fe4-S4] clusters in closed and open subunits of acetyl-CoA synthase/carbon monoxide dehydrogenase. Nat. Struct. Biol. **10**, 271–279 (2003). https://doi.org/10.1038/nsb912

Daumann, L.J.: Essential and ubiquitous: the emergence of lanthanide metallobiochemistry. Angew. Chem. Int. Ed. **58**(37), 12795–12802 (2019). https://doi.org/10.1002/anie.201904090

Draine, B.T.: Physics of the Interstellar and Intergalactic Medium. Princeton Series in Astrophysics. Princeton University, Princeton (2011)

Duley, W.W.: Redox reactions and the optical properties of interstellar grains. Astrophys. J. **240**, 950–955 (1980). https://doi.org/10.1086/158308

Elschenbroich, K.: Organometallics. Wiley-VCH, Weinheim (2006)

Fernández-García, C., Grefenstette, N.M., Powner, M.W.: Selective aqueous acetylation controls the photoanomerization of α-cytidine-5'-phosphate. Chem. Commun. **54**, 4850–4853 (2018). https://doi.org/10.1039/C8CC01929K

Fioroni, M.: Astrochemistry of transition metals? The selected cases of $[FeN]^+$ $[FeNH]^+$ and $[(CO)_2FeN]^+$: pathways toward CH_3NH_2 and HNCO. Phys. Chem. Chem. Phys. **16**, 24312–24322 (2014). https://doi.org/10.1039/C4CP03218G

Fioroni, M.: Astro-organometallics of Fe, Co, Ni: stability, IR fingerprints and possible locations. Comp. Theor. Chem. 1084, 196–212 (2016). https://doi.org/10.1016/j.comptc.2016.03.023

Fioroni, M., DeYonker, N.J.: H_2 formation on cosmic grain siliceous surfaces grafted with Fe^+: a silsesquioxanes-based computational model. ChemPhysChem **17**(21), 3390–3394 (2016). https://doi.org/10.1002/cphc.201600607

Fuchs, G.: Alternative pathways of carbon dioxide fixation: insights into the early evolution of Life? Ann. Rev. Microbi **65**(1), 631–658 (2011). https://doi.org/10.1146/annurev-micro-090110-102801

Furukawa, K., Ramesh, A., Zhou, Z., Weinberg, Z., Vallery, T., Winkler, W.C., Breaker, R.R.: Bacterial riboswitches cooperatively bind Ni(2+) or Co(2+) ions and control expression of heavy metal transporters. Mol. Cell. **57**(6), 1088–1098 (2015). https://doi.org/10.1016/j.molcel.2015.02.009

Furuya, R. S., Walmsley, C. M., Nakanishi, K., Schilke, P., Bachiller, R.: Interferometric observations of FeO towards Sagittarius B2. Astronom. Astrophys. **409**(2), L21–L24 (2003). https://doi.org/10.1051/0004-6361:20031304

Galli, D., Palla, F.: The dawn of chemistry. Ann. Rev. Astron. Astrophys. **51**(1), 163–206 (2013). https://doi.org/10.1146/annurev-astro-082812-141029

Gehrz, R.D., Truran, J.W., Williams, R.E., Starrfield, S.: Nucleosynthesis in classical novae and its contribution to the interstellar medium. Publ. Astron. Soc. Pac. **110**, 3–26 (1998). https://doi.org/10.1086/316107

Guth-Metzler, R., Bray, M.S., Moran, F.P., Suttapitugsakul, S., Montllor-Albalate, C., Bowman, J.C., Wu, R., Reddi, A.R., Okafor, C.D., Glass, J.B., Williams, L.D.: Cutting in-line with iron: ribosomal function and non-oxidative RNA cleavage. Nucleic Acid Res. **48**(15), 8663–8674 (2020). https://doi.org/10.1093/nar/gkaa586

Habing, H.J., Olofsson, H.: Asymptotic Giant Branch Stars. Springer, Heidelberg (2004)

Hadraoui, K., Cottin, H., Ivanovski, S. L., Zapf, P., Altwegg, K., Benilan, Y., Biver, N., Della Corte, V., Fray, N., Lasue, J., Merouane, S., Rotundi, A., Zakharov, V.: Distributed glycine in comet 67P/Churyumov-Gerasimenko. Astronom. Astrophys. **630**, A32 (2019). https://doi.org/10.1051/0004-6361/201935018

Halevy, I., Bachan, A.: The geologic history of seawater pH. Science **355**(6329), 1069–1071 (2017). https://doi.org/10.1126/science.aal4151

Halfen, D.T., Ziurys, L.M.: Laboratory detection of $[FeCO]^+$ by milllimeter/submillimeter velocity modulation spectroscopy. Astrophys. J. Lett. **657**, L61–L64 (2007). https://doi.org/10.1086/512853

Hamilton, V.E., Simon, A.A., Christensen, P.R., Reuter, D.C., Clark, B.E., Barucci, M.A., Bowles, N.E., Boynton, W.V., Brucato, J.R., Cloutis, E.A., Connolly Jr., H.C., Donaldson Hanna, K.L.,

Emery, J.P., L.Enos, H., Fornasier, S., Haberle, C.W., Hanna, R.D., Howell, E.S., Kaplan, H.H., Keller, L.P., Lantz, C., Li, J.Y., Lim, L.F., McCoy, T.J., Merlin, F., Nolan, M.C., Praet, A., Rozitis, B., Sandford, S.A., Schrader, D.L., Thomas, C.A., Zou, X.D., Lauretta, D.S., the OSIRIS-REx Team: Evidence for widespread hydrated minerals on asteroid (101955) Bennu. Nat. Astron. **3**, 332–340 (2019). https://doi.org/10.1038/s41550-019-0722-2

Hanwell, M.D., Curtis, D.E., Lonie, D.C., Vandermeersch, T., Zurek, E., Hutchison, G.R.: Avogadro: an advanced semantic chemical editor, visualization, and analysis platform. J. Cheminform. **4**, 1–17 (2012). https://doi.org/10.1186/1758-2946-4-17

Hao, J., Mokhtari, M., Pedreira-Segade, U., Michot, L.J., Daniel, I.: Transition metals enhance the adsorption of nucleotides onto clays: implications for the origin of Life. ACS Earth Space Chem. **3**(1), 109–119 (2019). https://doi.org/10.1021/acsearthspacechem.8b00145

Hartman, H.: Speculations on the origin and evolution of metabolism. J. Mol. Evol. **4**, 359–370 (1975). https://doi.org/10.1007/BF01732537

Hartshorn, R.M., Hellwich, K.H., Yerin, A., Damhus, T., Hutton, A.T.: Brief guide to the nomenclature of inorganic chemistry. Pure Appl. Chem. **87**(9–10), 1039–1049 (2015). https://doi.org/10.1515/pac-2014-0718

Hazen, R.: Paleomineralogy of the Hadean Eon: a preliminary species list. Am. J. Sci. **313**, 807–843 (2013). https://doi.org/10.2475/09.2013.01

Hazen, R.M.: An evolutionary system of mineralogy: proposal for a classification of planetary materials based on natural kind clustering. Am. Mineral. **104**(6), 810–816 (2019). https://doi.org/10.2138/am-2019-6709CCBYNCND

Hazen, R.M., Downs, R.T., Jones, A.P., Kah, L.: Carbon mineralogy and crystal chemistry. Rev. Mineral. Geochem. **75**(1), 7–46 (2013). https://doi.org/10.2138/rmg.2013.75.2

Henning, T., Semenov, D.: Chemistry in proto-planetary disks. Chem. Rev. **113**, 9016–9042 (2013). https://doi.org/10.1021/cr400128p

Herbst, E.: Three milieux for interstellar chemistry: gas, dust, and ice. Phys. Chem. Chem. Phys. **16**, 3344–3359 (2013). https://doi.org/10.1039/c3cp54065k

Herbst, E., van Dishoeck, E.F.: Complex organic interstellar molecules. Ann. Rev. Astron. Astrophys. **47**(1), 427–480 (2009). https://doi.org/10.1146/annurev-astro-082708-101654

Herbst, E., Vidali, G., Ceccarelli, C.: Complex organic molecules (COMs) in star-forming regions: a virtual special issue. ACS Earth Space Chem. **4**(4), 488–490 (2020). https://doi.org/10.1021/acsearthspacechem.0c00043

Herwig, F.: Evolution of asymptotic giant branch stars. Annu. Rev. Astron. Astrophys. **43**, 435–479 (2005). https://doi.org/10.1146/annurev.astro.43.072103.150600

Hsiao, C., Chou, I.C., Okafor, C.D., Bowman, J.C., O'Neill, E.B., Athavale, S.S., Petrov, A.S., Hud, N.V., Wartell, R.M., Harvey, S.C., Dean Williams, L.: RNA with iron(II) as a cofactor catalyses electron transfer. Nat. Chem. **5**(6), 525–528 (2013). https://doi.org/10.1038/nchem.1649

Huber, C., Wächtershäuser, G.: Activated acetic acid by carbon fixation on (Fe,Ni)S under primordial conditions. Science **276**(5310), 245–247 (1997). https://doi.org/10.1126/science.276.5310.245

Huber, C., Wächtershäuser, G.: Primordial reductive amination revisited. Tetrahedron Lett. **44**(8), 1695–1697 (2003). https://doi.org/10.1016/S0040-4039(02)02863-0

Irikura, K.K., Goddard, W.A., Beauchamp, J.L.: Prospects for the involvement of transition metals in the chemistry of diffuse interstellar clouds: formation of FeH^+ by radiative association. Int. J. Mass Spect. Ion Proc. **99**(3), 213–222 (1990). https://doi.org/10.1016/0168-1176(90)85031-V

Jack, S.W.: Systems chemistry on early Earth. Nature **459**, 171–172 (2009). https://doi.org/doi.org/10.1038/459171a

Jenkins, E.B., Savage, B.D., Spitzer L., J.: Abundances of interstellar atoms from UV absorption lines. Astrophys. J. **301**, 355 (1986). https://doi.org/10.1086/163906

Joalland, B., Simon, A., Marsden, C. J., Joblin, C.: Signature of $[SiPAH]^+$ π-complexes in the interstellar medium. Astronom. Astrophys. **494**(3), 969–976 (2009). https://doi.org/10.1051/0004-6361:200810863

Jones, A.P.: Dust evolution, a global view: III. core/mantle grains, organic nano-globules, comets and surface chemistry. R. Soc. Open Sci. **3**(12), 160224 (2016). https://doi.org/10.1098/rsos.160224

Jordi, J., Iliadis, C.: Nuclear astrophysics: the unfinished quest for the origin of the elements. Rep. Prog. Phys. **74**, 096901 (2011). https://doi.org/10.1088/0034-4885/74/9/096901

Joyce, G.F.: Forty years of *in-vitro* evolution. Angew. Chem. Int. Ed. **46**(34), 6420–6436 (2007). https://doi.org/10.1002/anie.200701369

Joyce, G.F., Szostak, J.W.: Protocells and RNA self-replication. Cold Spring Harb. Perspect. Biol. **10**, a034801 (2018). https://doi.org/10.1101/cshperspect.a034801

Junhyeok, S., Taylor, A.M., Michael, J.R.: Structural and functional synthetic model of mono-iron hydrogenase featuring an anthracene scaffold. Nat. Chem. **9**, 552–557 (2017). https://doi.org/10.1038/nchem.2707

Kamínski, T., Gottlieb, C. A., Menten, K. M., Patel, N. A., Young, K. H., Brünken, S., Müller, H. S. P., McCarthy, M. C., Winters, J. M., Decin, L.: Pure rotational spectra of TiO and TiO_2 in VY Canis Majoris. Astronom. Astrophys. **551**, A113 (2013). https://doi.org/10.1051/0004-6361/201220290

Kebukawa, Y., Kobayashi, H., Urayama, N., Baden, N., Kondo, M., Zolensky, M.E., Kobayashi, K.: Nanoscale infrared imaging analysis of carbonaceous chondrites to understand organic-mineral interactions during aqueous alteration. Proc. Nat. Ac. Sci. U.S.A. **116**(3), 753–758 (2019). https://doi.org/10.1073/pnas.1816265116

Kebukawa, Y., Kobayashi, H., Urayama, N., Baden, N., Kondo, M., Zolensky, M.E., Kobayashi, K.: Nanoscale infrared imaging analysis of carbonaceous chondrites to understand organic-mineral interactions during aqueous alteration. Proc. Nat. Ac. Sci. U.S.A. **116**(3), 753–758 (2019). https://doi.org/10.1073/pnas.1816265116

Keller, M.A., Kampjut, D., Harrison, S.A., Ralser, M.: Sulfate radicals enable a non-enzymatic Krebs cycle precursor. Nat. Ecol. Evol. **1**, 0083 (2017). https://doi.org/10.1038/s41559-017-0083

Keller, M.A., Turchyn, A.V., Ralser, M.: Non-enzymatic glycolysis and pentose phosphate pathway-like reactions in a plausible Archean ocean. Mol. Syst. Biol. **10**(4), 725–736 (2014). https://doi.org/10.1002/msb.20145228

Keller, M.A., Zylstra, A., Castro, C., Turchyn, A.V., Griffin, J.L., Ralser, M.: Conditional iron and pH-dependent activity of a non-enzymatic glycolysis and pentose phosphate pathway. Science Adv. **2**(1) (2016). https://doi.org/10.1126/sciadv.1501235

Kim, H.J., Furukawa, Y., Kakegawa, T., Bita, A., Scorei, R., Benner, S.A.: Evaporite borate-containing mineral ensembles make phosphate available and regiospecifically phosphorylate ribonucleosides: borate as a multifaceted problem solver in prebiotic chemistry. Angew. Chem. Int. Ed. **55**(51), 15816–15820 (2016). https://doi.org/10.1002/anie.201608001

Kimura, K., Nuth III, J.A., Ferguson, F.T.: Is the 21 micron feature observed in some post-AGB stars caused by the interaction between Ti atoms and fullerenes? Astrophys. J. **632**, L159–L162 (2005). https://doi.org/doi.org/10.1086/497987

Kitadai, N., Nakamura, R., Yamamoto, M., Takai, K., Yoshida, N., Oono, Y.: Metals likely promoted protometabolism in early ocean alkaline hydrothermal systems. Sci. Adv. **5**(6) (2019). https://doi.org/10.1126/sciadv.aav7848

Kitazato, K., Milliken, R.E., Iwata, T., Abe, M., Ohtake, M., Matsuura, S., Arai, T., Nakauchi, Y., Nakamura, T., Matsuoka, M., Senshu, H., Hirata, N., Hiroi, T., Pilorget, C., Brunetto, R., Poulet, F., Riu, L., Bibring, J.P., Takir, D., Domingue, D.L., Vilas, F., Barucci, M.A., Perna, D., Palomba, E., Galiano, A., Tsumura, K., Osawa, T., Komatsu, M., Nakato, A., Arai, T., Takato, N., Matsunaga, T., Takagi, Y., Matsumoto, K., Kouyama, T., Yokota, Y., Tatsumi, E., Sakatani, N., Yamamoto, Y., Okada, T., Sugita, S., Honda, R., Morota, T., Kameda, S., Sawada, H., Honda, C., Yamada, M., Suzuki, H., Yoshioka, K., Hayakawa, M., Ogawa, K., Cho, Y., Shirai, K., Shimaki, Y., Hirata, N., Yamaguchi, A., Ogawa, N., Terui, F., Yamaguchi, T., Takei, Y., Saiki, T., Nakazawa, S., Tanaka, S., Yoshikawa, M., Watanabe, S., Tsuda, Y.: The surface composition of asteroid 162173 Ryugu from Hayabusa2 near-infrared spectroscopy. Science **364**(6437), 272–275 (2019). https://doi.org/10.1126/science.aav7432

Kletzin, A., Adams, M.W.: Tungsten in biological systems. FEMS Microbiol. Rev. **18**(1), 5–63 (1996). https://doi.org/10.1111/j.1574-6976.1996.tb00226.x

Klotz, A., Marty, P., Boissel, P., de Caro, D., Serra, G., Mascetti, J., de Parseval, P., Derouault, J., Daudey, J.P., Chaudret, B.: Possible contribution of organometallic species in the solar system ices. Reactivity and spectroscopy. Planet. Space Sci. **44**, 957–965 (1996). https://doi.org/10.1016/0032-0633(96)00026-8

Klotz, A., Marty, P., Boissel, P., Serra, G., Chaudret, B., Daudey, J.P.: Contribution of metal bonded Polycyclic Aromatic Hydrocarbons to the interstellar depletion. Astron. Astrophys. **304**, 520–530 (1995)

Klotz, A., Ristorcelli, I., de Caro, D., Serra, S., Chaudret, B., Daudey, J.P., Giard, M., Barthelat, J.C., Marty, P.: Evaluation of the role of organometallic species in the chemistry of interstellar and circumstellar media. AIP Conf. Proc. **312**, 705 (1994). https://doi.org/10.1063/1.46628

Köhler, M., Jones, A., Ysard, N.: A hidden reservoir of Fe/FeS in interstellar silicates? Astronom. Astrophys. **565**, L9 (2014). https://doi.org/10.1051/0004-6361/201423985

Komskii, D.I.: Transition Metal Compounds. Cambridge University Press. (2014)

Krissansen-Totton, J., Arney, G.N., Catling, D.C.: Constraining the climate and ocean pH of the early Earth with a geological carbon cycle model. Proc. Nat. Ac. Sci. U.S.A. **115**(16), 4105–4110 (2018). https://doi.org/10.1073/pnas.1721296115

Kuljeet, K.M., Piyush, S.: Transition elements in supernova presolar grains: condensation versus implantation. Astrophys. J. **853**(1), 12 (2018). https://doi.org/10.3847/1538-4357/aa9e8e

Kwok, S.: Complex organics in space from solar system to distant galaxies. Astron. Astrophys. Rev. **24**, 8 (2016). https://doi.org/10.1007/s00159-016-0093-y

Lane, T.W., Morel, F.M.M.: A biological function for cadmium in marine diatoms. Proc. Nat. Ac. Sci. U.S.A. **97**(9), 4627–4631 (2000). https://doi.org/10.1073/pnas.090091397

Le Guillou, C., Bernard, S., Brearley, A.J., Remusat, L.: Evolution of organic matter in Orgueil, Murchison and Renazzo during parent body aqueous alteration: *in situ* investigations. Geochim. Cosmochim. Ac. **131**, 368–392 (2014). https://doi.org/10.1016/j.gca.2013.11.020

Leising, M.D., Share, G.H.: The gamma ray light curves of SN 1987A. Astrophys. J. **357**, 638–650 (1990). https://doi.org/10.1086/168952

Li, A., Liu, J.M., Jiang, B.W.: On iron monoxide nanoparticles as a carrier of the mysterious 21 μ emission features in post-asymptotic giant branch stars. Astrophys. J. **777**, 111 (2013). https://doi.org/10.1088/0004-637X/777/2/111

Li, C., Liu, Y.: Bridging Heterogeneous and Homogeneous Catalysis. Concepts, Strategies, and Applications. Wiley VCH, Weinheim (2014)

Li, M., Huang, S., Petaev, M.I., Zhu, Z., Steffen, J.H.: Dust condensation in evolving discs and the composition of planetary building blocks. Mon. Not. R. Astron. Soc. **495**(3), 2543–2553 (2020). https://doi.org/10.1093/mnras/staa1149

Limongi, M., Chieffi, A.: Explosive nucleosynthesis in massive stars. AIP Conf. Proc. 1269, 110–119 (2010). https://doi.org/10.1063/1.3485121

Lin, L.H., Slater, G.F., Sherwood Lollar, B., Lacrampe-Couloume, G., Onstott, T.: The yield and isotopic composition of radiolytic H_2, a potential energy source for the deep subsurface biosphere. Geochim. Cosmochim. Acta **69**(4), 893–903 (2005). https://doi.org/10.1016/j.gca.2004.07.032

Lin, S.Y., Wang, Y.C., Hsiao, C.: Prebiotic iron originates the peptidyl transfer origin. Mol. Biol. Evol. **36**(5), 999–1007 (2019). https://doi.org/10.1093/molbev/msz034

Lippard, S.K., Berg, J.M.: Principles of Bioinorganic Chemistry. University Science Books, Herndon (1994)

Lodders, K.: Solar system abundances and condensation temperatures of the elements. Astrophys. J. **591**, 1220–1247 (2003). https://doi.org/10.1086/375492

Lodders, K., Fegley, B.: Asymptotic giant branch stars. In: Proc. IAU Symp., vol. 191, pp. 279–289. Cambridge University Press, Cambridge (1999)

Mann, J.B., Meek, T.L., Knight, E.T., Capitani, J.F., Allen, L.C.: Configuration energies of the d-block elements. J. Am. Chem. Soc. **122**(21), 5132–5137 (2000). https://doi.org/10.1021/ja9928677

Mariani, A., Russell, D.A., Javelle, T., Sutherland, J.D.: A light-releasable potentially prebiotic nucleotide activating agent. J. Am. Chem. Soc. **140**(28), 8657–8661 (2018). https://doi.org/10.1021/jacs.8b05189

Marty, P., de Parseval, P., Klotz, A., Serra, G., Boissel, P.: Organometallic reaction in the gas-phase: measurement of the formation and dissociation rates of organometallic complexes. Astron. Astrophys. **316**, 270–274 (1996)

McGuire, B.A.: 2018 census of interstellar, circumstellar, extragalactic, protoplanetary disk, and exoplanetary molecules. Astrophys. J. Suppl. S. **239**(2), 17 (2018). https://doi.org/10.3847/1538-4365/aae5d2

McMurry, J.E., Begley, T.P.: The organic chemistry of biological pathways, 2nd edn. Roberts and Company, USA (2016)

de Meijere, A., Bräse, S., Oestreich, M.: Metal–Catalyzed Cross–Coupling Reactions and More. Wiley VCH, Weinheim (2013)

Meisch, H.U., Hoffmann, H., Reinle, W.: Vanadium catalysis in the nonenzymatic transamination of δ-aminolevulinic acid. Z. Naturforsch. C **33**(9–10), 623–628 (1978). https://doi.org/10.1515/znc-1978-9-1003

Mestdagh, H., Rolando, C., Sablier, M., Billy, N., Gouedard, G., Vigue, J.: Gas-phase reaction of iron carbonyl cations with atomic hydrogen and atomic nitrogen. J. Am. Chem. Soc. **114**(2), 771–773 (1992). https://doi.org/10.1021/ja00028a060

Methikkalam, R.R.J., Ghosh, J., Bhuin, R.G., Bag, S., Ragupathy, G., Pradeep, T.: Iron assisted formation of CO_2 over condensed CO and its relevance to interstellar chemistry. Phys. Chem. Chem. Phys. **22**, 8491–8498 (2020). https://doi.org/10.1039/C9CP06983F

Miessler, G.L., Fischer, P.J., Tarr, D.A.: Inorganic Chemistry. Pearson, London (2014)

Mingen, P.: Kinetic condensation of metals in the early solar system: unveiling the cooling history of solar nebula by refractory metal nuggets. Icarus **350**, 113851 (2020). https://doi.org/10.1016/j.icarus.2020.113851

Morowitz, H.J., Srinivasan, V., Smith, E.: Ligand Field Theory and the origin of Life as an emergent feature of the periodic table of elements. Bio. Bull. US **219**, 1–6 (2010). https://doi.org/10.1086/BBLv219n1p1

Morrison, S.M., Runyon, S.E., Hazen, R.: The paleomineralogy of the Hadean Eon revisited. Life **8**, 64 (2018). https://doi.org/10.3390/life8040064

Moss, G.P., Smith, P.A.S., Tavernier, D.: Glossary of class names of organic compounds and reactivity intermediates based on structure (IUPAC Recommendations 1995). Pure Appl. Chem. **67**(8–9), 1307–1375 (1995). https://doi.org/10.1351/pac199567081307

Muchowska, K.B., Varma, S.J., Chevallot-Beroux, E., Lethuillier-Karl, L., Li, G., Moran, J.: Metals promote sequences of the reverse Krebs cycle. Nat. Ecol. Evol. **1**, 1716–1721 (2017). https://doi.org/10.1038/s41559-017-0311-7

Muchowska, K.B., Varma, S.J., Moran, J.: Synthesis and breakdown of universal metabolic precursors promoted by iron. Nature **569**, 104–107 (2019). https://doi.org/10.1038/s41586-019-1151-1

Muchowska, K.B., Varma, S.J., Moran, J.: Nonenzymatic metabolic reactions and Life's origins. Chem. Rev. **120**(15), 7708–7744 (2020). https://doi.org/10.1021/acs.chemrev.0c00191

Nakajima, T., Yabushita, Y., Tabushi, I.: Amino acid synthesis through biogenetic-type CO_2 fixation. Nature **256**, 60–61 (1975). https://doi.org/10.1038/256060a0

Nielsen, F.H.: Modern Nutrition in Health and Disease, pp. 283–303. Williams and Wilkins, Philadelphia (1999)

Nomoto, K., Kobayashi, C., Tominaga, N.: Nucleosynthesis in stars and the chemical enrichment of galaxies. Ann. Revi. Astron. Astrophys. **51**(1), 457–509 (2013). https://doi.org/10.1146/annurev-astro-082812-140956

Noun, M., Baklouti, D., Brunetto, R., Borondics, F., Calligaro, T., Dionnet, Z., d'Hendecourt L., L.S., Nsouli, B., Ribaud, I., Roumie, M., Della-Negra, S.: A mineralogical context for the organic matter in the Paris meteorite determined by a multi-technique analysis. Life **9**, 44 (2019). https://doi.org/10.3390/life9020044

Pasek M.A., Dworkin, J.P., Lauretta, D.S.: A radical pathway for organic phosphorylation during schreibersite corrosion with implications for the origin of life. Geochimi. Cosmochim. Acta **71**(7), 1721–1736 (2007). https://doi.org/https://doi.org/10.1016/j.gca.2006.12.018

Patel, B.H., Percivalle, C., Ritson, D.J., Duffy, C.D., Sutherland, J.D.: Common origins of RNA, protein and lipid precursors in a cyanosulfidic protometabolism. Nat. Chem. **7**, 301–307 (2015). https://doi.org/10.1038/nchem.2202

Pearce, B.K.D., Tupper, A.S., Pudritz, R.E., Higgs, P.G.: Constraining the time interval for the origin of Life on Earth. Astrobiology **18**(3), 343–364 (2018). https://doi.org/10.1089/ast.2017

Pearson, V.K., Sephton, M.A., Kearsley, A.T., Bland, P.A., Franchi, I.A., Gilmour, I.: Clay mineral-organic matter relationships in the early solar system. Meteorit. Planet. Sci. **37**(12), 1829–1833 (2002). https://doi.org/10.1111/j.1945-5100.2002.tb01166.x

Peretó, J.: Out of fuzzy chemistry: from prebiotic chemistry to metabolic networks. Chem. Soc. Rev. **41**, 5394–5403 (2012). https://doi.org/10.1039/C2CS35054H

Perry, R.S., Mcloughlin, N., Lynne, B.Y., Sephton, M.A., Oliver, J.D., Perry, C.C., Campbell, K., Engel, M.H., Farmer, J.D., Brasier, M.D., Staley, J.T.: Defining biominerals and organominerals: direct and indirect indicators of Life. Sediment. Geol. **201**(1), 157–179 (2007). https://doi.org/10.1016/j.sedgeo.2007.05.014

Peters, J.W., Cody, G., Russell, M., Ferry, J.G., Schoonen, M.A.A.: Evolution and life: surviving catastrophes and extremes on Earth and beyond. In: Prebiotic Organometallic Catalysis, 1538, p. 5566. Astrobiology Science Conference 2010, League City, Texas (2010)

Peters, J.W., Schut, G.J., Boyd, E.S., Mulder, D.W., Shepard, E.M., Broderick, J.B., King, P.W., Adams, M.W.: [FeFe]- and [NiFe]-hydrogenase diversity, mechanism, and maturation. Biochim. Biophys. Acta Mol. Cell Res. **1853**(6), 1350–1369 (2015). https://doi.org/10.1016/j.bbamcr.2014.11.021

Pirajno, F.: Hydrothermal Processes and Mineral Systems. Springer, Amsterdam (2009)

Pitsch, S., Eschenmoser, A., Gedulin, B., Hui, S., Arrhenius, G.: Mineral induced formation of sugar phosphates. Orig. Life Evol. Biosph. **25**, 297–334 (1995). https://doi.org/10.1007/BF01581773

Pizzarello, S., Shock, E.: The organic composition of carbonaceous meteorites: the evolutionary story ahead of biochemistry. Cold Spring Harb. Perspect. Biol. **2**, a002105 (2010). https://doi.org/10.1101/cshperspect.a002105

Potapov, A., Jäger, C., Henning, T.: Ice coverage of dust grains in cold astrophysical environments. Phys. Rev. Lett. **124**, 221103 (2020). https://doi.org/10.1103/PhysRevLett.124.221103

Prasad, S.S., Huntress W. T., J.: A model for gas phase chemistry in interstellar clouds: I. The basic model, library of chemical reactions, and chemistry among C, N, and O compounds. Astrophys. J. Suppl. **43**, 1–35 (1980). https://doi.org/10.1086/190665

Preiner, M., Igarashi, K., Muchowska, K.B., Yu, M., Varma, S.J., Kleinermanns, K., Nobu, M.K., Kamagata, Y., Tüysüz, H., Moran, J., Martin, W.F.: A hydrogen-dependent geochemical analogue of primordial carbon and energy metabolism. Nat. Ecol. Evo. **4**, 534–542 (2020). https://doi.org/10.1038/s41559-020-1125-6

Pross, A.: What is Life? How Chemistry Becomes Biology. Oxford University Press, Oxford (2012)

Pulliam, R.L., Savage, C., Agúndez, M., Cernicharo, J., Guélin, M., Ziurys, L.M.: Identification of KCN in IRC+10216: evidence for selective cyanide chemistry. Astrophys. J. Lett. **725**, L181 (2010). https://doi.org/10.1088/2041-8205/725/2/l181

Puzzarini, C.: Grand challenges in astrochemistry. Front. Astron. Space Sci. **7**, 19 (2020). https://doi.org/10.3389/fspas.2020.00019

R., K., Pitsch, S., Arrhenius, G.: Mineral induced formation of pentose-2,4-bisphosphates. Orig. Life Evol. Biosph. **29**(2), 139–152 (1999). https://doi.org/10.1023/a:1006540518676

R., S., Crestini, C., Costanzo, G., Negri, R., Di Mauro, E.: A possible prebiotic synthesis of purine, adenine, cytosine, and 4(3H)-pyrimidinone from formamide: implications for the origin of Life. Bioorg. Med. Chem. **9**(5), 1249–1253 (2001). https://doi.org/10.1016/s0968-0896(00)00340-0

Ragsdale, S.W.: Enzymology of the Wood–Ljungdahl pathway of acetogenesis. Ann. N. Y. Acad. Sci. **1125**, 129–136 (2011). https://doi.org/10.1196/annals.1419.015

Ramaty, R., Kozlovsky, B.D., Lingenfelter, R.E., Reeves, H.: Light elements and cosmic rays in the early galaxy. Astrophys. J. **488**, 730–748 (1997). https://doi.org/10.1086/304744

Ricardo, A., Carrigan, M.A., Olcott, A.N., Benner, S.A.: Borate minerals stabilize ribose. Science **303**(5655), 196–196 (2004). https://doi.org/10.1126/science.1092464

Rich, A.: On the Problems of Evolution and Biochemical Information. Academic Press, New York (1962)

Rich, R.L.: Periodic Correlations. W. A. Benjamin, Reading (1965)

Ritson, D.J., Battilocchio, C., Ley, S.V., Sutherland, J.D.: Mimicking the surface and prebiotic chemistry of early Earth using flow chemistry. Nat. Commun. **9**, 1821 (2018). https://doi.org/10.1038/s41467-018-04147-2

Ritson, D.J., Sutherland, J.D.: Synthesis of aldehydic ribonucleotide and amino acid precursors by photoredox chemistry. Ang. Che. Int. Ed. **52**(22), 5845–5847 (2013). https://doi.org/10.1002/anie.201300321

Roberts, S.J., Szabla, R., Todd, Z.R., Stairs, S., Bučar, D.K., Šponer, J., Sasselov, D.D., Powner, M.W.: Selective prebiotic conversion of pyrimidine and purine anhydronucleosides into Watson-Crick base-pairing *arabino*-furanosyl nucleosides in water. Nat. Commun. **9**, 4073 (2018). https://doi.org/10.1038/s41467-018-06374-z

Robertson, M.P., Joycek, G.F.: The origins of the RNA world. Cold Spring Harb. Perspect. Biol. **4**, a003608 (2012). https://doi.org/10.1101/cshperspect.a003608

Rose, W.K.: Advanced Stellar Astrophysics. Cambridge University Press, Cambridge (1998). https://doi.org/10.1017/CBO9781139171205

Ruf, A., D'Hendecourt, L., Schmitt-Kopplin, P.: Data-driven astrochemistry: one step further within the origin of life puzzle. Life **8**, 18 (2018). https://doi.org/10.3390/life8020018

Ruf, A., Kanawati, B., Hertkorn, N., Yin, Q.Z., Moritz, F., Harir, M., Lucio, M., Michalke, B., Wimpenny, J., Shilobreeva, S., Bronsky, B., Saraykin, V., Gabelica, Z., Gougeon, R.D., Quirico, E., Ralew, S., Jakubowski, T., Haack, H., Gonsior, M., Jenniskens, P., Hinman, N.W., Schmitt-Kopplin, P.: Previously unknown class of metalorganic compounds revealed in meteorites. Proc. Nat. Ac. Sci. U.S.A. **114**(11), 2819–2824 (2017). https://doi.org/10.1073/pnas.1616019114

Saito, M.A., Sigman, D.M., Morel, F.M.M.: The bioinorganic chemistry of the ancient ocean: the co-evolution of cyanobacterial metal requirements and biogeochemical cycles at the Archean Proterozoic boundary? Inorg. Chim. Acta **356**, 308–318 (2003). https://doi.org/0.1016/S0020-1693(03)00442-0

Saladino, R., Ciambecchini, U., Crestini, C., Costanzo, G., Negri, R., Di Mauro, E.: One-pot TiO_2-catalyzed synthesis of nucleic bases and acyclonucleosides from formamide: implications for the origin of Life. ChemBioChem **4**(6), 514–521 (2003). https://doi.org/10.1002/cbic.200300567

Saladino, R., Crestini, C., Ciambecchini, U., Ciciriello, F., Costanzo, G., Di Mauro, E.: Synthesis and degradation of nucleobases and nucleic acids by formamide in the presence of montmorillonites. ChemBioChem **5**(11), 1558–1566 (2004). https://doi.org/10.1002/cbic.200400119

Saladino, R., Crestini, C., Neri, V., Brucato, J.R., Colangeli, L., Ciciriello, F., Di Mauro, E., Costanzo, G.: Synthesis and degradation of nucleic acid components by formamide and cosmic dust analogues. ChemBioChem **6**(8), 1368–1374 (2005). https://doi.org/10.1002/cbic.200500035

Saladino, R., Neri, V., Crestini, C., Costanzo, G., Graciotti, M., Di Mauro, E.: Synthesis and degradation of nucleic acid components by formamide and iron sulfur minerals. J. Am. Chem. Soc. **130**(46), 15512–15518 (2008). https://doi.org/10.1021/ja804782e

Sandford, S.A., Nuevo, M., Bera, P.P., Lee, T.J.: Prebiotic astrochemistry and the formation of molecules of astrobiological interest in interstellar clouds and protostellar disks. Chem. Rev. **120**(11), 4616–4659 (2020). https://doi.org/10.1021/acs.chemrev.9b00560

Savin, D.W., Brickhouse, N.S., Cowan, J.J., Drake, R.P., Federman, S.R., Ferland, G.J., Frank, A., Gudipati, M.S., Haxton, W.C., Herbst, E., Profumo, S., Salama, F., Ziurys, L.M., Zweibel, E.G.: The impact of recent advances in laboratory astrophysics on our understanding of the cosmos. Rep. Prog. Phys. **75**(3), 036901 (2012). https://doi.org/10.1088/0034-4885/75/3/036901

Savin, D.W., Brickhouse, N.S., Cowan, J.J., Drake, R.P., Federman, S.R., Ferland, G.J., Frank, A., Gudipati, M.S., Haxton, W.C., Herbst, E., Profumo, S., Salama, F., Ziurys, L.M., Zweibel, E.G.: The impact of recent advances in laboratory astrophysics on our understanding of the cosmos. Rep. Prog. Phys. **75**, 036901 (2012). https://doi.org/10.1088/0034-4885/75/3/036901

Sawai, H.: Catalysis of internucleotide bond formation by divalent metal ions. J. Am. Chem. Soc. **98**(22), 7037–7039 (1976). https://doi.org/10.1021/ja00438a050

Sawai, H., Higa, K., Kuroda, K.: Synthesis of cyclic and acyclic oligocytidylates by uranyl ion catalyst in aqueous solution. J. Chem. Soc. Perkin Trans. **1**, 505–508 (1992). https://doi.org/10.1039/P19920000505

Sawai, H., Kuroda, K., Hojo, T.: Uranyl ion as a highly effective catalyst for internucleotide bond formation. Bull. Chem. Soc. Jpn. **62**(6), 2018–2023 (1989). https://doi.org/10.1246/bcsj.62.2018

Sawai, H., Orgel, L.E.: Oligonucleotide synthesis catalyzed by the zinc(2+) ion. J. Am. Chem. Soc. **97**(12), 3532–3533 (1975). https://doi.org/10.1021/ja00845a050

Sawai, H., Yamamoto, K.: Lanthanide ion as a catalyst for internucleotide bond formation. Bull. Chem. Soc. Japan **69**(6), 1701–1704 (1996). https://doi.org/10.1246/bcsj.69.1701

Schilter, D., Camara, J.M., Huynh, M.T., Hammes-Schiffer, S., Rauchfuss, T.B.: Hydrogenase enzymes and their synthetic models: the role of metal hydrides. Chem. Rev. **116**(15), 8693–8749 (2016). https://doi.org/10.1021/acs.chemrev.6b00180

Schmitt-Kopplina, P., Gabelicab, Z., Gougeonc, R.D., Feketea, A., Kanawatia, B., Harira, M., Gebefuegia, I., Eckeld, G., Hertkorna, N.: High molecular diversity of extraterrestrial organic matter in Murchison meteorite revealed 40 years after its fall. Proc. Nat. Acad. Sci. U.S.A. **107**, 2763–2768 (2010). https://doi.org/10.1073/pnas.0912157107

Schöier, F.L., Maercker, M., Justtanont, K., Olofsson, H., Black, J.H., Decin, L., deKoter, A., Waters, R.: A chemical inventory of the S-type AGB star X Cygni based on Herschel/HIFI observations of circumstellar line emission. Astron. Astrophys. **530**, A83 (2011). https://doi.org/10.1051/0004-6361/201116597

Schwander, D., Berg, T., Schönhense, G., Ott, U.: Condensation of refractory metals in asymptotic giant branch and other stellar environments. Astrophys. J. **793**(1), 20 (2014). https://doi.org/10.1088/0004-637x/793/1/20

Schwander, D., Berg, T., Schönhense, G., Ott, U.: Condensation of refractory metals in asymptotic giant branch and other stellar environments. Astrophys. J. **793**, 20 (2014). https://doi.org/10.1088/0004-637X/793/1/20

Schwarz, G., Mendel, R., Ribbe, M.: Molybdenum cofactors, enzymes and pathways. Nature **460**, 839–847 (2009). https://doi.org/10.1038/nature08302

Seeger, P.A., Fowler, W.A., Clayton, D.D.: Nucleosynthesis of heavy elements by neutron capture. Astrophys. J. Suppl. **11**, 121–166 (1965). https://doi.org/10.1086/190111

Seoane, L.F., Solé, R.V.: Information theory, predictability and the emergence of complex Life. R. Soc. Open Sci. **5**(2), 172221 (2018). https://doi.org/10.1098/rsos.172221

Serra, G., Chaudret, B., Saillard, Y., Le Beuze, A., Rabaa, H., Ristorcelli, I., Klotz, A.: Organometallic chemistry in the interstellar medium. I—are organometallic reactions efficient processes in astrochemistry? Astron. Astrophys. **260**, 489–493 (1992)

Sheridan, P.M., Brewster, M.A., Ziurys, L.M.: Rotational rest frequencies for CrO and CrN. Astrophys. J. **576**, 1108–1114 (2002). https://doi.org/10.1086/341813

Sheridan, P.M., Ziurys, L.M., Hirano, T.: Rotational rest frequencies for FeN (X2Δi) and revised spectroscopic constants for FeC (X3Δi). Astrophys. J. **593**, L141–L144 (2003). https://doi.org/10.1086/378177

Shock, E., Canovas, P.: The potential for abiotic organic synthesis and biosynthesis at seafloor hydrothermal systems. Geofluids **10**(1–2), 161–192 (2010). https://doi.org/10.1111/j.1468-8123.2010.00277.x

Sleeper, H., Orgel, L.: The catalysis of nucleotide polymerization by compounds of divalent lead. J. Mol. Evol. **12**, 357–364 (1979). https://doi.org/10.1007/BF01732030

Smith, E., Morowitz, H.: The Origin and Nature of Life on Earth: The Emergence of the Fourth Geosphere. Cambridge University Press, Cambridge (2016)

Smith, E., Morowitz, H.J.: Universality in intermediary metabolism. Proc. Nat. Ac. Sci. U.S.A. **101**(36), 13168–13173 (2004). https://doi.org/10.1073/pnas.0404922101

Smith, K.E., House, C.H., Jr., R.D.A., Dworkin, J.P., Callahan, M.P.: Organometallic compounds as carriers of extraterrestrial cyanide in primitive meteorites. Nat. Commun. **10**, 2777 (2019). https://doi.org/10.1038/s41467-019-10866-x

Smith, K.T.: Molecules in interstellar space. Science **363**(6428), 704–705 (2019). https://doi.org/10.1126/science.363.6428.704-g

Snow, T.P., Rachford, B.L., Figoski, L.: Gas-phase iron abundances and depletions in translucent interstellar lines of sight from far UV spectroscopic Explorer Observations of FeII Lines. Astrophys. J. **573**(2), 662–669 (2002). https://doi.org/10.1086/340661

Sousa, F.L., Martin, W.F.: Biochemical fossils of the ancient transition from geoenergetics to bioenergetics in prokaryotic one carbon compound metabolism. Biochim. Biophys. Acta (BBA)-Bioenergetics **1837**(7), 964–981 (2014). https://doi.org/10.1016/j.bbabio.2014.02.001

Springsteen, G., Yerabolu, J.R., Nelson, J., Rhea, C.J., Krishnamurthy, R.: Linked cycles of oxidative decarboxylation of glyoxylate as protometabolic analogs of the citric acid cycle. Nat. Commun. **9**, 91 (2018). https://doi.org/10.1038/s41467-017-02591-0

Stairs, S., Nikmal, A., Bučar, D.K., Zheng, S.L., Szostak, J.W., Powner, M.W.: Divergent prebiotic synthesis of pyrimidine and 8-oxo-purine ribonucleotides. Nat. Commun. **8**, 15270 (2017). https://doi.org/10.1038/ncomms15270

Stockbridge, R.B., Lewis, C.A., Yuan, Y., Wolfenden, R.: Impact of temperature on the time required for the establishment of primordial biochemistry, and for the evolution of enzymes. Proc. Nat. Ac. Sci. U.S.A. **107**(51), 22102–22105 (2010). https://doi.org/10.1073/pnas.1013647107

Stritzinger, M., Leibundgut, B., Walch, S., Contardo, G.: Constraints on the progenitor systems of type Ia supernovae. Astron. Astrophys. **450**, 241–251 (2006). https://doi.org/10.1051/0004-6361:20053652

Sutherland, J.D.: The origin of Life–out of the blue. Angew. Chem. Int. Ed. **55**(1), 104–121 (2016). https://doi.org/10.1002/anie.201506585

Szczepanski, J., Wang, H., Tielens, A.G.G.M., Eyler, J., Oomens, J.: Infrared spectroscopy of gas-phase complexes of Fe^{+} and PAH molecules. Astrophys. J. **646**, 666–680 (2006). https://doi.org/10.1086/504867

Tarakeshwar, P., Buseck, P.R., Timmes, F.X.: On the structure, magnetic properties, and infrared spectra of iron pseudocarbynes in the interstellar medium. Astrophys. J. **879**(1), 2 (2019). https://doi.org/10.3847/1538-4357/ab22b7

Tenenbaum, E.D., Dodd, J.L., Milam, S.N., Woolf, N.J., Ziurys, L.M.: Arizona radio observatory 1 mm spectral survey of IRC+10216 and VY Canis Majoris (215–285 GHz). Astrophys. J., Suppl. Ser. **190**, 348–417 (2010a). https://doi.org/10.1088/0067-0049/190/2/348

Tenenbaum, E.D., Dodd, J.L., Milam, S.N., Woolf, N.J., Ziurys, L.M.: Comparative spectra of oxygen-rich versus carbon-rich circumstellar shells: VY Canis Majoris and IRC+10216 at 215–285 GHz. Astrophys. J. Lett. **720**, L102 (2010b). https://doi.org/10.1088/2041-8205/720/1/l102

Tenenbaum, E.D., Ziurys, L.M.: Exotic metal molecules in oxygen-rich envelopes: detection of AlOH ($X1\Sigma+$) in VY Canis Majoris. Astrophys. J. Lett. **712**, L93 (2010). https://doi.org/10.1088/2041-8205/712/1/L93

Thiabaud, A., Marboeuf, U., Alibert, J., Cabral, N., Leya, I., Mezger, K.: From stellar nebula to planets: the refractory components. Astron. Astrophys. **562**, A27 (2014). https://doi.org/10.1051/0004-6361/201322208

Thiabaud, Amaury, Marboeuf, Ulysse, Alibert, Yann, Cabral, Nahuel, Leya, Ingo, Mezger, Klaus: from stellar nebula to planets: the refractory components. Astronom. Astrophys. **562**, A27 (2014). https://doi.org/10.1051/0004-6361/201322208

Tielens, A.G.G.M.: Molecular and chemical processes and interactions, atomic molecular chemical and grain processes. Rev. Mod. Phys. **85**, 1021–1081 (2013). https://doi.org/10.1103/RevModPhys.85.1021

Tsuchida, R.: Absorption spectra of co-ordination compounds. I. Bull. Chem. Soc. Jpn **13**(5), 388–400 (1938). https://doi.org/10.1246/bcsj.13.388

van de Loosdrecht, J., Botes, F., Ciobica, I., Ferreira, A., Gibson, P., Moodley, D., Saib, A., Visagie, J., Weststrate, C., Niemantsverdriet, J.: Fischer–Tropsch synthesis: catalysts and chemistry. In: J. Reedijk, K. Poeppelmeier (eds.) Comprehensive Inorganic Chemistry II (Second Edition), 2nd edn., pp. 525–557. Elsevier, Amsterdam (2013)

Varma, S.J., Muchowska, K.B., Chatelain, P., Moran, J.N.: Iron reduces CO_2 to intermediates and end-products of the Acetyl-CoA pathway. Nat. Ecol. Evol. **2**, 1019–1024 (2018). https://doi.org/10.1038/s41559-018-0542-2

Vogiatzis, K.D., Polynski, M.V., Kirkland, J.K., Townsend, J., Hashemi, A., Liu, C., Pidko, E.A.: Computational approach to molecular catalysis by 3d transition metals: challenges and opportunities. Chem. Rev. **119**(4), 2453–2523 (2019). https://doi.org/10.1021/acs.chemrev.8b00361

Volbeda, A., Garcin, E., Piras, C., de Lacey, A.L., Fernandez, V.M., Hatchikian, E.C., Frey, M., Fontecilla-Camps, J.C.: Structure of the [NiFe] hydrogenase active site: evidence for biologically uncommon Fe ligands. J. Am. Chem. Soc. **118**(51), 12989–12996 (1996). https://doi.org/10.1021/ja962270g

Wächtershäuser, G.: Evolution of the first metabolic cycles. Proc. Nat. Ac. Sci. U.S.A. **87**(1), 200–204 (1990). https://doi.org/10.1073/pnas.87.1.200

Wilkins, R.G.: Kinetics and Mechanism of Reactions of Transition Metal Complexes. Wiley VCH, Weinheim (2002). https://doi.org/10.1002/3527600825

Wood, B.J., Smythe, D.J., Harrison, T.: The condensation temperatures of the elements: a reappraisal. Am. Mineral. **104**(6), 844–856 (2019). https://doi.org/10.2138/am-2019-6852CCBY

Yadav, M., Kumar, R., Krishnamurthy, R.: Chemistry of abiotic nucleotide synthesis. Chem. Rev. **120**(11), 4766–4805 (2020). https://doi.org/10.1021/acs.chemrev.9b00546

Yesiltas, M., Kebukawa, Y.: Associations of organic matter with minerals in Tagish lake meteorite via high spatial resolution synchrotron-based FTIR microspectroscopy. Meteorit. Planet. Sci. **51**(3), 584–595 (2016). https://doi.org/10.1111/maps.12609

Yi, R., Hongo, Y., Fahrenbach, A.C.: Synthesis of imidazole-activated ribonucleotides using cyanogen chloride. Chem. Commun. **54**, 511–514 (2018). https://doi.org/10.1039/C7CC08489G

Zack, L.N., Halfen, D.T., Ziurys, L.M.: Detection of FeCN (X^4 Δ_i) in IRC+10216: a new interstellar molecule. Astrophys. J. **733**(2), L36 (2011). https://doi.org/10.1088/2041-8205/733/2/l36

Ziegler, E.W., Kim, H.J., Benner, S.A.: Molybdenum(VI)-catalyzed rearrangement of prebiotic carbohydrates in formamide, a candidate prebiotic solvent. Astrobiology **18**(9), 1159–1170 (2018). https://doi.org/doi.org/10.1089/ast.2017.1742

Chapter 2
Mineralogical Environments of the Hadean Eon: Rare Elements Were Ubiquitous in Surface Sites of Rock-Forming Minerals

Rare Elements Ubiquitous on Rock-Forming Mineral Surfaces in the Hadean

Robert M. Hazen and Shaunna M. Morrison

Abstract Surfaces of condensed inorganic phases, including minerals and volcanic glass, are often invoked in origins-of-life models as potential catalysts, templates, reactants, or protective environments for critical prebiotic reactions. Consequently, mineralogists have compiled inventories of early Earth minerals and their reactive surface sites. These lists of probable near-surface minerals present during Earth's Hadean Eon (>4.0 Ga) collate more than 400 species, including phases in meteorites, the igneous minerals of Earth's earliest differentiated crust, magma and fumarole (hot vapor deposited) minerals of volcanic deposits, and a host of minerals that formed through interactions with Earth's emerging hydrosphere. These varied minerals incorporate more than two dozen chemical elements that are critical to biochemistry. However, minerals with high concentrations of several elements thought to be essential for the chemistry of life's origins and/or subsequent metabolism, including B, P, V, Co, Ni, Cu, Zn, and Mo, were volumetrically trivial, if not completely absent, from that Hadean mineral inventory. Consequently, it may not be reasonable to invoke minerals such as borates or molybdates in origins-of-life scenarios. We propose an alternative possibility: common rock-forming minerals with modest concentrations (typically 10s to 1000s of parts-per-million levels) of trace elements may have been critical to origins-of-life chemistry. Accordingly, to complement lists of Hadean mineral species we catalog trace and minor elements in such ubiquitous minerals as olivine, high-Ca pyroxene, plagioclase feldspars, and magnetite, as well as volcanic glass. We conclude that varied reactive surface sites

---

R. M. Hazen (✉) · S. M. Morrison
Earth and Planets Laboratory, Carnegie Institution for Science, Washington, DC, USA
e-mail: rhazen@carnegiescience.edu

A. Neubeck, S. McMahon (eds.), *Prebiotic Chemistry and the Origin of Life*, Advances in Astrobiology and Biogeophysics,
https://doi.org/10.1007/978-3-030-81039-9_2

with all elements essential to life's origins were abundant in Hadean geochemical environments.

2.1 Introduction

The origin of life was a geochemical process, dependent on a wide range of chemical building blocks. While the first lifeforms probably did not require all of the 38 different chemical elements employed by the diversity of life on Earth today (Fraústo da Silva and Williams 2001), it likely relied on several trace and minor elements, including B, P, Mo, and a number of first-row transition elements that played idiosyncratic roles in biogenesis (e.g., Benner and Hutter 2002; Ricardo et al. 2004; Kim et al. 2016; Ziegler et al. 2018). To some extent, biological uses of minor elements have changed through deep time—an important facet of the co-evolving geosphere and biosphere, presumably in response to variations in bioavailability of those elements (Lyons et al. 2015; Jelen et al. 2016; Moore et al. 2017). Nevertheless, a number of relatively rare elements must have played critical chemical roles in life's origins.

In this context, minerals and their chemically diverse surfaces have long been invoked as critical players in origins-of-life chemistry in such varied roles as catalysts for biomolecular production (Wächtershäuser 1988a; Cody 2004); templates for assembly of molecular components, especially polymers (Ferris et al. 1996; Hill et al. 1998; Orgel 1998); surfaces for selective adsorption, notably chiral selection (Hazen and Sholl 2003; Hazen 2006); and protection of those adsorbed molecules in otherwise destructive environments (Parsons et al. 1998; Smith 1998). Since Bernal (1951) and Goldschmidt (1952) speculated on such varied roles of common rock-forming minerals in the origins of life, scores of researchers have explored fundamental principles, as well as idiosyncratic scenarios, for mineral-mediated biogenesis (for example, see references in Ferris et al. 1996; Lahav 1999; Hazen 2005; Cleaves et al. 2012; Deamer 2011).

Several rock-forming mineral groups have become central to hypotheses related to the origin of life. Clay minerals, notably smectite group species such as montmorillonite (Hazen et al. 2013a), have been widely invoked in origins scenarios (Lahav et al. 1978; Cairns-Smith and Hartman 1986; Cairns-Smith 1977, 1982, 2005; Ferris and Ertem 1992, 1993; Ferris 1993, 2005; Heinen and Lauwers 1996; Ertem and Ferris 1996, 1997; Hanczyc et al. 2003). The catalytic potential of transition metal sulfides with Fe, Ni, Co, Zn, and/or Cu have also been subject to significant experimental and theoretical focus (Wächtershäuser 1988a, b, 1990a, b, 1993; Blöchl et al. 1992; Russell et al. 1994; Russell and Hall 1997; Huber and Wächtershäuser 1997, 1998; Brandes et al. 1998; Bebié and Schoonen 2000; Cody et al. 2000, 2001, 2004; Huber et al. 2003, 2012; Cody 2004, 2005; Mulkidjanian 2009; Mulkidjanian and Galperin 2009). Among the other common minerals and mineral groups considered in origins scenarios are quartz (SiO_2; Bonner et al. 1974, 1975; Soai et al. 1999; Evgenii and Wolfram 2000), feldspar group (Parsons et al.

1998; Smith 1998), zeolite group (Smith 1998; Smith et al. 1999), olivine group (Berndt et al. 1996; McCollom and Seewald 2001), mica group (Hansma 2010), hydroxides (Holm et al. 1993; Weber 1995; Pitsch et al. 1995), hydroxylapatite ($Ca_5(PO_4)_3OH$; Weber 1982; Acevedo and Orgel 1986; Benner and Hutter 2002), carbonates (Hazen et al. 2001; Kim et al. 2013), and rutile (TiO_2; Summers and Chang 1990; Jonsson et al. 2009). In addition, less common minerals have received consideration for specific prebiotic roles: e.g., ferrous metals alloys, primarily the meteoritic Fe-Ni alloys kamacite and taenite, as catalysts for organic synthesis (Smirnov et al. 2008); phosphides, notably meteoritic schreibersite [$(Fe,Ni)_3P$], as a reactive source of P (Pasek and Lauretta 2005; Pasek et al. 2007; Pasek 2008); borates from evaporite deposits, such as kernite [$Na_2B_4O_6(OH)_2 \cdot 3H_2O$] or colemanite [$CaB_3O_4(OH)_3 \cdot H_2O$], for specific steps in RNA synthesis (Ricardo et al. 2004; Grew et al. 2011; Kim et al. 2016); and molybdates, such as powellite ($CaMoO_4$), for catalysis of conformational changes of carbohydrates (Ziegler et al. 2018).

The reliance of origins-of-life models on this diversity of minerals presupposes that these species were readily available on the Hadean Earth. Nevertheless, few authors have attempted to document Earth's Hadean mineral inventory (see, however, Schoonen et al. 2004; Papineau 2010; Hazen 2013; Morrison et al. 2018). A common theme of these contributions is the assumption that if a mineral species was rare or lacking, then it is unlikely to have played significant roles in the origins of life. In this chapter we consider the possible contributions of minerals from a slightly different perspective—that of the ubiquitous trace and minor elements in common rock-forming minerals that offer a remarkable diversity of potentially reactive surface sites (Morrison et al. 2018). First, however, we revisit the question of what minerals were likely present on Earth at the time of life's origins.

2.2 An Evolutionary System of Mineralogy

"Mineral evolution" is defined as the study of the emerging diversity and distribution of minerals in the context of deep time (Zhabin 1981; Hazen et al. 2008, 2011; Hazen and Ferry 2010; Hazen 2012). We contend that the condensed phases that form planets and moons display congruent complexification of the mineral realm as new physical, chemical, and (in the case of Earth) biological processes create new mineral-forming environments of temperature, pressure, and composition. Initially, contributions to mineral evolution focused on qualitative overviews of individual elements, including U (Hazen et al. 2009), B (Grew et al. 2011), Hg (Hazen et al. 2012), C (Hazen et al. 2013b), Mo (Golden et al. 2013), Be (Grew and Hazen 2014), and Li (Grew et al. 2019).

More recently, we have attempted to place mineral evolution on a more rigorous quantitative footing by applying methods of data analysis and visualization, including network analysis, cluster analysis, affinity analysis, and models of the statistical distributions of common versus rare species (Hazen 2014, 2017; Hazen et al. 2015, 2017a, 2017b; Hystad et al. 2015, 2019; Krivovichev et al. 2017; Liu et al. 2017a, 2018; Morrison et al. 2017, 2020; Moore et al. 2018; Hazen et al. 2019). We have also introduced an "Evolutionary System" of mineral classification (Hazen 2019), based on the concept of historical natural kinds (Cleland 2011; Ereshefsky 2014; Godman 2019; Cleland et al. 2020) and the application of cluster analysis to identifying mineral historical natural kinds (e.g., Everitt 2011).

This effort, which builds on and complements the official mineral classification protocols of the International Mineralogical Association's Commission on New Minerals, Nomenclature and Classification (IMA-CNMNC; e.g., Burke 2006; Mills et al. 2009; Schertl et al. 2018), is being published sequentially as chronological parts, the first five of which cover pre-terrestrial minerals, most of which are preserved in meteorites. Part I cataloged stellar minerals (literally "stardust") that predate our solar nebula, i.e., prior to 4.567 Ga (Hazen and Morrison 2020). Subsequently, in Part II we explored primary interstellar and nebular condensates commencing ~4.567 Ga (Morrison and Hazen 2020), while the primary mineralogy of chondrules from ~4.566 to 4.561 Ga was the focus of Part III (Hazen et al. 2021). Part IV summarized the primary asteroidal mineralogy of non-chondritic meteorites from ~4.566 to 4.560 Ga, as well as high-pressure impact mineralogy preserved in meteorites (Morrison and Hazen 2021), and Part V reviews the diverse secondary meteorite minerals formed by aqueous alteration and/or thermal metamorphism (Hazen and Morrison 2021). All told, these contributions catalogued more than 290 different pre-terrestrial minerals, including a number of as yet unnamed crystalline and amorphous solids not approved by the IMA-CNMNC.

When considering paleomineralogy of the Hadean Eon (prior to 4 billion years ago), it is important to evaluate both meteorite minerals and those that arose during the earliest stages of Earth's crustal evolution (Hazen 2013; Morrison et al. 2018). Today, the IMA-CNMNC has approved more than 5700 mineral "species," each of which represents a unique combination of end-member composition and idealized crystal structure (rruff.info/ima; accessed 19 November 2020). Of these varied minerals, the great majority postdate the origins of life—indeed, an estimated 50% of known mineral species are directly or indirectly the consequence of biological modifications to the near-surface environment (Hazen et al. 2008, 2017b; Hazen 2010; Sverjensky and Lee 2010; Hazen and Morrison 2022; Hazen and Papineau 2012). Furthermore, minerals incorporating a number of rare chemical elements as essential constituents (i.e., as an integral part of the idealized chemical formula), including Li, B, Be, Hg, Se, As, Bi, Sb, U, and Th, as well as many others, may have required a billion years or more of repeated fluid-rock interactions to achieve sufficient concentration to form macroscopic minerals (Hazen et al. 2009, 2012; Grew et al. 2011, 2019; Grew and Hazen 2014). Consequently, the majority of minerals—possibly more than 4000 of the 5700 known species today—were not available to contribute to the origins of life.

Nevertheless, prebiotic physical and chemical processes at or near Earth's surface significantly increased the diversity of minerals beyond the meteorite inventory. At least three formation mechanisms played important roles in the new terrestrial minerals that likely emerged during the Hadean Eon: (1) igneous activity that produced minerals in Earth's earliest differentiated crust; (2) volcanic eruptions, especially those associated with tectonic activity, that generated magma and fumaroles (vents of hot vapor) that formed mineral deposits; and (3) interactions with Earth's emerging hydrosphere, which generated a host of new H-bearing minerals. These three processes are considered below.

Igneous Processes of the Differentiated Crust On Earth, mineralogical diversity increased by the sequential evolution of igneous rocks, including the generation of basaltic magmas from partially molten peridotite and the subsequent generation of granitic melts by partial melting of hydrated basalt (e.g., Bowen 1928; Yoder 1976; London 2008). Many primary igneous phases, notably members of the olivine, pyroxene, and feldspar groups, had already been produced abundantly in differentiated planetesimals and brought to Earth in achondrite meteorites (Morrison and Hazen 2021). However, new mineralogical richness emerged on Earth through the production of hydrous igneous silicates, including diverse amphibole group (e.g., actinolite, cummingtonite, edenite, hornblende, tremolite), mica group (annite, muscovite, phlogopite), and other "biopyribole" minerals (Deer et al. 1962; Veblen and Buseck 1979).

A critical mineralogical innovation on Earth was the appearance of silica-rich igneous rocks, collectively called "granite." These buoyant lithologies [by virtue of their relatively high Si/(Mg + Fe) compared to basalt] were essential to the formation of the first extensive landmasses and, ultimately, continents. Granite arises from partial melting of wet basaltic crust—a process that could only occur on a water-rich planet like Earth that was hot enough (i.e., large enough) to partially re-melt basalt. That partial melt was not only rich in Si, but it also concentrated many "incompatible" elements such as Be, Li, Cs, Y, Nb, Sn, Ta, Th, and U. As primary granite minerals such as quartz, feldspar, and mica crystallized, the remaining eutectic fluid became ever more enriched in these and other rare elements, which eventually reached concentrations high enough to form their own minerals. The result was hundreds of new "pegmatite" minerals (London 2008). However, the cyclic process of element concentration and mineralogical diversification must have required significant time; London (2008) estimates that a billion years elapsed before the generation of the earliest "complex" pegmatites, which hosted many of the rare species that are unique to these remarkable deposits.

Volcanism and Hadean Tectonics Mountain building and other consequences of tectonic activity must have played important roles in the diversification of Hadean minerals. However, the mechanisms and rates of tectonic processes and associated volcanism were likely different from modern plate tectonics (e.g., Shirey and Richardson 2011; Van Kranendonk 2019). In particular, Hadean volcanism may have been driven more by vertical processes of rising magma plumes, as opposed to the lateral motions of divergence, convergence, and subduction associated with plate tectonics. Consequently, the question of when plate tectonics began (and, by association, the rate and extent of Hadean continental crust formation) remains a matter of significant debate.

If Hadean volcanism was primarily a consequence of plume-driven vertical tectonics, as suggested by some field studies on Earth's oldest rocks (Van Kranendonk 2019), then volcanic rocks were likely associated with S-rich fluids that generated a significant number of new phases, notably transition metal sulfides. Furthermore, if volcanoes breached the ocean's surface then a wide variety of vapor-deposited fumarolic minerals incorporating uncommon elements such B, As, V, Se, Sb, Hg, Bi, Tl, and Pb, may have appeared for the first time. As today, many of those minerals would have been ephemeral, for example sublimating under sunlight or dissolving after each significant rain storm.

On the other hand, if modern-style subduction and associated arc volcanism were established in the Hadean Eon, then it is likely that hundreds of new mineral species appeared, for example those found in massive sulfide deposits that contain more than 100 exotic "sulfosalt" minerals (metal sulfide minerals with one or more other chalcogenide elements, for example, Se, As, Sb, or Bi). However, recent studies suggest that subduction-driven plate tectonics, and associated continent formation and arc volcanism, did not become a significant process on Earth until approximately 3 billion years ago (Shirey and Richardson 2011). Therefore, the diversity of minerals associated with volcanic activity may have been significantly less in the Hadean Eon than today.

Earth's Hydrosphere Surface weathering and aqueous activity must have led to numerous additional new hydrous phases, including hydroxide, clay, and zeolite group minerals. Of special note are the varied layer silicates with sub-micrometer-scale grains collectively called clay minerals. These minerals, especially from the smectite group of "expandable clays" that can reversibly incorporate water, alkali, and/or alkaline earth cations between rigid structural silicate layers, have long been employed in origins-of-life models because they possess exceptional high surface areas (typically $>10^6$ cm^2/cm^3; Hazen 2017) and because some varieties have the ability to absorb organic molecules. However, the nomenclature and identities of clay minerals often lead to confusion, as there are more than 50 distinct approved species, as well as many mixed phase assemblages that are sometimes also characterized as "clays" (e.g., bauxite, bentonite, limonite, steatite). Clay minerals have been subdivided into seven groups, which are characterized by: (1) their particular layered atomic structure, (2) their ranges of interlayer constituents, and (3) their ability to expand in an aqueous environment (Deer et al. 1962; Hazen et al. 2013a).

Hazen et al. (2013a) enumerated five common mechanisms by which clay minerals form: (1) alteration of prior silicates in a subsurface aqueous or hydrothermal environment; (2) precipitation from solution, notably in marine sediments ("authigenesis"); (3) transformation by low-grade metamorphism at depth in the crust, with subsequent exposure by uplift related to plate tectonics; (4) weathering reactions of surface rocks, especially under oxic and/or acidic conditions, which is a process that often produces "soil"; and (5) the advent of soil-forming microbes, fungi, and plant roots and associated biological weathering following the rise of the terrestrial biosphere. Of these five clay-forming mechanisms, only the first (through the process of "serpentinization"; Schrenk et al. 2013) and the fourth (anoxic near-surface weathering) could have played significant roles before about 3.5 billion years ago. Prior to that time, the second mechanism (authigenesis) would have been limited by a relative paucity of Hadean and Paleoarchean marine sediments. Clay minerals formed by the third mechanism, low-grade metamorphism, would have rarely been exposed at Earth's surface prior to plate tectonics. Furthermore, before the origins of life neither oxic weathering (mechanism 4) or biotic terrestrial soil formation (mechanism 5) could have occurred. Therefore, most clay minerals found in modern sediments were not present during the Hadean Eon. Indeed, the only volumetrically major clay mineral species prior to life's origins were probably members of the serpentine group (e.g., chrysotile and greenalite) and smectite group (saponite and ferrosaponite) as a consequence of the alteration of olivine, pyroxene, and other ferromagnesian silicates in the abundant mafic/ultramafic rocks of the early crust (Arndt and Nisbet 1982). Aqueous alteration of Ca-Al silicates, notably the feldspar anorthite ($CaAl_2Si_2O_8$), undoubtedly resulted in production of some aluminous clay minerals, including montmorillonite, kaolinite, and halloysite, with the likely generation of mixed clay assemblages. In addition, significant clay mineral production, notably the smectite group minerals montmorillonite and nontronite, was likely associated with weathering of volcanic rocks in their typical acidic surface environments.

Earth's dynamic hydrosphere also increased mineral diversity through the generation of at least three other major mineral groups. Hydroxides of magnesium, iron, and aluminum, such as brucite [$Mg(OH)_2$], goethite [$FeO(OH)$], and gibbsite [$Al(OH)_3$], would have been common weathering products. Zeolite group minerals, including chabazite [$(Na_3K)Al_4Si_8O_{24}\cdot 11H_2O$], stilbite [$NaCa_4(Al_9Si_{27})O_{72}\cdot 28H_2O$], and natrolite ($Na_2Si_3Al_2O_{10}\cdot 2H_2O$), would have formed, as they do today, when hydrothermal fluids altered porous basaltic lava or near-surface basalt intrusive environments. Finally, the Hadean Earth may have hosted localized deposits of evaporite minerals, for example halite (NaCl) and sylvite (KCl). However, more exotic evaporite minerals such as borates must be preceded by three events, none of which are certain to have occurred during the Hadean Eon: (1) the formation of extensive arid landmasses, (2) the incursion of inland seas of significant extent, and (3) stranding and evaporation of those extensive bodies of saline water to precipitate a variety of salts. Rare evaporite minerals such as borates only appear during the final stages of such a sequence (Grew et al. 2011).

Other Mechanisms of Mineralization Morrison et al. (2018) suggested that Earth's Hadean mineral inventory could have expanded by three additional localized phenomena.

1. Prior to widespread predation by cellular life, organic molecules such as amino acids, polycyclic aromatic hydrocarbons, or nucleobases could have become selected and concentrated to form crystals. An intriguing possibility in this context is local formation of chiral molecular solids.

2. Large bolide impacts could have generated new minerals, not only shock phases (included in the inventory of meteorite minerals; Morrison and Hazen 2021), but also associated with dynamic zones of hydrothermal mineralization in the circumferential fractured zones of impact sites. Such mineralization, which has led to a number of productive ore bodies on Earth today (Pirajno 2009), may have been more prevalent during the Hadean Eon.
3. A number of phenomena can produce significant local redox excursions. Such effects include lightning, notably volcanic lightning (Essene and Fisher 1986; Pasek and Block 2009; Wadsworth et al. 2017); photo-oxidation, for example of Fe^{2+} compounds to produce Fe^{3+} and hydrogen (Braterman et al. 1983; Borowska and Mauzerall 1987; Kim et al. 2013); crystal settling of Fe^{2+}-rich or other reduced minerals from a magma that consequently becomes progressively more oxidized (Wager and Brown 1968; Weinstein et al. 1988); or immiscibility in a magma, especially separation of an S-rich or C-rich fluid from a silicate magma (Veksler et al. 2012; Lester et al. 2013; Savelyev et al. 2018). The associated local redox excursions can lead to a variety of new phases, both oxidized and reduced.

In summary, many of the minerals considered in origins-of-life scenarios would have been present abundantly on the Hadean Earth. All of the major rock-forming minerals, including quartz, olivine, pyroxene, feldspar, mica and clay layer silicates, zeolites, hydroxides, carbonates, and sulfide groups were present in near-surface environments more than 4 billion years ago. However, other potentially critical phases, including those with B, P, Co, Cu, V, and Mo, were either absent or were present only as volumetrically minor phases. Furthermore, a number of these minerals, including borates, molybdates, and phosphides, would have occurred only in restricted, idiosyncratic localities—borates at a mid-content evaporite deposit, for example, or phosphides at an iron meteorite fall. An origin of life scenario requiring the juxtaposition of such rare and isolated minerals seems unlikely. However, there is an obvious, though long overlooked, solution to this dilemma.

2.3 Trace and Minor Elements in Common Rock-Forming Minerals

Minerals, though traditionally described with an idealized chemical formula such as Mg_2SiO_4 (the olivine group mineral forsterite), invariably contain a wealth of minor and trace elements. For example, an individual forsterite sample might be better represented by a complex chemical formula such as $[(Mg_{0.91}Fe_{0.04}Mn_{0.02}Ca_{0.01}Ni_{0.008}Co_{0.003}V_{0.001})_2(Si_{0.99}P_{0.01})O_4]$. Indeed, Earth's major crustal repositories for most rare elements are as trace elements in common rock-forming minerals. For example, while cobalt is an essential element in 70 different mineral species (rruff.info/ima; accessed 16 August 2020), the cumulative total mass of all known Co-minerals in Earth's ore deposits represents less than

Table 2.1 Trace element concentrations (parts-per-million) in common rock-forming minerals

Source	B	P	V	Co	Ni	Cu	Zn	Mo	References
Average crust	10	1050	120	25	84	60	70	1.2	1
Continental crust	17	1200	97	17	47	28	67	1.1	2
Ocean crust	1.8	1000	299	44	100	81	87	~1	3
Ocean basalt glass (average)	–	800	350	45	125	75	100	0.8	4
Ocean basalt glass (max)	–	2100	500	55	200	160	165	9	4
Olivine (average igneous)	–	1000	150	100	2000	15	50	–	5–9
Olivine (maximum)	10	4000	260	220	4800	60	330	–	5–10
Augite (average igneous)	70	90	250	75	150	10	70	–	9, 11, 12
Augite (maximum)	120	600	6000	250	5000	3100	110	–	9, 11
Plagioclase (average igneous)	2	500	–	–	10	10	5	–	13–15
Plagioclase (maximum)	15	1400	20	5	100	50	25	–	13–15
Magnetite (maximum)	–	–	4200	225	2100	1600	725	0.5	16, 17

References: 1 = Haynes (2016); 2 = Rudnick and Gao (2005); 3 = White and Klein (2014); 4 = Jenner and O'Neill (2012); 5 = Deer et al. (1982); 6 = Herzberg et al. (2016); 7 = Canil and Fedortchouk (2001); 8 = O'Reilly et al. (1997); 9 = Liu et al. (2014); 10 = Kent and Rossman (2002); 11 = Deer et al. (1997); 12 = Koritnig (1965); 13 = Storm and Holland (1957); 14 = Deer et al. (2001); 15 = Sun et al. (2017); 16 = Paster et al. (1974); 17 = Deer et al. (2011)

5×10^{-8} of the Co incorporated in forsterite and other minerals at an average concentration of 48 parts-per-million (ppm) in the basalt that forms most of Earth's ocean crust (Herzberg et al. 2016; Hazen et al. 2017a; Koritnig 1965; see Table 2.1). A similar situation applies to all but about a dozen of the 72 elements that are essential to the definition of one or more mineral species (i.e., H, C, O, Na, Mg, Al, Si, S, K, Ca, Ti, and Fe). In other words, idealized mineral formulas do not adequately represent the distribution and availability of most chemical elements and, therefore, do not accurately reflect the availability of trace and minor elements for catalyzing biochemical reactions related to life's origins.

Consider eight minor and trace elements that have been invoked in life's origins and/or biochemistry: B, P, V, Co, Ni, Cu, Zn, and Mo. Average crustal compositions, as well as measured concentrations of these elements in various crustal reservoirs, are tabulated in Table 2.1. Note that the average crustal composition of at least some of these key minor elements may have been significantly greater during the Hadean Eon. For example, Liu et al. (2017b; their Fig. 2) compiled ~80,000 bulk igneous rock analyses for samples of known age and found that the average Ni and Co contents were approximately twice the modern values. In spite of such secular variations, the observation that some minerals concentrate key trace and minor elements holds throughout geological time.

Boron Boron, notably in the form of evaporite borate minerals (Ricardo et al. 2004; Kim et al. 2016), has been invoked in key steps in biogenesis related to the assembly of RNA. However, boron-containing minerals, including borates and the borosilicate tourmaline group, were undoubtedly volumetrically trivial phases on the Hadean Earth. Nevertheless, boron was abundantly available on the Hadean Earth: the average crust contains 10 parts per million (ppm) B, with enhanced concentrations as high as 120 ppm in Al-bearing minerals such as the clinopyroxene augite [$Ca(Mg,Fe,Al)Si_2O_6$] and the feldspar plagioclase [$(Na,Ca)(Al,Si)_4O_8$]. Note that since at least the Paleoarchean Eon (~4 Ga), augite and plagioclase have been volumetrically the dominant minerals in ocean floor basalt (e.g., Rollinson 2007), which blankets ~70% of Earth's surface today.

Phosphorus Phosphorus is a minor element, accounting for ~1000 ppm in the crust. In this instance, the P-bearing mineral hydroxylapatite [$Ca(PO_4)_3OH$] is a common, if volumetrically minor, phase in basalt and other crustal igneous rocks. Less common, but potentially significant for origins-of-life scenarios, is the reactive iron-nickel phosphide mineral schreibersite (Pasek et al. 2007). However, P is also a ubiquitous and significant component of several common mineral phases, averaging 2100 ppm in basaltic glass from ocean floor volcanoes, and as much as 4000 ppm in some igneous olivine [$(Mg,Fe)_2SiO_4$] and 1400 ppm in plagioclase. Note, however, that P in plagioclase is often present as exsolved hydroxylapatite (Deer et al. 2001).

Vanadium The first-row transition element vanadium, which plays a role as an essential micronutrient, is present at ~100 ppm in Earth's crust. No vanadium minerals of any significance were present during Earth's Hadean Eon (e.g., Liu et al. 2018). However, V can be significantly concentrated in mineral phases with up to 6000 ppm in augite and 4200 ppm in the common iron oxide magnetite (Fe_3O_4).

Cobalt Cobalt is a scarce transitional metal (25 ppm) that plays important roles in industry and biochemistry (Hazen et al. 2017a; Moore et al. 2018). Cobalt minerals were rare and volumetrically insignificant during the Hadean Eon. Meteoritic iron can contain up to ~1 wt % Co (Brearley and Jones 1998), while secondary thermal processing can result in exsolution of the Co-rich alloy, wairauite (CoFe; Hua et al. 1995). Far more abundant is Co available as a trace element in the common rock-forming Mg-Fe minerals, including olivine (220 ppm), augite (225 ppm), and magnetite (225 ppm).

Nickel Nickel, at 84 ppm in the average crust, plays important biological roles as one of several transition metals in oxidoreductase enzymes that mediate electron transfer reactions. Indeed, nickel active centers have been proposed to occur in some of the most ancient metalloenzymes (Nitschke et al. 2013). With the exception of falls of meteorites with Fe-Ni alloys and other Ni-bearing phases, and localized concentrations of Ni-enrich iron sulfides at some seafloor hydrothermal vents, Ni tends to be dispersed in the crust as a minor/trace element. Highest concentrations occur in olivine (to 4800 ppm), augite (5000 ppm), and magnetite (2100 ppm). As with the other transition metals listed here, element partitioning often leads to concentration factors of 5 or more.

Copper At 60 ppm in the average crust, copper is a minor element that has nevertheless come to play a major role in biology in many metabolic enzymes. (However, to date it has received little attention in origins-of-life scenarios.) A number of Cu minerals, including native copper and a variety of copper sulfides, have been identified as extremely minor phases in meteorites (Hazen and Morrison 2021). In addition, it is plausible that chalcopyrite ($CuFeS_2$) was present during the Hadean Eon in seafloor hydrothermal systems. Nevertheless, Cu was far more widespread as a minor element in basaltic glass (average 160 ppm), augite (to 3100 ppm), and magnetite (1600 ppm).

Zinc Zinc plays many roles in biochemistry today, and it has been invoked in origins-of-life scenarios (see, however, Mulkidjanian 2009; Mulkidjanian and Galperin 2009). Minerals with essential zinc were probably scarce during the Hadean, possibly limited to highly localized grains of sphalerite (ZnS). Nevertheless, zinc was widely available as a minor constituent of basaltic glass (to 165 ppm), olivine (330 ppm), augite (110 ppm), and magnetite (725 ppm)—all significantly enhanced above the crustal average of 70 ppm.

Molybdenum At 1.2 ppm, molybdenum is a rare element in the average crust. In spite of that scarcity, it plays crucial biological roles as an electron carrier, for example in nitrogenase, which facilitates reduction of di-nitrogen to ammonium. Molybdenum minerals have also been invoked by Ziegler et al. (2018), who suggest that molybdates, such as powellite ($CaMoO_4$), catalyze critical conformational changes of carbohydrates. Here the difficulty of resorting to minerals with essential Mo is especially acute, as Mo minerals would have been vanishingly rare on the Hadean Earth. For example, decades of meticulous study of meteorites have resulted in the discovery of a few micron-scale grains of Mo-rich alloys and secondary molybdates. Reliance on such sparse, submicroscopic phases for essential steps in life's origins seems risky, at best. However, as with the other trace elements noted above, Mo is significantly enriched, to a maximum 9 ppm in basaltic glass, which blankets much of the ocean floor. Given that ocean floor basalt also holds phases with significant concentrations of B, P, V, Co, Ni, Cu, and Zn, we suggest that the innumerable potentially reactive Mo-bearing sites associated with seafloor volcanism represent the more likely loci for prebiotic reactions.

This observation has important implications for any origin-of-life model that invokes reactive mineral surface sites. If one requires a catalytic site with B, P, Co, Mo, or any other scarce element, such sites are many orders of magnitude more prevalent as dispersed "impurity" atoms at the surfaces of the commonest minerals compared with the more concentrated atoms of extremely rare, widely dispersed minerals. While detailed information on all of the trace and minor elements in rock-forming minerals is not yet widely available, enough data are published to establish general trends (Table 2.1).

2.4 Conclusions

Understanding the diversity and distribution of Hadean minerals provides an important environmental context for studies of life's geochemical origins. However, lists of plausible prebiotic mineral species can be misleading, especially when plausible prebiotic catalytic roles for relatively rare elements, such as boron, molybdenum, and phosphorus, are invoked. Minerals of these elements that are relatively rare on Earth today would have likely been significantly rarer, if present at all, during Earth's first 500 million years. Indeed, the total volumes of minerals with all but the most common 15 or 20 chemical elements would have been trivial in the global context. This circumstance has led to challenges to what otherwise appear to be elegant chemical solutions to key steps in life's origins (Grew et al. 2011).

The ubiquity of rare trace and minor elements in common rock-forming minerals provides a largely overlooked solution to this dilemma. The most common rock-forming minerals feature an abundance of potentially reactive surface sites of varied chemistry and geometry. Not only are there individual molybdate, phosphate, borate, or any other desired surface sites, but there are also vast numbers of sites where two or more atoms—boron and phosphorus, or calcium and manganese, for example—are juxtaposed in various geometries.

* * *

An unresolved debate in origins-of-life research is whether biogenesis is a rare event limited by finicky reactions and quirky chemistry or, rather, if life is a cosmic imperative (Monod 1970; Schoffeniels 1976; De Duve 1995; Davies 2016; Hazen 2017). For a time, mineralogical investigations of the Hadean Eon would seem to have favored the former camp, as many rare chemical elements and their exotic minerals would seem to have been volumetrically trivial, if not entirely absent, from Earth's earliest near-surface environments. However, the realization that the most common minerals host a rich diversity of reactive surface sites removes at least one impediment to hypotheses of a robust, deterministic prebiotic chemical milieu.

Acknowledgments Studies of mineral evolution and mineral ecology have been supported by the Alfred P. Sloan Foundation, the W. M. Keck Foundation, the John Templeton Foundation, the NASA Astrobiology Institute ENIGMA team, a private foundation, and the Carnegie Institution for Science. Any opinions, findings, or recommendations expressed herein are those of the authors and do not necessarily reflect the views of the National Aeronautics and Space Administration.

References

Acevedo, O.L., Orgel, L.E.: Template-directed oligonucleotide ligation on hydroxylapatite. Nature. **321**, 790–792 (1986)

Arndt, N.T., Nisbet, E.G. (eds.): Komatiites. George Allen & Unwin, London (1982)

Bebié, J., Schoonen, M.A.A.: Pyrite surface interaction with selected organic aqueous species under anoxic conditions. Geochem. Trans. **1**, 47 (2000)

Benner, S.A., Hutter, D.: Phosphates, DNA, and the search for nonterrean life: a second generation model for genetic molecules. Bioorg. Chem. **30**, 62–80 (2002)

Bernal, J.D.: The Physical Basis of Life. Routledge and Kegan Paul, London (1951)

Berndt, M.E., Allen, D.E., Seyfried Jr., W.E.: Reduction of CO_2 during serpentization of olivine at 300°C and 500 bar. Geology. **24**, 351–354 (1996)

Blöchl, E., Keller, M., Wächtershäuser, G., Stetter, K.O.: Reactions depending on iron sulfide and linking geochemistry with biochemistry. Proc. Natl. Acad. Sci. USA. **89**, 8117–8120 (1992)

Bonner, W.A., Kavasmaneck, P.R., Martin, F.S., Flores, J.J.: Asymmetric adsorption of alanine by quartz. Science. **186**, 143–144 (1974)

Bonner, W.A., Kavasmaneck, P.R., Martin, F.S., Flores, J.J.: Asymmetric adsorption by quartz: a model for the prebiotic origin of optical activity. Orig. Life. **6**, 367–376 (1975)

Borowska, Z.K., Mauzerall, D.C.: Efficient near ultraviolet light induced formation of hydrogen by ferrous hydroxide. Orig. Life Evol. Biosph. **17**, 251–259 (1987)

Bowen, N.L.: The Evolution of the Igneous Rocks. Princeton University Press, Princeton, NJ (1928)

Brandes, J.A., Boctor, N.Z., Cody, G.D., Cooper, B.A., Hazen, R.M., Yoder Jr., H.S.: Abiotic nitrogen reduction on the early earth. Nature. **395**, 365–367 (1998)

Braterman, P.S., Cairns-Smith, A.G., Sloper, R.W.: Photo-oxidation of hydrated Fe^{2+}—significance for banded iron formations. Nature. **303**, 163–164 (1983)

Brearley, A.J., Jones, R.H.: Chondritic meteorites. Rev. Mineral. **36**, 3.01–3.398 (1998)

Burke, E.A.J.: The end of CNMMN and CCM—long live the CNMNC! Elements. **2**, 388 (2006)

Cairns-Smith, A.G.: Takeover mechanisms and early biochemical evolution. Biosystems. **9**, 105–109 (1977)

Cairns-Smith, A.G.: Genetic Takeover and the Mineral Origins of Life. Cambridge University Press, U.K. (1982)

Cairns-Smith, A.G.: Sketches for a mineral genetic system. Elements. **1**, 157–161 (2005)

Cairns-Smith, A.G., Hartman, H.: Clay Minerals and the Origin of Life. Cambridge University Press, U.K. (1986)

Canil, D., Fedortchouk, Y.: Olivine-liquid partitioning of vanadium and other trace elements, with applications to modern and ancient picrites. Can. Mineral. **39**, 319–330 (2001)

Cleaves II, H.J., Scott, A.M., Hill, F.C., Leszczynski, J., Sahai, N., Hazen, R.M.: Mineral-organic interfacial processes: potential roles in the origins of life. Chem. Soc. Rev. **41**, 5502–5525 (2012)

Cleland, C.E.: Prediction and explanation in historical natural science. Br. J. Philos. Sci. **62**, 551–582 (2011)

Cleland, C.E., Hazen, R.M., Morrison, S.M.: Historical natural kinds and mineralogy: systematizing contingency in the context of necessity. Proc. Natl. Acad. Sci. **118**(1), e2015370118 (2020)

Cody, G.D.: Transition metal sulfides and the origins of metabolism. Annu. Rev. Earth Planet. Sci. **32**, 569–599 (2004)

Cody, G.D.: Geochemical connections to primitive metabolism. Elements. **1**, 139–143 (2005)

Cody, G.D., Boctor, N.Z., Filley, T.R., Hazen, R.M., Scott, J.H., Yoder Jr., H.S.: The primordial synthesis of carbonylated iron-sulfur clusters and the synthesis of pyruvate. Science. **289**, 1339–1342 (2000)

Cody, G.D., Boctor, N.Z., Hazen, R.M., Brandes, J.A., Morowitz, H.J., Yoder Jr., H.S.: Geochemical roots of autotrophic carbon fixation: hydrothermal experiments in the system citric acid, H2O-(±FeS)-(±NiS). Geochim. Cosmochim. Acta. **65**, 3557–3576 (2001)

Cody, G.D., Boctor, N.Z., Brandes, J.A., Filley, T.L., Hazen, R.M., Yoder Jr., H.S.: Assaying the catalytic potential of transition metal sulfides for abiotic carbon fixation. Geochim. Cosmochim. Acta. **68**, 2185–2196 (2004)

Davies, P.: Many planets, not much life. Sci. Am. (2016). https://doi.org/10.1038/scientificamerican0916-8

De Duve, C.: Vital Dust: Life as a Cosmic Imperative. Basic Books, New York (1995)

Deamer, D.: First Life: Discovering the Connections Between Stars, Cells, and How Life Began. University of California Press, Berkeley, CA (2011)

Deer, W.A., Howie, R.A., Zussman, J. (eds.): Rock-Forming Minerals: Sheet Silicates, vol. 3. Longmans, Green & Co, London (1962)

Deer, W.A., Howie, R.A., Zussman, J.: Rock-Forming Minerals: Orthosilicates, vol. 1A. London, Geological Society of London (1982)

Deer, W.A., Howie, R.A., Zussman, J.: Rock-Forming Minerals: Single-Chain Silicates, vol. 2A. London, Geological Society of London (1997)
Deer, W.A., Howie, R.A., Zussman, J.: Rock-Forming Minerals: Feldspars, vol. 4A. London, Geological Society of London (2001)
Deer, W.A., Howie, R.A., Zussman, J.: Rock-Forming Minerals: Non-Silicates Oxides, Hydroxides and Sulfides, vol. 5A. London, Geological Society of London (2011)
Ereshefsky, M.: Species, historicity, and path dependency. Philos. Sci. **81**, 714–726 (2014)
Ertem, G., Ferris, J.P.: Synthesis of RNA oligomers on heterogeneous templates. Nature. **379**, 238–240 (1996)
Ertem, G., Ferris, J.P.: Template-directed synthesis using the heterogeneous templates produced by montmorillonite catalysis. A possible bridge between the prebiotic and RNA worlds. J. Am. Chem. Soc. **119**, 7197–7201 (1997)
Essene, E.J., Fisher, D.C.: Lightning strike fusion: extreme reduction and metal-silicate liquid immiscibility. Science. **234**, 189–193 (1986)
Everitt, B.: Cluster Analysis. Wiley, Chichester (2011)
Evgenii, K., Wolfram, T.: The role of quartz in the origin of optical activity on earth. Orig. Life Evol. Biosph. **30**, 431–434 (2000)
Ferris, J.P.: Catalysis and prebiotic synthesis. Orig. Life Evol. Biosph. **23**, 307–315 (1993)
Ferris, J.P.: Mineral catalysis and prebiotic synthesis: montmorillonite-catalyzed formation of RNA. Elements. **1**, 145–149 (2005)
Ferris, J.P., Ertem, G.: Oligomerization of ribonucleotides on montmorillonite—reaction of the 5′-phosphorimidazolide of adenosine. Science. **257**, 1387–1389 (1992)
Ferris, J.P., Ertem, G.: Montmorillonite catalysis of RNA oligomer formation in aqueous solution—a model for the prebiotic formation of RNA. J. Am. Chem. Soc. **115**, 12270–12275 (1993)
Ferris, J.P., Hill, A.R., Liu, R., Orgel, L.E.: Synthesis of long prebiotic oligomers on mineral surfaces. Nature. **381**, 59–61 (1996)
Fraústo da Silva, J.J.R., Williams, R.J.P.: The Biological Chemistry of the Elements. The Inorganic Chemistry of Life, 2nd edn. Oxford University Press, Oxford (2001)
Godman, M.: Scientific realism with historical essences: the case of species. Synthese. (2019). https://doi.org/10.1007/s11229-018-02034-3
Golden, J., McMillan, M., Downs, R.T., Hystad, G., Stein, H.J., Zimmerman, A., Sverjensky, D.A., Armstrong, J., Hazen, R.M.: Rhenium variations in molybdenite (MoS_2): evidence for progressive subsurface oxidation. Earth Planet. Sci. Lett. **366**, 1–5 (2013)
Goldschmidt, V.M.: Geochemical aspects of the origin of complex organic molecules on the earth, as precursors to organic life. New Biol. **12**, 97–105 (1952)
Grew, E.S., Hazen, R.M.: Beryllium mineral evolution. Am. Mineral. **99**, 999–1021 (2014)
Grew, E.S., Bada, J.L., Hazen, R.M.: Borate minerals and origin of the RNA world. Orig. Life Evol. Biosph. **41**, 307–316 (2011)
Grew, E.S., Hystad, G., Toapanta, M., Eleish, A., Ostroverkhova, A., Golden, J., Hazen, R.M.: Lithium mineral evolution and ecology: comparison with boron and beryllium. Eur. J. Mineral. **31**, 755–774 (2019). https://doi.org/10.1127/ejm/2019/0031-2862
Hanczyc, M.M., Fujikawa, S.M., Szostak, J.W.: Experimental models of primitive cellular compartments: encapsulation, growth and division. Science. **302**, 618–622 (2003)
Hansma, H.A.: Possible origin of life between mica sheets: does life imitate mica? J. Biomol. Struct. Dyn. **31**, 888–895 (2010)
Haynes, W.M.: CRC Handbook of Chemistry and Physics, 97th edn, pp. 14–17. CRC Press, Boca Raton, FL (2016)
Hazen, R.M.: Genesis: The Scientific Quest for Life's Origin. Joseph Henry Press, Washington, DC (2005)
Hazen, R.M.: Mineral surfaces and the prebiotic selection and organization of biomolecules. Am. Mineral. **91**, 1715–1729 (2006)
Hazen, R.M.: The evolution of minerals. Sci. Am. **303**, 58–65 (2010)

Hazen, R.M.: The Story of Earth: The First 4.5 Billion Years, from Stardust to Living Planet. Viking, New York (2012)
Hazen, R.M.: Paleomineralogy of the hadean eon: a preliminary list. Am. J. Sci. **313**, 807–843 (2013)
Hazen, R.M.: Data-driven abductive discovery in mineralogy. Am. Mineral. **99**, 2165–2170 (2014)
Hazen, R.M.: Chance, necessity, and the origins of life: a physical sciences perspective. Philos. Trans. R. Soc. **A375**, 20160353 (2017). https://doi.org/10.1098/rsta.2016.0353
Hazen, R.M.: An evolutionary system of mineralogy: proposal for a classification based on natural kind clustering. Am. Mineral. **104**, 810–816 (2019). https://doi.org/10.2138/am-2019-6709
Hazen, R.M., Ferry, J.M.: Mineral evolution: mineralogy in the fourth dimension. Elements. **6**, 9–12 (2010)
Hazen, R.M., Morrison, S.M.: An evolutionary system of mineralogy, part I: stellar mineralogy (>13 to 4.6 Ga). Am. Mineral. **105**, 627–651 (2020)
Hazen, R.M., Morrison, S.M.: An evolutionary system of mineralogy, Part V: planetesimal aqueous and thermal alteration of planetesimals (4.565 to 4.550 Ga). Am Mineral. (2021). https://doi.org/10.2138/am-2021-7760
Hazen, R.M., Morrison, S.M.: On the paragenetic modes of minerals: A mineral evolution perspective. American Mineralogist, in press (2022). https://doi.org/10.2138/am-2021-8099
Hazen, R.M., Papineau, D.: Mineralogical co-evolution of the geosphere and biosphere. In: Knoll, A.H., Canfield, D.E., Konhauser, K.O. (eds.) Fundamentals of Geobiology, pp. 333–350. Oxford, UK, Wiley-Blackwell (2012)
Hazen, R.M., Sholl, D.S.: Chiral selection on inorganic crystalline surfaces. Nat. Mater. **2**, 367–374 (2003)
Hazen, R.M., Filley, T.R., Goodfriend, G.A.: Selective adsorption of L- and D-amino acids on calcite: implications for biochemical homochirality. Proc. Natl. Acad. Sci. USA. **98**, 5487–5490 (2001)
Hazen, R.M., Papineau, D., Bleeker, W., Downs, R.T., Ferry, J.M., McCoy, T.J., Sverjensky, D.A., Yang, H.: Mineral evolution. Am. Mineral. **93**, 1693–1720 (2008)
Hazen, R.M., Ewing, R.C., Sverjensky, D.A.: Evolution of uranium and thorium minerals. Am. Mineral. **94**, 1293–1311 (2009)
Hazen, R.M., Bekker, A., Bish, D.L., Bleeker, W., Downs, R.T., Farquhar, J., Ferry, J.M., Grew, E.S., Knoll, A.H., Papineau, D.F., Ralph, J.P., Sverjensky, D.A., Valley, J.W.: Needs and opportunities in mineral evolution research. Am. Mineral. **96**, 953–963 (2011)
Hazen, R.M., Downs, R.T., Golden, J., Hystad, G., Grew, E.S., Azzolini, D., Sverjensky, D.A.: Mercury (Hg) mineral evolution: a mineralogical record of supercontinent assembly, changing ocean geochemistry, and the emerging terrestrial biosphere. Am. Mineral. **97**, 1013–1042 (2012)
Hazen, R.M., Sverjensky, D.A., Azzolini, D., Bish, D.L., Elmore, S., Hinnov, L., Milliken, R.E.: Clay mineral evolution. Am. Mineral. **98**, 2007–2029 (2013a)
Hazen, R.M., Jones, A.P., Kah, L., Sverjensky, D.A.: Carbon mineral evolution. In: Hazen, R.M., Jones, A.P., Baross, J. (eds.) Carbon in Earth, pp. 79–107. Mineralogical Society of America, Washington, DC (2013b)
Hazen, R.M., Grew, E.S., Downs, R.T., Golden, J., Hystad, G.: Mineral ecology: chance and necessity in the mineral diversity of terrestrial planets. Can. Mineral. **53**, 295–323 (2015). https://doi.org/10.3749/canmin.1400086
Hazen, R.M., Hystad, G., Golden, J.J., Hummer, D.R., Liu, C., Downs, R.T., Morrison, S.M., Ralph, J., Grew, E.S.: Cobalt mineral ecology. Am. Mineral. **102**, 108–116 (2017a)
Hazen, R.M., Grew, E.S., Origlieri, M., Downs, R.T.: On the mineralogy of the "Anthropocene Epoch.". Am. Mineral. **102**, 595–611 (2017b)
Hazen, R.M., Downs, R.T., Elesish, A., Fox, P., Gagné, O., Golden, J.J., Grew, E.S., Hummer, D.R., Hystad, G., Krivovichev, S.V., Li, C., Liu, C., Ma, X., Morrison, S.M., Pan, F., Pires, A.J., Prabhu, A., Ralph, J., Runyon, S.E., Zhong, H.: Data-driven discovery in mineralogy: recent advances in data resources, analysis, and visualization. China Eng. **5**, 397–405 (2019). https://doi.org/10.1016/j.eng.2019.03.006

Hazen, R.M., Morrison, S.M., Prabhu, A.: An evolutionary system of mineralogy, part III: primary chondrule mineralogy (4.566 to 4.561 Ga). Am. Mineral. (2021). https://doi.org/10.2138/am-2020-7564

Heinen, W., Lauwers, A.M.: Organic sulfur compounds resulting from interaction of iron sulfide, hydrogen sulfide and carbon dioxide in an aerobic aqueous environment. Orig. Life Evol. Biosph. **26**, 131–150 (1996)

Herzberg, C., Vidito, C., Starkey, N.A.: Nickel–cobalt contents of olivine record origins of mantle peridotite and related rocks. Am. Mineral. **101**, 1952–1966 (2016)

Hill, A.R., Böhler, C., Orgel, L.E.: Polymerization on the rocks: negatively-charged α-amino acids. Orig. Life Evol. Biosph. **28**, 235–243 (1998)

Holm, N.G., Ertem, G., Ferris, J.P.: The binding and reactions of nucleotides and polynucleotides on iron oxide hydroxide polymorphs. Orig. Life Evol. Biosph. **23**, 195–215 (1993)

Hua, X., Eisenhour, D.D., Buseck, P.R.: Cobalt-rich, nickel-poor metal (wairauite) in the Ningqiang carbonaceous chondrite. Meteoritics. **30**, 106–109 (1995)

Huber, C., Wächtershäuser, G.: Activated acetic acid by carbon fixation on (Fe,Ni)S under primordial conditions. Science. **276**, 245–247 (1997)

Huber, C., Wächtershäuser, G.: Peptides by activation of amino acids with CO on (Ni,Fe)S surfaces: Implications for the origin of life. Science. **281**, 670–672 (1998)

Huber, C., Eisenreich, W., Hecht, S., Wächtershäuser: A possible primordial peptide cycle. Science. **301**, 938–940 (2003)

Huber, C., Kraus, F., Hanzlik, M., Eisenreich, W., Wächtershäuser, G.: Elements of metabolic evolution. Chem. Eur. J. (2012). https://doi.org/10.1002/chem.201102914

Hystad, G., Downs, R.T., Hazen, R.M.: Mineral frequency distribution data conform to a LNRE model: prediction of Earth's "missing" minerals. Math. Geosci. **47**, 647–661 (2015)

Hystad, G., Eleish, A., Downs, R.T., Morrison, S.M., Hazen, R.M.: Bayesian estimation of Earth's undiscovered mineralogical diversity using noninformative priors. Math. Geosci. **51**, 401–417 (2019). https://doi.org/10.1007/s11004-019-09795-8

Jelen, B.I., Giovannelli, D., Falkowski, P.G.: The role of microbial electron transfer in the coevolution of the biosphere and geosphere. Annu. Rev. Microbiol. **70**, 45–62 (2016)

Jenner, F.E., O'Neill, H.S.C.: Analysis of 60 elements in 616 ocean floor basaltic glasses. Geochem. Geophys. Geosyst. **13** (2012). https://doi.org/10.1029/2011GC004009

Jonsson, C.M., Jonsson, C.L., Sverjensky, D.A., Cleaves, H.J., Hazen, R.M.: Attachment of L-glutamate to rutile (TiO_2): a potentiometric, adsorption, and surface complexation study. Langmuir. **25**, 12127–12135 (2009)

Kent, A.J.R., Rossman, G.R.: Hydrogen, lithium and boron in mantle-derived olivine: the role of coupled substitutions. Am. Mineral. **87**, 1432–1436 (2002)

Kim, J.D., Yee, N., Nanda, V., Falkowski, P.G.: Anoxic photochemical oxidation of siderite generates molecular hydrogen and iron oxides. Proc. Natl. Acad. Sci. USA. **110**, 10073–10077 (2013)

Kim, H., Furukawa, Y., Kakegawa, T., Bita, A., Scorei, R., Benner, S.A.: Evaporite borate-containing mineral ensembles make phosphate available and regiospecifically phosphorylate ribonucleosides: borate as a multifaceted problem solver in prebiotic chemistry. Angew. Chem. Int. Ed. **55**, 15816–15820 (2016)

Koritnig, S.: Geochemistry of phosphorus. I. The replacement of Si^{4+} by P^{5+} in rock-forming silicate minerals. Geochimica et Cosmochimica Acta, **29**, 361–371. (1965)

Krivovichev, S.V., Krivovichev, V.G., Hazen, R.M.: Structural and chemical complexity of minerals: correlations and time evolution. Eur. J. Mineral. **18**, 231–236 (2017). https://doi.org/10.1127/ejm/2018/0030-2694

Lahav, N.: Biogenesis: Theories of Life's Origin. Oxford University Press, New York (1999)

Lahav, N., White, D., Chang, S.: Peptide formation in the prebiotic era: thermal condensation of glycine in fluctuating clay environments. Science. **201**, 67–69 (1978)

Lester, G.W., Clark, A.H., Kyser, T.K., Naslund, H.R.: Experiments on liquid immiscibility in silicate melts with H_2O, P, S, F and Cl: implications for natural magmas. Contrib. Mineral. Petrol. **166**, 329–349 (2013)

Liu, X., Xiong, X., Audetat, A., Li, Y., Song, M., Li, L., Sun, W., Ding, X.: Partitioning of copper between olivine, orthopyroxene, clinopyroxene, spinel, garnet and silicate melts at upper mantle conditions. Geochim. Cosmochim. Acta. **125**, 1–22 (2014)

Liu, C., Hystad, G., Golden, J.J., Hummer, D.R., Downs, R.T., Morrison, S.M., Grew, E.S., Hazen, R.M.: Chromium mineral ecology. Am. Mineral. **102**, 612–619 (2017a)

Liu, C., Knoll, A.H., Hazen, R.M.: Geochemical and mineralogical evidence that Rodinian assembly was unique. Nat. Commun. **8**, 1950 (2017b)

Liu, C., Eliesh, A., Hystad, G., Golden, J.J., Downs, R.T., Morrison, S.M., Hummer, D.R., Ralph, J.P., Fox, P., Hazen, R.M.: Analysis and visualization of vanadium mineral diversity and distribution. Am. Mineral. **103**, 1080–1086 (2018). https://doi.org/10.2138/am-2018-6274

London, D.: Pegmatites, vol. 10. Mineralogical Association of Canada, Quebec City (2008)

Lyons, T.W., Fike, D.A., Zerkle, A.: Emerging biogeochemical views of Earth's ancient microbial worlds. Elements. **11**, 415–421 (2015)

McCollom, T.M., Seewald, J.S.: A reassessment of the potential for reduction of dissolved CO_2 to hydrocarbons during serpentinization of olivine. Geochim. Cosmochim. Acta. **65**, 3769–3778 (2001)

Mills, S.J., Hatert, F., Nickel, E.H., Ferrais, G.: The standardization of mineral group hierarchies: application to recent nomenclature proposals. Eur. J. Mineral. **21**, 1073–1080 (2009)

Monod, J.: Chance and Necessity: An Essay on the Natural Philosophy of Modern Biology. Knopf, New York (1970)

Moore, E.K., Jelen, B.I., Giovannelli, D., Raanan, H., Falkowski, P.G.: Metal availability and the expanding network of microbial metabolisms in the Archaean eon. Nat. Geosci. **10**, 629–636 (2017)

Moore, E.K., Hao, J., Sverjensky, D.A., Jelen, B.I., Meyer, M., Hazen, R.M., Falkowski, P.G.: Geological and chemical factors that impacted the biological utilization of cobalt in the Archean eon. J. Geophys. Res. Biogeosci. **123**, 743–759 (2018). https://doi.org/10.1002/2017G004067

Morrison, S.M., Hazen, R.M.: An evolutionary system of mineralogy, part II: interstellar and solar nebula primary condensation mineralogy (> 4.565 Ga). Am. Mineral. (2020). https://doi.org/10.2138/am-2020-7447

Morrison, S.M., Hazen, R.M.: An evolutionary system of mineralogy, part IV: planetesimal differentiation and impact mineralization (4.566 to 4.560 Ga). Am. Mineral. (2021). https://doi.org/10.2138/am-2021-7632

Morrison, S.M., Liu, C., Eleish, A., Prabhu, A., Li, C., Ralph, J., Downs, R.T., Golden, J.J., Fox, P., Hummer, D.R., Meyer, M.B., Hazen, R.M.: Network analysis of mineralogical systems. Am. Mineral. **102**, 1588–1596 (2017). https://doi.org/10.2138/am-2017-6104

Morrison, S.M., Runyon, S.E., Hazen, R.M.: The paleomineralogy of the Hadean Eon revisited. Life. **8**, 64 (2018). https://doi.org/10.3390/life8040064

Morrison, S.M., Buongiorno, J., Downs, R.T., Eleish, A., Fox, P., Giovannelli, D., Golden, J.J., Hummer, D.R., Hystad, G., Kellogg, L.H., Kreylos, O., Krivovichev, S.V., Liu, C., Prabhu, A., Ralph, J., Runyon, S.E., Zahirovic, S., Hazen, R.M.: Visualizing carbon mineral systems: recent advances in C mineral evolution, mineral ecology, and network analysis. Front. Earth Sci. (2020). https://doi.org/10.3389/feart.2020.00208

Mulkidjanian, A.Y.: On the origin of life in the zinc world: 1. Photosynthesizing, porous edifices built of hydrothermally precipitated zinc sulfide as cradles of life on Earth. Biol. Direct. **4**, 26 (2009)

Mulkidjanian, A.Y., Galperin, M.Y.: On the origin of life in the zinc world: 2. Validation of the hypothesis on the photosynthesizing zinc sulfide edifices as cradles of life on Earth. Biol. Direct. **4**, 27 (2009)

Nitschke, W., McGlynn, S.E., Milner-White, E.J., Russell, M.J.: On the antiquity of metalloenzymes and their substrates in bioenergetics. Biochim. Biophys. Acta. **1827**, 871–881 (2013)

O'Reilly, S.Y., Chen, D., Griffin, W.L., Ryan, C.G.: Minor elements in olivine from spinel lherzolite xenoliths: implications for thermobarometry. Mineral. Mag. **61**, 257–269 (1997)

Orgel, L.E.: Polymerization on the rocks: theoretical introduction. Orig. Life Evol. Biosph. **28**, 227–234 (1998)

Papineau, D.: Mineral environments of the earliest Earth. Elements. **6**, 25–30 (2010)

Parsons, I., Lee, M.R., Smith, J.V.: Biochemical evolution II: origin of life in tubular microstructures in weathered feldspar surfaces. Proc. Natl. Acad. Sci. USA. **95**, 15173–15176 (1998)

Pasek, M.A.: Rethinking early Earth phosphorus geochemistry. Proc. Natl. Acad. Sci. USA. **105**, 853–858 (2008)

Pasek, M., Block, K.: Lightning-induced reduction of phosphorus oxidation state. Nat. Geosci. **2**, 553–556 (2009)

Pasek, M.A., Lauretta, D.S.: Aqueous corrosion of phosphide minerals from iron meteorites: a highly reactive source of prebiotic phosphorus on the surface of the early earth. Astrobiology. **5**, 515–535 (2005)

Pasek, M.A., Dworkin, J.P., Lauretta, D.S.: A radical pathway for organic phosphorylation during schreibersite corrosion with implications for the origin of life. Geochim. Comsochim. Acta. **71**, 1721–1736 (2007)

Paster, T.P., Schauwecker, D.S., Haskin, L.A.: The behavior of some trace elements during solidification of the Skaergaard layered series. Geochim. Cosmochim. Acta. **38**, 1549–1577 (1974)

Pirajno, F.: Hydrothermal Processes and Mineral Systems. Springer, Amsterdam, The Netherlands (2009)

Pitsch, S., Eschenmoser, A., Gedulin, B., Hui, S., Arrhenius, G.: Mineral induced formation of sugar phosphates. Orig. Life Evol. Biosph. **25**, 297–334 (1995)

Ricardo, A., Carrigan, M.A., Olcott, A.N., Benner, S.A.: Borate minerals stabilize ribose. Science. **303**, 196 (2004)

Rollinson, H.: Early Earth Systems: A Geochemical Approach. Blackwell, Malden, MA (2007)

Rudnick, R.L., Gao, S.: Composition of the continental crust. In: Rudnick, R.L., Holland, H.D., Turekian, K.K. (eds.) The Crust, vol. 3, pp. 1–64. Elsevier, Amsterdam, The Netherlands (2005)

Russell, M.J., Hall, A.J.: The emergence of life from iron monosulphide bubbles at a submarine hydrothermal redox and pH front. J. Geol. Soc. Lond. **154**, 377–402 (1997)

Russell, M.J., Daniel, R.M., Hall, A.J., Sherringham, J.: A hydrothermally precipitated catalytic iron-sulfide membrane as a first step toward life. J. Mol. Evol. **39**, 231–243 (1994)

Savelyev, D.P., Kamenetsky, V.S., Danyushevsky, L.V., Botcharnikov, R.E., Kamenetsky, M.B., Park, J.-W., Portnyagin, M.V., Olin, P., Krasheninnikov, S.P., Hauff, F.: Immiscible sulfide melts in primitive oceanic magmas: evidence and implications from picrite lavas (Eastern Kamchatka, Russia). Am. Mineral. **103**, 886–898 (2018)

Schertl, H.-P., Mills, S.J., Maresch, W.V.: A Compendium of IMA-Approved Mineral Nomenclature. International Mineralogical Association, Melbourne (2018)

Schoffeniels, E.: Anti-Chance: a Reply to Monod's Chance and Necessity. Pergamon Press, Oxford, UK (1976)

Schoonen, M.A.A., Smirnov, A., Cohn, C.: A perspective on the role of minerals in prebiotic synthesis. Ambio. **33**, 539–551 (2004)

Schrenk, M.O., Brazelton, W.J., Lang, S.Q.: Serpentinization, carbon, and deep life. Rev. Mineral. Geochem. **75**, 575–606 (2013)

Shirey, S.B., Richardson, S.H.: Start of the Wilson cycle at 3 Ga shown by diamonds from the subcontinental mantle. Science. **333**, 434–436 (2011). https://doi.org/10.1126/science.1206275

Smirnov, A., Hausner, D., Laffers, R., Strongin, D.R., Schoonen, M.A.A.: Abiotic ammonium formation in the presence of Ni-Fe metals and alloys and its implications for the Hadean nitrogen cycle. Geochem. Trans. **9**, 5 (2008)

Smith, J.V.: Biochemical evolution. I. Polymerization on internal, organophilic silica surfaces of dealuminated zeolites and feldspars. Proc. Natl. Acad. Sci. USA. **95**, 3370–3375 (1998)

Smith, J.V., Arnold Jr., F.P., Parsons, I., Lee, M.R.: Biochemical evolution III: Polymerization on organophilic silica-rich surfaces, crystal-chemical modeling, formation of first cells, and geological clues. Proc. Natl. Acad. Sci. USA. **96**, 3479–3485 (1999)

Soai, K., Osanai, S., Kadowaki, K., Yonekubo, S., Shibata, S., Sato, I.: D- and L-quartz-promoted highly enantioselective synthesis of a chiral organic compound. J. Am. Chem. Soc. **121**, 11235–11236 (1999)

Storm, T.W., Holland, H.D.: The distribution of nickel in the Lambertville diabase. Geochim. Cosmochim. Acta. **11**, 335–347 (1957)

Summers, D.P., Chang, S.: Prebiotic ammonia from reduction of nitrite by iron (II) on the early Earth. Nature. **365**, 630–632 (1990)

Sun, C., Graff, M., Liang, Y.: Trace element partitioning between plagioclase and silicate melt: the importance of temperature and plagioclase composition, with implications for terrestrial and lunar magmatism. Geochim. Cosmochim. Acta. **206**, 273–295 (2017)

Sverjensky, D.A., Lee, N.: The great oxidation event and mineral diversification. Elements. **6**, 31–36 (2010)

Van Kranendonk, M.: Chapter 19 - Paleoarchean development of a continental nucleus: the east Pilbara terrane of the Pilbara Craton, Western Australia. In: Van Kranendonk, M., Bennett, V., Hoffman, E. (eds.) Earth's Oldest Rocks, pp. 437–462. Elsevier, Dordrecht (2019). https://doi.org/10.1016/B978-0-444-63901-1.00019-8

Veblen, D.R., Buseck, P.R.: Serpentine minerals: intergrowths and new combination structures. Science. **206**, 1398–1400 (1979)

Veksler, I.V., Dorfman, A.M., Dulski, P., Kamenetsky, V.S., Danyushevsky, L.V., Jeffries, T., Dingwell, D.B.: Partitioning of elements between silicate melt and immiscible fluoride, chloride, carbonate, phosphate and sulfate melts, with implications to the origin of natrocarbonatite. Geochim. Cosmochim. Acta. **79**, 20–40 (2012)

Wächtershäuser, G.: Before enzymes and templates: theory of surface metabolism. Microbiol. Rev. **52**, 452–484 (1988a)

Wächtershäuser, G.: Pyrite formation, the first energy source for life: a hypothesis. Syst. Appl. Microbiol. **10**, 207–210 (1988b)

Wächtershäuser, G.: Evolution of the first metabolic cycles. Proc. Natl. Acad. Sci. USA. **87**, 200–204 (1990a)

Wächtershäuser, G.: The case for the chemoautotrophic origin of life in an iron-sulfur world. Orig. Life Evol. Biosph. **20**, 173–176 (1990b)

Wächtershäuser, G.: The cradle chemistry of life: on the origin of natural products in a pyrite-pulled chemoautotrophic origin of life. Pure Appl. Chem. **65**, 1343–1348 (1993)

Wadsworth, F.B., Vasseur, J., Llewellin, E.W., Genareau, K., Cimarelli, C., Dingwell, D.B.: Size limits for rounding of volcanic ash particles heated by lightning. J. Geophys. Res. Solid Earth. **122**, 1977–1989 (2017)

Wager, L.R., Brown, G.M.: Layered Igneous Rocks. W. H. Freeman and Company, New York (1968)

Weber, A.: Formation of pyrophosphate on hydroxyapatite with thioesters as condensing agents. Biosystems. **15**, 183–189 (1982)

Weber, A.: Prebiotic polymerization: oxidative polymerization of 2,3-dimercapto-1-propanol on the surface of iron(III) hydroxide oxide. Orig. Life Evol. Biosph. **25**, 53–60 (1995)

Weinstein, S.A., Yuen, D.A., Olson, P.L.: Evolution of crystal-settling in magma-chamber convection. Earth Planet. Sci. Lett. **87**, 237–248 (1988)

White, W.M., Klein, E.M.: Composition of the oceanic crust. In: Treatise on Geochemistry, vol. 4, 2nd edn, pp. 457–496. Elsevier, Oxford (2014)

Yoder Jr., H.S.: Generation of Basaltic Magma. National Academy of Sciences Press, Washington, DC (1976)

Zhabin, A.G.: Is there evolution of mineral speciation on Earth? Doklady Earth Sci. Sect. **247**, 142–144 (1981)

Ziegler, E.W., Kim, H.-J., Benner, S.A.: Molybdenum (VI)-catalyzed rearrangement of prebiotic carbohydrates in formamide, a candidate prebiotic solvent. Astrobiology. **18**, 1159–1170 (2018)

Chapter 3
The Geological Platform for the Origin of Life on Earth

Eva E. Stüeken and Nicholas J. Gardiner

Abstract Resolving the geochemical steps which led to the first living organisms on Earth is perhaps one of the most challenging tasks in science. A prime reason for this challenge is that these steps cannot be reproduced on laboratory benchtops; numerous environmental parameters (including minerals, physicochemical conditions and mixing processes) were present while prebiotic chemistry unfolded, and these factors undoubtedly played significant roles in the final outcome. In this chapter, we argue that any proposed chemical pathway that leads to the formation of a particular biomolecule therefore needs to be placed into the geological context of the early Earth as a whole, a context which was likely very different from the modern Earth in ways that we outline. Further advances in origin of life research will require: (a) identifying key organic synthesis reactions whose requirements can be met by realistic environmental settings; and (b) identifying utility in geochemical reactions that are known to occur abundantly in natural environmental settings. This chapter will show how evidence from Hadean zircons, Archean crustal remnants, computational geodynamic models and observations of other planets can be used to reconstruct the hydrologically and volcanically dynamic geological platform on which prebiotic reaction networks were operating. Within this framework, several geochemical reactions will be highlighted that were likely common and that may have contributed productively to life's origin. Ultimately, this geological vantage point may help us understand what makes other planets not just habitable but also suitable for independent biogenesis.

E. E. Stüeken (✉)
School of Earth & Environmental Sciences, University of St Andrews, St Andrews, UK
e-mail: ees4@st-andrews.ac.uk

N. J. Gardiner
School of Earth & Environmental Sciences, University of St Andrews, St Andrews, UK

School of Earth, Atmosphere and Environment, Monash University, Clayton, VIC, Australia

A. Neubeck, S. McMahon (eds.), *Prebiotic Chemistry and the Origin of Life*, Advances in Astrobiology and Biogeophysics,
https://doi.org/10.1007/978-3-030-81039-9_3

3.1 Introduction: Life Did Not Emerge in a Beaker

Reconstructing the origin of life on Earth has challenged scientists for centuries. Even the simplest living cell is composed of myriads of complex molecules with very specific arrangements and interactions, making it difficult to understand how organisms capable of metabolism and reproduction emerged from abiotic materials on the early Earth. Traditionally, origin-of-life research has been driven by organic chemists who have provided key insights into the properties and functions of biomolecules. For example, it is now known that all life as we know it shares a set of essential amino acids and nucleobases which therefore likely date back to life's origin. It is also well known that DNA (deoxyribonucleic acid) and RNA (ribonucleic acid)—the two information-carriers in living cells—contain a specific set of nucleobases, ribose sugar and a backbone of phosphate ions, and that cell membranes are made of lipids. Sooner or later, any origin-of-life scenario therefore needs to lead to the formation of these basic constituents.

More recently, with the discovery of exoplanets and potentially habitable conditions on other celestial bodies in our solar system, the origin of life has moved into the focus of planetary science. It is likely that certain microbial organisms, in particular extremophiles, would be capable of surviving in the deep oceans of the moons Europa or Enceladus (Chyba 2000; McKay et al. 2014; Glein et al. 2015) or in crater lakes on early Mars (Grotzinger et al. 2014), but this recognition has brought with it the question of which properties a planetary body needs to have for an independent origin of life. The ability of a planet to support life (habitability) does not necessarily imply that this planet can also enable biogenesis. Further, it may be possible to create genomes artificially within an organic chemistry lab (Gibson et al. 2010), but that does not mean that we have uncovered the processes that led to our own origin on the early Earth. The organic chemical reactions that lead to the formation and assembly of key biomolecules must be embedded in a plausible planetary environment.

Although this requirement may be seen as a limitation to organic synthesis reactions, which are most easily carried out and understood under well-defined conditions, it also provides an opportunity, because no organic chemistry laboratory can ever be as diverse as the natural laboratory of the Earth. Environmental conditions where organic molecules and other key ingredients for life can persist range from the interior of the crust to the upper reaches of the atmosphere. Numerous gradients in temperature, pressure, pH, salinity, composition, irradiation and water activity create a nearly infinite number of niches that may each have nurtured very distinct prebiotic reactions, and mixing processes exist that exchange products and reactants between these niches. We know that life managed to emerge from this milieu and persist for billions of years, and it is conceivable that such environmental diversity was in fact an essential requirement for our own success (Stüeken et al. 2013). It is therefore important to place the origin of life within this global network and consider the possibility that those planetary processes that were the most dominant within each setting were also the most productive towards the emergence of a resilient biosphere. In the following, we will revisit the fundamental ingredients that were probably needed during biogenesis (Sect. 3.2), review our understanding of early Earth environments (Sect. 3.3), and conclude with a synthesis of planetary provisions and prebiotic chemistry.

3.2 A Geologist's Recipe for Life

Before we interrogate the Earth's rock record for suitable prebiotic settings, we need to remind ourselves of the essential processes or properties that we are looking for. These are defined largely from the vantage point of modern biology, but they are so fundamental that we regard them as universal requirements:

3.2.1 CHNOPS Sources and Activation Energy for Organic Synthesis

The major elements of life are carbon, hydrogen, nitrogen, oxygen, phosphorus and sulphur (CHNOPS). The major forms of these elements in the Earth's upper crust and atmosphere today are CO_2 or CH_4 gas or carbonate rock, liquid H_2O or H_2 gas, N_2 gas, phosphate minerals, and SO_2 or H_2S gas or dissolved SO_4^{2-}. As discussed below (Sect. 3.2), these molecules were probably also the most abundant CHNOPS species on the early Earth, though in different proportions than today. Hence prebiotic reactions needed to occur which converted these naturally occurring inorganic molecules into organic compounds. It remains possible that some organic matter was brought to Earth by meteorites. For example, carbonaceous chondrites have been shown to contain amino acids, nucleobases and even sugars which survived the impact (Pizzarello and Cronin 2000; Callahan et al. 2011; Furukawa et al. 2019). It is not well known how exactly those molecules were produced within the meteorite parent body. However, given that appropriate conditions evidently existed within multiple meteorite parent bodies, it is likely that also on Earth organic molecules were synthesised abiotically from inorganic reactants. Such synthesis reactions would have required the reduction of CO_2, N_2 and SO_2. These reduction reactions are thermodynamically feasibly in the presence of H_2. For example, CO_2 and H_2 are in chemical disequilibrium with each other and would tend to form reduced carbon such as CH_4 or longer-chain hydrocarbons. However, the activation energy of these reactions is high (e.g., Lane et al. 2010). The origin of life therefore probably required not just provisions of the CHNOPS elements but also large sources of energy to overcome this activation barrier. Activation energy can therefore be considered a requirement for life in order to convert the most abundant CHNOPS species that occur in the environment into bioavailable forms. Plausible sources of energy include short-wave electromagnetic radiation, lightning, geothermal heat and impact shock, and we will further discuss these below within the context of the early Earth (Sects. 3.3 and 3.4). These energy sources likely led to the production of a range of organic molecules including formaldehyde, cyanide, simple amino acids, some nucleobases, and hydrocarbons (e.g., Miller and Urey 1959; McCollom and Seewald 2007; Cleaves 2008; Tian et al. 2011).

3.2.2 Polymerisation

Chemical synthesis of organic molecules produces monomers, which needed to be joined into polymers or into more complex monomeric compounds. One example of polymerisation is the formation of peptides from amino acids, which are important building blocks of enzymes in living cells. Reactions between amino acids involve the loss of H_2O, which supports the idea that at least some polymerisation reactions and other synthesis reactions may have proceeded in dry conditions, perhaps on the surfaces of clay minerals (Ferris 2006; Becker et al. 2018). Clays have been shown to bind nucleobases by adsorption and could have acted as templates for the formation of RNA precursors. Similarly, it has been shown that borate minerals can catalyse the formation of ribose sugar from formaldehyde under alkaline conditions (Ricardo et al. 2004). However, polymerisation has also been documented from experiments mimicking deep-sea hydrothermal vents where it may be templated by sulphide minerals (Burcar et al. 2015). The greatest challenge in polymerisation reactions, which has to our knowledge not yet been resolved, is selectivity of specific monomers from a diverse mixture. As discussed below, envisioning the origin of life in a global context may help overcome this problem, because it would allow for the formation of different monomers and polymers in different settings.

3.2.3 Encapsulation

All life as we know it is encapsulated within a cell membrane. Although cell sizes of extant organisms range over several orders of magnitude, some form of encapsulation is probably required to maintain all interacting biomolecules within close proximity to each other and to protect the cellular machinery from external influences, such as changes in pH or salinity. Cell membranes are made of lipids, and because of their hydrophobic nature, these can arrange spontaneously into vesicles, if they are suspended in freshwater (Black et al. 2013; Deamer 2017). In saline media, the spontaneous vesicular organisation breaks down, but some form of coagulation is still possible (Black et al. 2013). An alternative to encapsulation that has been proposed for the prebiotic world is an organic coating of mineral surfaces in the ocean that was perhaps created from such coagulated lipids and sugars. Such a 'ribofilm' (Baross and Martin 2015) could effectively have acted as a capsule for mineral-bound synthesis reactions. Both vesicles and fragments of organic films could have acted as transport agents for organic material. In fact, if vesicles and/or 'ribofilms' were created abiotically on the early Earth, then it is likely that they were exchanged between environmental settings. Their presence therefore needs to be considered in any origin-of-life model.

3.2.4 Redox Gradients and Mineral Catalysis

Replication and metabolism are two defining characteristics of life, and they can be related to each other through the equation $\Delta G = \Delta H - T\Delta S$ (Schoepp-Cothenet et al. 2012). Here ΔS (entropy) is a quantification of the number of different ways the energy can be distributed within a system and therefore the inverse of cellular and genomic organisation that are created during replication. ΔS is therefore negative. ΔH (enthalpy) is the energy that is either released or absorbed during metabolic reactions, such as the energy that humans gain from the reaction between inhaled O_2 and consumed organic carbon. ΔG can be thought of as the progression of life. Thermodynamics dictates that a reaction is favourable, *i.e.* that life progresses, if $\Delta G < 0$. In other words, replication (*i.e.* the creation of negative entropy $\Delta S < 0$) is for the most part (with a small temperature (T) dependence) only possible if it is counterbalanced by a negative change in entropy ($\Delta H < 0$). Life therefore depends strictly on metabolic reactions that are exergonic, *i.e.* energy-releasing. All life as we know it uses exergonic redox reactions where an oxidant and a reductant react with each other, and the energy that is released in the process is channelled into the production of new molecular bonds. Replication is thus driven by electrochemical (redox) energy, using redox couples provided by the environment. Importantly, many of these redox reactions require catalysts (enzymes) to proceed. This makes sense, because if the metabolites were reacting spontaneously in the environment, the flow of energy released during the reaction would be much more difficult to control. Many of these catalysts frequently contain transition metals in their structure, because metals are ideal for the rapid exchange of electrons and ligands (Da Silva and Williams 2001). Some of these enzymatic metal clusters resemble minerals that also form abiotically in some surface environments, in particular around hydrothermal vents (Cody 2004; Russell and Martin 2004). Fundamentally, life can thus be thought of as a catalytic machinery that is built around metallic centres. The ubiquity of this machinery across the tree of life suggests that it is ancient and an inherent property of our biosphere. Any plausible origin-of-life scenario therefore needs to include a constant supply of oxidants and reductants as well as metal catalysts. These components not only help in generating organic monomers (Sect. 3.2.1), but they would have been essential for the emergence of autotrophic metabolisms and for the development of other cellular functions.

This list of requirements highlights that the difficulty of envisioning the origin of life within a single environmental setting. Based on current knowledge, numerous environmental conditions need to be fulfilled to create the chemical complexity that defines biology. At the same time, this list illustrates that numerous prebiotic processes were probably tightly linked to planetary properties. Understanding the origin of life therefore demands a more thorough understanding of the early Earth.

3.3 What Did the Earth Look Like When Life Emerged?

The Earth has evolved over its 4.56 billion-year-long history, and when exactly life emerged is not well known, making it difficult to know what state the Earth was in at the time. However, we can place a minimum age on the origin of life based on the oldest known biosignatures. The most widely accepted oldest evidence for life dates back to 3.49 Ga, and it comes from the Dresser Formation in the Pilbara region of Western Australia. These rocks contain stromatolites—microbially induced carbonate mounds that are now silicified (Walter et al. 1980)—and they show isotopic evidence of microbial sulphate reduction (Shen et al. 2001; Ueno et al. 2008). From this point onwards, many sedimentary rocks of low metamorphic grade from Australia, South Africa and other parts of the world show diverse relics of microbial life in the form of diagnostic carbon isotope fractionation, stromatolites, or even microfossils (e.g., Schidlowski 2001; Allwood et al. 2006; Sugitani et al. 2015). Older sedimentary rocks are preserved in Greenland and Canada, and they have experienced much higher grades of metamorphism and metasomatism. Reports of biosignatures based on carbon isotope ratios (Mojzsis et al. 1996; Rosing 1999; Tashiro et al. 2017), stromatolites (Nutman et al. 2016) and microfossils (Dodd et al. 2017) dating back to at least 3.8 Ga are therefore contested (van Zuilen et al. 2002; Allwood et al. 2018; Whitehouse et al. 2019). In the earlier Hadean, heavy meteorite bombardment may have inhibited the long-term establishment of a biosphere (Ryder 2003). In the following, we will therefore assume that the origin of life falls around the time of the Hadean-Archean transition, somewhere between 4.0 and 3.5 Ga. The rock record that is preserved from this time to inform us about environmental conditions is very limited, but we can make inferences from slightly younger rocks, from modern analogues and from computational models.

3.3.1 The Solid Earth

Much of life as we know it resides on continental platforms, including marine shelves and dry land. While it is conceivable that life could originate in a water world without exposed land masses, the presence of continents offers a much wider range of environmental niches and may be crucial for the formation of anhydrous biopolymers (Sect. 3.2.2). Understanding the tectonic regime of the early Earth is therefore crucial for identifying plausible settings for prebiotic chemistry. The early (>3.0 Ga) Earth was a very different planet to today. It was hotter, where higher rates of internal radiogenic heat production resulted in mantle potential temperatures of perhaps up to 200 °C above today (Herzberg et al. 2010). A hotter mantle means higher degrees of mantle melting, resulting in more magnesian and less siliceous mantle melts (e.g., komatiites), and greater degrees of volcanic activity, which is the main conduit for planetary heat loss. In the context of the origin of life, intense volcanism would have replenished the supplies of atmospheric gases (CO_2, SO_2, N_2) that form the basis of organic matter (Sect. 3.2.1).

The early Earth was geochemically primitive and less differentiated; the emergence of the first, preserved, stable continental crust only occurred during the Eoarchean (3.6–4.0 Ga), some 600 million years after planetary accretion, and thus possibly coincided with life's origins. Archean continental crust is largely of the tonalite-trondhjemite-granodiorite (TTG) variety, a form of granite which is compositionally distinct to continental crust produced on the modern Earth. This compositional difference reflects both a different source and distinct processes of partial melting, fractionation and crystallization. TTGs were formed via a chain of partial melting: melting mantle to create mafic crust such as basalt, and then remelting this (hydrated) mafic crust to create evolved granitic crust. Many TTG terranes are today associated with Archean cratons, which are underpinned by a deep, refractory lithospheric root. The simultaneous growth of the overlying continental crust and the underlying depleted lithosphere potentially allowed Earth's early continental lithosphere to remain buoyant and resist being recycled back into the mantle. However, Earth's early continental lithosphere was likely rheologically weaker than modern continental lithosphere due to the elevated geothermal gradients, and hence less able to support horizontal orogenic (mountain-building) processes. Instead of the plate tectonic geodynamic paradigm prevalent today, it is probable that different prevailing geodynamics dominated on the early Earth, most likely so-called stagnant-lid or "vertical" type tectonics rather than stable subduction processes (Brown et al. 2020). However, the preserved rock record from the early Earth is highly fragmentary and has typically undergone extensive reworking and metamorphism since its initial growth; trace element and radiogenic isotope geochemistry are thus some of the best approaches to decipher the geodynamic nature of the early Earth and the potential surficial conditions therein.

3.3.1.1 The Hadean to Eoarchean: The Earliest Fragments

Very little is known about the Earth during the Hadean Eon (4.6–4.0 Ga), the period immediately after planetary accretion and the ca. 4.5 Ga moon-forming impact. The known surviving terrestrial fragments from this period are microscopic (~100 μm) zircon crystals, now found within Neoarchean sandstones of the Jack Hills and Mount Narryer localities in northern Western Australia, and rare ca. 4.0 Ga evolved components of the Acasta Gneisses and potentially older mafic components of the Nuvvuagittuq Greenstone belt, both of northern Canada. Zircons are common accessory minerals in granitic rocks but rare in mafic crust. The Jack Hills zircons therefore provide an important portal into the nature of Earth's earliest crust, and open questions remain whether they represent relic Hadean continental lithosphere or alternatively highly fractionated mafic melts. U–Pb geochronology of the Jack Hills grains has proved their antiquity, with a dominant age peak at 4.2 Ga and contested ages up to 4.4 Ga (reviewed in Harrison 2009). Hafnium isotopic systematics of the zircons most likely point towards a Hadean Earth surfaced in a predominately mafic (anorthositic) crust as a result of a crystallized magma ocean, such as is found on the Moon, which underwent episodic magmatic reworking (Kemp et al. 2010). Similarly, the oldest components of the Acasta Gneisses, the ca. 4.0 Ga Idiwhaa Gneisses, are likely the product of internal differentiation in a mantle upwell, similar to a modern mid-ocean ridge analogue (Reimink et al. 2016), and do not represent relic continental lithosphere. Frequent resurfacing of the early Hadean Earth may have created uninhabitable conditions, meaning that productive prebiotic reactions probably began only after some degree of mantle cooling.

The oldest preserved stable TTG crust is now found as components of the ca. 3.9 Ga Itsaq Gneiss Complex of south west Greenland (Nutman et al. 1996), and it remains an open question why essentially no crust appears to have survived the Hadean-Eoarchean boundary. The Itsaq Gneisses represent relic Archean mid-crust now exhumed. In contrast, Archean upper crust is now mainly represented by the Paleoarchean granite-greenstone terranes, such as the East Pilbara Terrane of Western Australia, and the Barberton Greenstone Belt of southern Africa. These granite-greenstone terranes are typified by granitic domes some 50–200 km in diameter, surrounded by steeply-dipping supracrustal sequences (greenstones), packages of mafic-ultramafic and felsic volcanic rocks and siliciclastic sedimentary units. Due to their low metamorphic grade and relatively little deformation, coupled with their exceptional preservation, these analogues of Archean continental near-surface crust have become highly studied areas for evidence of both the presence and conditions for the development of earliest life.

3.3.1.2 The 3.5–3.2 Ga East Pilbara Terrane: Geodynamics and Continental Emergence

The East Pilbara Terrane of the Pilbara Craton, Western Australia is perhaps the archetypal granite-greenstone terrane. A well-developed "vertical tectonic" geodynamic model for its Paleoarchean development (Fig. 3.1) describes the growth of a volcanic plateau over the span of roughly 500 million years through episodic volcanic resurfacing, likely driven through mantle plume activity (Van Kranendonk et al. 2007; Smithies et al. 2009). This resurfacing built and thickened a plateau which in its early form was dominated by successions of mafic-ultramafic crust. The plateau eventually reached a critical crustal thickness (>50 km) where high geothermal gradients drove internal partial melting of the buried and hydrated mafic rocks to form TTG melts. The TTG domes of the East Pilbara Terrane can be geochemically divided into contemporaneous "High Al" and "Low Al" series whose trace element characteristics imply simultaneous TTG generation at different depths of melting (Champion and Smithies 2019), adding support for an infracrustal melting scenario. The TTG melts were more buoyant than the surrounding mafic crust, and the terrane developed over time via so-called "sagduction", where gravitational instability caused the denser greenstones to sink around rising TTG melts to form the domes (Johnson et al. 2014), and where repeated pulses of melting drove overturn events generating composite granitic domal complexes. This process

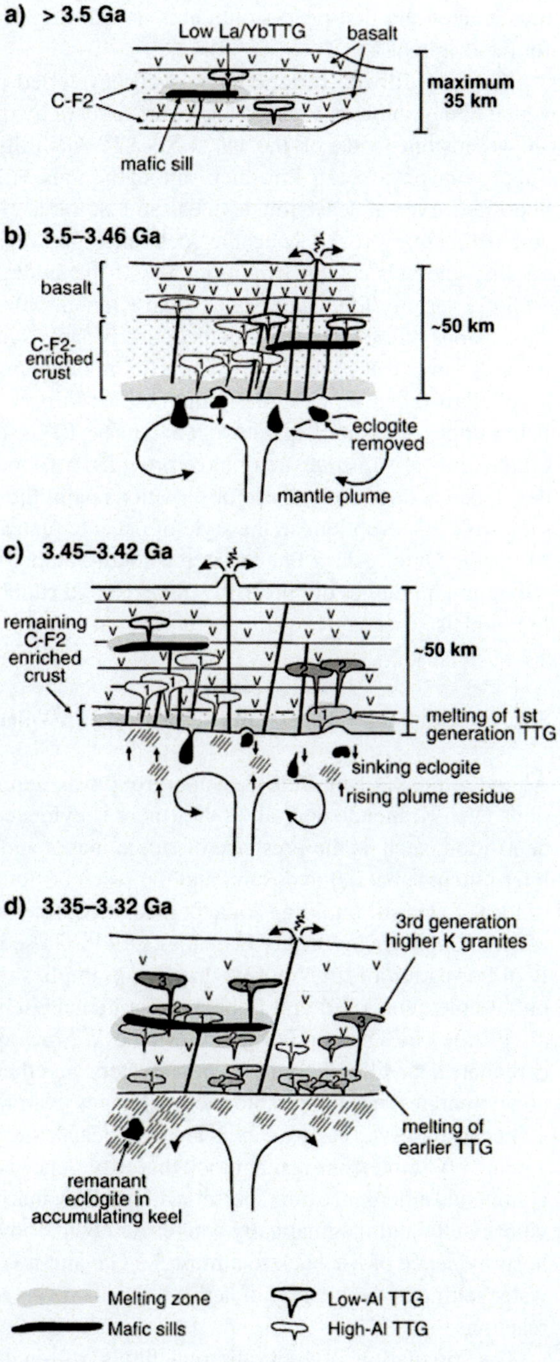

Fig. 3.1 Cartoon showing the development of the Paleoarchean crust of the Pilbara Craton. From Smithies et al. (2009)

thus created the first proto-continents, which may have offered exposed platforms for the origin of life.

The East Pilbara Terrane has a well-developed dome-and-keel architecture in which the granite complexes are surrounded by narrow, steeply-dipping arcuate greenstone belts of the ca. 3.53–3.17 Ga Pilbara Supergroup. The Pilbara Supergroup is a > 15 km-thick autochthonous volcano-sedimentary succession deposited over at least four cycles, and dominated by metamorphosed volcanic and sedimentary rocks. Volcanic rocks include komatiites, tholeiitic and enriched basalts, and felsic (e.g., rhyolitic) units, the latter which can be age-correlated with the various TTG supersuites (Van Kranendonk et al. 2007). The stratigraphic successions of the greenstone belts define broadly synclinal structures which young inwards towards the domes. Evidence for an older, possibly Eoarchean in age, now-cryptic crustal substrate underlying the East Pilbara Terrane is implied via Hf and Nd isotopes measured in the Paleoarchean TTGs (Gardiner et al. 2017). These cryptic crustal substrates may have provided a foundation upon which the plateau developed. Collectively, these observations paint the picture of a planet with large-scale volcanic eruptions in the style of oceanic plateaus, perhaps comparable to the Mesozoic Ontong-Java Plateau in the south Pacific. Horizontal plate tectonics and tall mountain ranges did probably not exist and cannot be relied upon for origin-of-life models.

3.3.1.3 Surficial Expressions of an Archean Volcanic Plateau

Although much of the Pilbara Supergroup was deposited under submarine conditions (Van Kranendonk et al. 2007), there is evidence for periods of shallow water deposition, such as the presence of ripple marks and desiccation cracks, reflecting intermittent subaerial exposure, and the siliciclastic units in part reflect the erosion of local TTGs, all implying some degree of continental emergence. The recognition of an angular unconformity beneath ca. 3.46 Ga sedimentary units of the Strelley Pool Formation of the Warrawoona Group implies that these units were deposited on an emergent, rigid and buoyant continental shelf previously eroded (Buick et al. 1995). The Dresser Formation of the Warrawoona Group yields evidence of carbonate deposition in shallow, quiet waters, and these are interspersed with coarse conglomerates, implying rapid facies changes (Van Kranendonk et al. 2008). Thus, it appears that even early on in its Paleoarchean history, the East Pilbara Terrane was a relatively stable continental block that experienced degrees of at least intermittent continental emergence interspersed with shallow marine conditions whilst still being volcanically and magmatically active. As noted above, the Dresser Formation also hosts evidence of life back to almost 3.5 Ga, and it is conceivable that this shallow-water setting in the vicinity of active volcanism was important for organic synthesis reactions.

The circulation of hydrothermal fluids within the Paleoarchean units of the Pilbara Supergroup is evident from a variety of hydrothermal-related mineral deposit types, including Volcanogenic-Hosted Massive Sulfide (VHMS), porphyry, epithermal, and chert-barite (Hutson et al. 2019). The felsic units of the Warrawoona Group are host to some of Earth's oldest recognized VHMS-type deposits, including the Big Stubby VHMS deposit of the 3.47 Ga Duffer Formation. The Dresser Formation of the North Pole Dome consists of lower and upper chert units either side of a central basalt unit including well-developed pillows. Here, massive barite veins and lenses are found within the chert layers, notably within the lower North Pole Chert. The stromatolites that have been described from the Dresser Formation in this area are interpreted to have formed alongside the barite lenses in a low temperature (100–75 °C,) near-neutral pH, hot spring environment within a terrestrial caldera setting (Harris et al. 2009). In summary, the East Pilbara Terrane is perhaps the type example of Paleoarchean vertical tectonics. Evidence points to pulsed volcanic resurfacing over the period 3.5–3.2 Ga, with concomitant TTG magmatism, and the early (3.5 Ga) formation of a stable continental block. This block experienced intermittent continental emergence leading to its surficial environment experiencing periodic, but sustained, periods of shallow marine to subaerial emergence, with evidence for the circulation of warm hydrothermal fluids allowing the development of habitable conditions in shallow warm ponds.

3.3.2 *The Atmospheric Envelope*

As noted above, geological evidence suggests that the early Earth was a volcanically active planet, and this volcanism almost certainly impacted the composition and density of Earth's atmosphere. The atmosphere must play an important role in any origin-of-life scenario for at least two reasons: First, it stabilises planetary climate and therefore controls habitability, and second it is a large repository of the gaseous elements (CHNOS) that make up life as we know it.

3.3.2.1 The Composition of Earth's Atmosphere

Life requires liquid water, but the Sun was about 30% less luminous in the early Archean compared to today (Gough 1981). Therefore, at the time of life's origin, the Earth must have had a sufficiently dense atmosphere with high abundances of greenhouse gases to maintain a liquid water ocean at the surface. This atmosphere was probably created by volcanic outgassing and its composition controlled by that of the upper mantle. Today, the upper mantle contains minor amounts of Fe^{3+} (in addition to a large reservoir of Fe^{2+}), which may have formed during core formation and disproportionation of some Fe^{2+} into Fe^{3+} and Fe^0 (Frost and McCammon 2008). The major gases in equilibrium with a mantle of this composition would likely have been N_2 and CO_2 with only trace amounts of CH_4 and H_2 (Catling and Zahnle 2020). Under the 'faint young Sun', three-dimensional global circulation models suggest that CO_2 pressures of around 60–100 mbar were needed to keep the Earth above the freezing point of liquid water (Charnay et al.

2013; Wolf and Toon 2013), but these estimates are contingent upon the distribution of land masses (Rosing et al. 2010) and will therefore require further geological constraints. Besides higher levels of CO_2, also the pressure of N_2 gas may have been higher than today (up to 2–3 bars) and contributed to greenhouse warming by pressure broadening (Goldblatt et al. 2009). This term describes a phenomenon where collisions between N_2 and CO_2 enhance the absorption of infrared radiation by CO_2. On the other hand, strong XUV radiation from the Sun may have eroded N_2 from the upper atmosphere and limited the total N_2 pressure (Lammer et al. 2019). Whether or not N_2 contributed to greenhouse warming on the early Earth is therefore unresolved.

While the abundances of the various gases are still debated, it is now widely accepted that the prebiotic atmosphere was only weakly reducing, *i.e.* dominated by N_2 and CO_2, unlike earlier models which assumed an atmosphere dominated by NH_3, CH_4 and H_2 (Miller 1953). Such a reducing atmosphere with free NH_3, CH_4 and H_2 would be difficult to reconcile with the inferred composition of Earth's upper mantle (Trail et al. 2011). Further, NH_3 is rapidly photolyzed under UV radiation and could therefore not have built up to significant levels (Kuhn and Atreya 1979). Highly reducing conditions may nevertheless have been created locally or temporarily during impacts (Zahnle et al. 2020). Most asteroids have a much more reducing composition compared to the Earth's upper mantle, and this material could therefore have favoured the temporary production of reduced carbon and nitrogen species shortly after an impact event. Activation energy from the impact itself may even have led to the formation of organic matter (Yuto et al. 2020; Zahnle et al. 2020). But outside of this impact scenario, the development of life would likely have proceeded under an atmosphere dominated by N_2 and CO_2.

3.3.2.2 Lightning and Photochemistry

The composition of the atmosphere matters for prebiotic chemistry because it determines the products of photochemical and lightning-induced reactions. Both lightning and electromagnetic radiation are energy sources that can help overcome activation energies for the formation of organic monomers or for creating new chemical disequilibria (Sect. 3.2). For this reason, the atmosphere has long played a pivotal role in origin-of-life models. In 1953, Harald Urey and Stanley Miller applied an electric discharge to a gas mixture of NH_3, CH_4 and H_2S and found it to generate a suite of organic compounds, including amino acids (Miller 1953). Subsequent experiments with gas mixtures of N_2 and CO_2 that more closely capture our modern understanding of the early atmosphere were less productive (Cleaves et al. 2008), but still small amounts of amino acids were produced. Rainout of these reaction products onto land surfaces and into the surface ocean could therefore have been an important ingredient for life's origin. Other minor products of lightning reactions would have been nitrogen oxides (Navarro-González et al. 1998), which could have acted as a strong oxidizing agent in proto-metabolic reactions. Indeed, phylogenetic data suggest that metabolic NO utilisation is ancient, possibly reflecting that early life was taking advantage of lightning products as metabolites (Ducluzeau et al. 2008).

In addition to lightning, also electromagnetic radiation could have played an important role in prebiotic chemistry, because the Earth did not have an ozone shield at the time and therefore UV radiation was able to penetrate much deeper into the atmosphere. Mass-independent fractionation (MIF) of sulphur isotopes—a phenomenon that occurs during photolysis of volcanogenic SO_2 to H_2SO_4 and S_8—is direct evidence that UV photolysis was widespread in Earth's atmosphere until about 2.4 Ga (Johnston 2011). These MIF signatures are abundantly preserved in Archean sedimentary rocks. $SO_4{}^{2-}$ can act as an oxidant and probably fuelled the first sulphate- reducing communities recorded in the 3.5 Ga Dresser Formation (Shen et al. 2001; Ueno et al. 2008). Evidence of this atmospheric sulphate flux is also preserved in the form of barite deposits that show MIF signatures (Roerdink et al. 2012). For other elements, the effect of UV photolysis is not documented, but it is conceivable that UV radiation also induced some oxidation of Fe^{2+} and Mn^{2+} to Fe^{3+} and Mn^{4+}, respectively (Anbar and Holland 1992). The Sun's UV flux would thus have created a source of oxidants at the Earth's surface. In addition, theoretical models suggest that UV photolysis in the upper atmosphere could have generated formaldehyde (CH_2O) and hydrogen cyanide (HCN) (Cleaves II 2008; Tian et al. 2011), both of which can act as precursors for the formation of amino acids, nucleobases and sugars. The exact fluxes of any of these reaction products are highly uncertain, but the MIF-S record, which currently extends back to 4.0 Ga (Thomassot et al. 2015), gives us confidence that UV radiation had a notable effect on geochemical cycles in the Hadean and Archean eons.

3.3.2.3 Aerosols

Besides producing reagents that rained out onto the Earth's surface, the atmosphere may also have hosted more complex reactions that took place within or on the surfaces of aerosols. Water droplets ejected from the sea surface may have been enriched in hydrophobic organic matter that was concentrated at the water-air interface (Dobson et al. 2000; Tuck 2002) (see also Sect. 3.3.2). Once suspended in the air, these droplets may have acted as reactors in which organic matter became progressively more concentrated as water evaporated. An important example of such a reaction is the phosphorylation of organic molecules as a key step towards the formation of RNA (Vaida 2017; Castañeda et al. 2019). Furthermore, aerosols would have been a transport pathway of material from the ocean to land surfaces.

3.3.3 The Ocean

The geological record tells us that much of the early Earth was submerged under oceanic waters. Although continental land masses probably existed and may have hosted crucial prebiotic reactions, perhaps in 'warm little ponds' fuelled by volcanism, continents were likely smaller and less extensive than today. The most productive origin-of-life scenario may therefore be one that takes advantage of processes occurring within the ocean.

3.3.3.1 Seawater Composition

Similar to the atmosphere, which was buffered by the redox state of the upper mantle, the early ocean was buffered by the composition of the crust. Most O_2 in the ocean and atmosphere is biologically produced with only very minor photochemical sources (Haqq-Misra et al. 2011), and so on a lifeless world the composition of seawater would have been anoxic. The pH was probably buffered to a value around 7 (compared to today's value of 8.1) by a combination of authigenic clay formation in sediments (Isson and Planavsky 2018), hydrothermal activity, volcanic CO_2 outgassing and carbonatization of mafic crust (Halevy and Bachan 2017; Krissansen-Totton et al. 2018). Salinity of the early ocean is much less well constrained, but it could have been 1–2 times higher than today (Knauth 2005). Under these anoxic conditions, Fe^{2+}, derived from weathering and hydrothermal alteration of mafic crust, was much more soluble than it is today. Dissolved Fe^{2+} concentrations have been estimated to around 0.1–1 mM, buffered by the precipitation of greenalite and/or siderite on the ocean floor (Tosca et al. 2016). Direct evidence for high levels of dissolved iron come from thick packages of banded iron formations composed of interlayered chert and iron-rich precipitates that are thought to have formed on (proto-)continental shelves as a result of Fe^{2+} upwelling from the deep ocean (Konhauser et al. 2017). Within banded iron formations, previous studies have documented relatively high levels of Ni and Co (Konhauser et al. 2009; Swanner et al. 2014), which were probably enriched in the early ocean due to leaching from a relatively more mafic (komatiitic) oceanic crust. In contrast, Zn and Cu levels were probably similar to or lower than today (Chi Fru et al. 2016; Robbins et al. 2016). These metals are important for origin-of-life models, because they act as catalysts in many enzymes and are therefore important for overcoming activation energies within cells.

Another interesting parameter to consider is the concentration of dissolved organic matter, which would have been a balance between abiotic sources and sinks. The modern deep sea contains on average 40 μM of dissolved organic carbon (Dittmar and Stubbins 2014), which is produced from degrading biomass and consumed mostly during aerobic respiration. On a prebiotic world, the major sources of dissolved organic matter would have been high-energy reactions in the atmosphere (Sects. 3.3.2.2 and 3.3.2.3), on beaches (Sect. 3.3.3.3) and in hydrothermal vents (Sect. 3.3.3.4); sinks of organic matter would have been photochemical and hydrothermal destruction, coagulation and sinking to the seafloor, followed by thermal degradation during burial. Earlier estimates derived a value of 3–70 mM (Miller and Orgel 1974; Bar-Nun and Shaviv 1975), *i.e.* 2–3 orders of magnitude above today's levels, but these results have been questioned (Nissenbaum 1976) though not rectified. More detailed estimates on the speciation of prebiotic organic matter are needed.

3.3.3.2 The Ocean-Atmosphere Interface

A phenomenon observable on the modern ocean is the formation of a thin coating of polymeric organic matter at the sea surface (Hunter and Liss 1981). In particular hydrophobic molecules tend to concentrate in this film, but it can also contain proteins and complex organic polymers. It is likely that hydrocarbons produced abiotically on the early Earth formed a similar sea surface coating, which has been termed the 'primordial oil slick' (Lasaga et al. 1971). This oil slick may have hosted additional synthesis reactions, such as polymerisation of HCN into sugars or the formation of tetrapyrroles (Nilson 2002; Soares et al. 2012), and it may have protected the underlying water column from UV radiation (Cleaves and Miller 1998). Furthermore, it is likely that some of these organic molecules adsorbed to floating mineral particles and pumice rafts. Pumice—a frothy and glassy rock type that is produced during explosive volcanism—floats in water and is known to form extensive rafts following volcanic eruptions. The oldest known pumice dates back to 3.5 Ga (Brasier et al. 2011), indicating that the same process took place on the early Earth. The glassy material that makes up pumice quickly alters to zeolite minerals, which themselves have hydrophobic surfaces and are therefore particularly susceptible for accumulating hydrophobic organic matter. Within these floating rock pores, desiccation may lead to further polymerisation reactions. Eventually, pumice gets water-logged and sinks to the seafloor. It is thus an efficient mechanism of transporting material from the sea surface to the bottom of the ocean.

3.3.3.3 Beaches and Shorelines

The interface between proto-continents and the ocean could itself have been an important setting for prebiotic reactions. Today, estuaries are among the most productive habitats because organisms thrive on nutrients delivered by rivers, and at the same time beaches around the world trap plastic waste materials. In a similar style, prebiotic beaches and shorelines, such as that preserved by the Dresser Formation in the Pilbara craton (Sect. 3.1), would have received input of organic material that was washed down from lakes and land surfaces as well as material that was brought in from the sea surface. Beach sand itself displays gradients in water activity, temperature and irradiation, which may have fostered new suites of chemical reactions. Another important aspect of beach sand is the accumulation of heavy minerals, including pyrite, monazite and uraninite in the Archean (Rasmussen and Buick 1999). Uraninite is radioactive, and experiments show that this form of radiation can induce polymerisation reactions as well as liberate phosphate from minerals such as monazite (Adam 2007). Materials produced in beach settings would ultimately have been washed into the open ocean and thus participated in further reactions.

3.3.3.4 Sea Ice

The presence of sea ice on the early Earth is one of the biggest unknowns, largely because global climate is not well constrained. One the one hand, the young Sun was fainter, but on the other hand, geological evidence of glaciation is sparse throughout the Archean (de Wit and Furnes 2016). If sea ice existed, locally and/or temporarily at high latitudes, it would have been part of the global prebiotic network and may have contributed important building blocks. As seawater freezes, the salt content is expelled, resulting in a highly porous ice structure rich in saline brine pockets. These brine pockets also concentrate organic matter and host phsychrophilic (cold-loving) microorganisms today (Thomas and Dieckmann 2002). Experiments suggest that prebiotic organic matter could have been stabilised in this cold saline environment, because the low temperature would have slowed down degradation while the concentrating mechanism of freezing may have induced the formation of new organic linkages (Trinks et al. 2005; Feller 2017). And any new products that were generated within sea ice pockets could later have been released during melting.

3.3.3.5 Deep-Sea Hydrothermal Vents

While the presence of sea ice is uncertain, we have direct evidence from the geological record that the early ocean hosted areas of intense hydrothermal activity (Sect. 3.3.1.3). Shortly after the discovery of submarine 'black smokers' in the 1970s, Baross and Hoffman (1985) were the first to propose that hydrothermal vents may have played a pivotal role in the origin of life because they offer a range of physicochemical gradients, catalytic minerals, and energy sources. Black smokers are so-called *magma-driven* hydrothermal vents where a deep crustal magma chamber acts as a heat source for seawater that seeped into the oceanic crust through faults and fractures. The water attains temperatures of over 400 °C (Koschinsky et al. 2008) and drastically changes its chemical composition (von Damm 1990). The O_2 is completely consumed in reactions with ferrous iron minerals such as pyroxene and olivine. Some SO_4^{2-} is precipitated as anhydrite, but much of it is reduced to H_2S. Additional sulphide is leached from basalts. Heated water alters plagioclase feldpar and pyroxene into new mineral assemblages dominated by sericite, chlorite and quartz. Feldspar alteration generates protons, and this reaction is the main reason why the vent effluents are highly acidic with pH down to 3. Under these conditions, numerous transition metals are leached into the fluid. As the fluid rises to the seafloor, sulphide ore deposits precipitate as chimneys and in breccias. These sulphide minerals can act as catalysts for the abiotic reduction of CO_2 to CH_4 and longer-chain hydrocarbons (Cody 2004). Also organic thiols have been described, which may have acted as important proto-metabolites (Russell and Martin 2004).

A key reactant in the production of organics from CO_2 is molecular hydrogen (H_2), which forms during aqueous alteration of olivine and pyroxene. H_2 is enriched in magma-hosted vent fluids, but even more so in so-called *serpentinization-driven* systems that form in ultramafic rocks. Serpentinization describes the aqueous alteration of olivine to serpentine. Ferrous iron contained in the fayalite endmember of olivine is partially oxidized to magnetite while H_2O is reduced to H_2 (Kelley et al. 2001). This reaction raises the pH to >10 and it creates highly reducing conditions within the crust, which can locally favour the formation of elemental iron-nickel alloys (Preiner et al. 2018). These alloys have strong catalytic capabilities. Besides CO_2 reduction, also N_2 and NO_3^- reduction to NH_3 has been documented in experiments (Brandes et al. 1998). Serpentinizing sites could therefore have been important sources of ammonium to prebiotic chemistry. Any nitrogen oxides produced in the atmosphere during lightning reactions (Sect. 3.2) would have been converted to ammonium or organic amines in these settings. More recently it has become apparent that conditions in hydrothermal systems may even lead to the reduction of phosphate (P(V)) to phosphite (P(III)) (Herschy et al. 2018). Phosphite is readily soluble in seawater, and more reactive towards organic matter than phosphate (Pasek et al. 2017). It could therefore have been an important precursor for phosphorylated nucleotides and RNA. Lastly, the copious amounts of H_2 generated during serpentinization would have favoured the development of H_2-based metabolisms such as methanogenesis or acetogenesis, which are both present in hydrothermal settings today and known to be deeply rooted in the tree of life (Russell et al. 2010). Several enzymes within these metabolic pathways contain metal co-factors, which may be a relic of prebiotic geochemical processes.

On the modern Earth, magma-hosted hydrothermal vents are primarily concentrated along tectonic plate boundaries, in particular along mid-ocean ridges. The only purely serpentinization-driven system, Lost City, is located off-axis along a fault that exposed ultramafic rocks (Kelley et al. 2001). Also mixed systems, such as the Rainbow vent field, exist along mid-ocean ridges where both serpentinization and magmatic heating occur. On the early Earth, hydrothermal activity may have been more widespread on the seafloor, if the Earth was hotter and more volcanically active (Sect. 3.1). Trace element data from marine sedimentary rocks confirm a much stronger hydrothermal influence on seawater chemistry in the early Archean than thereafter (Viehmann et al. 2015). These data are thus direct evidence that the products of hydrothermal reactions were widely dispersed throughout the early ocean. Today, hydrothermal plumes travel for hundreds of kilometers and are thought to even drive biological productivity in the photic zone (Fitzsimmons et al. 2014). Hence it is likely that also on a prebiotic Earth, organic and inorganic materials contained in vent fluids were transported to the surface ocean and mixed with continental runoff.

3.4 Summary and Conclusions: The Earth Was a Global Chemical Reactor

As geologists, our number-one constraint on the origin of life on Earth is the geological record, which delineates boundary conditions for the composition of ocean and atmosphere, the size of land masses and the style of volcanic and hydrothermal activity. Some of the key insights gained from this record reveal that: (1) the early Earth experienced plateau-style mafic volcanic eruptions on the seafloor; (2) proto-continents grew by remelting of thick, hydrated mafic crust. These proto-continents were likely not as geographically extensive as modern continents but offered shallow and subaerial settings; (3) Earth's mantle was mildly oxidizing, leading to volcanic outgassing of CO_2, N_2 and H_2O rather than CH_4, NH_3 and H_2 as once thought; (4) submarine volcanism resulted in intense hydrothermal circulation, which strongly influenced seawater chemistry and likely created local niches for prebiotic chemistry; and (5) intense UV radiation from the 'faint young Sun' induced photochemical reactions in the atmosphere that likely led to the formation of simple organic monomers such as HCN and CH_2O. Importantly, we have geological evidence not only for the simultaneous existence of various geological settings but also for their interaction (Fig. 3.2). For example, some hydrothermal barite deposited on the Archean seafloor contains evidence of sulphur MIF produced in the atmosphere (Golding et al. 2011). Conversely, banded iron formations deposited on continental shelves preserve signatures of hydrothermal

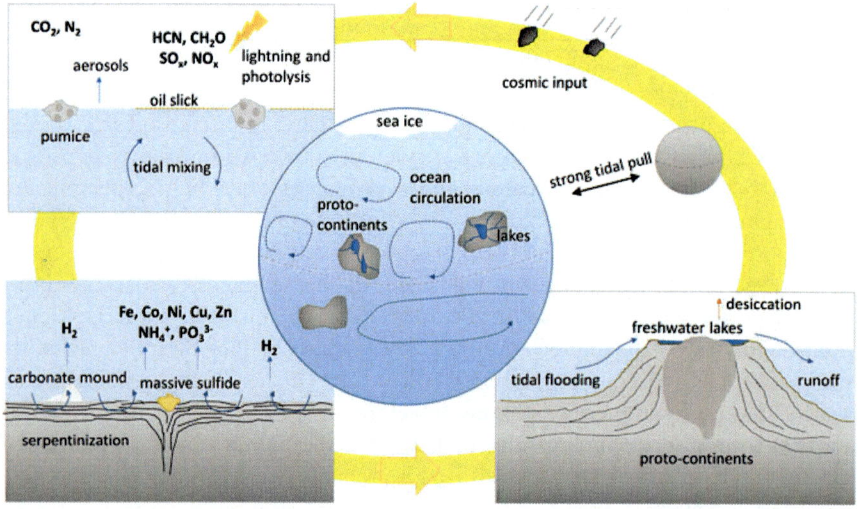

Fig. 3.2 The early Earth was a global chemical reactor with numerous interacting environmental settings and extra-terrestrial input by impacts and cosmic dust (top right). Displayed here are processes occurring near the ocean-atmosphere interface (top left), in hydrothermal settings on the seafloor (bottom left) and on small exposed land masses (bottom right)

vents (Viehmann et al. 2015). These data are thus direct evidence for physical mixing of material between different reservoirs, including the atmosphere, the deep sea and ocean margins. These mixing processes were probably driven by tidal currents and seismic activity. It is therefore imperative to place any origin-of-life settings into the global context of the early Earth. A disadvantage of large-scale mixing is the risk of dilution during transport; however, concentration mechanisms, such as on mineral surfaces or withing coagulated 'ribofilms' (Sect. 3.2) could have circumvented this problem and may in fact have allowed chemical reactions to be more selective. Furthermore, if prebiotic chemistry is firmly embedded within an environmentally diverse context, it opens up a myriad of niches for a much greater diversity of reactions than would be possible within an isolated setting. Lastly, if life originated from a global chemical reactor it may help explain the resilience and adaptability of Earth's biosphere as a whole. If we use Earth as an analogue for life on other worlds, it may therefore be necessary to target planets that are geologically diverse.

Acknowledgements We thank Sean McMahon and Anna Neubeck for inviting us to contribute to this book and for leading the editorial efforts. We also thank Professor Peter Clark who was a strong supporter of astrobiological research at St Andrews and would have enjoyed reading this compilation.

References

Adam, Z.: Actinides and the origin of life. Astrobiology. **7**, 852–872 (2007)

Allwood, A.C., Walter, M.R., Kamber, B.S., Marshall, C.P., Burch, I.W.: Stromatolite reef from the early Archaean era of Australia. Nature. **441**, 714–718 (2006)

Allwood, A.C., Rosing, M.T., Flannery, D.T., Hurowitz, J.A., Heirwegh, C.M.: Reassessing evidence of life in 3,700-million-year-old rocks of Greenland. Nature. **563**, 241–244 (2018)

Anbar, A.D., Holland, H.D.: The photochemistry of manganese and the origin of banded iron formations. Geochim. Cosmochim. Acta. **56**, 2595–2603 (1992)

Bar-Nun, A., Shaviv, A.: Dynamics of the chemical evolution of Earth's primitive atmosphere. Icarus. **24**, 197–210 (1975)

Baross, J., Hoffman, S.E.: Submarine hydrothermal vents and associated gradient environments as sites for the origin and evolution of life. Origins Life Evol. Biosph. **15**, 327–345 (1985)

Baross, J.A., Martin, W.F.: The ribofilm as a concept for life's origins. Cell. **162**, 13–15 (2015)

Becker, S., Schneider, C., Okamura, H., Crisp, A., Amatov, T., Dejmek, M., Carell, T.: Wet-dry cycles enable the parallel origin of canonical and non-canonical nucleosides by continuous synthesis. Nat. Commun. **9**, 1–9 (2018)

Black, R.A., Blosser, M.C., Stottrup, B.L., Tavakley, R., Deamer, D.W., Keller, S.L.: Nucleobases bind to and stabilize aggregates of a prebiotic amphiphile, providing a viable mechanism for the emergence of protocells. Proc. Natl. Acad. Sci. **110**, 13272–13276 (2013)

Brandes, J.A., Boctor, N.Z., Cody, G.D., Cooper, B.A., Hazen, R.M., Yoder Jr., H.S.: Abiotic nitrogen reduction on the early Earth. Nature. **395**, 365–367 (1998)

Brasier, M.D., Matthewman, R., McMahon, S., Wacey, D.: Pumice as a remarkable substrate for the origin of life. Astrobiology. **11**, 725–735 (2011)

Brown, M., Johnson, T.E., Gardiner, N.J.: Plate tectonics and the Archean Earth. Annu. Rev. Earth Planet. Sci. **48**, 291–320 (2020)

Buick, R., Thornett, J.R., McNaughton, N.J., Smith, J.B., Barley, M.E., Savage, M.: Record of emergent continental crust ~3.5 billion years ago in the Pilbara craton of Australia. Nature. **375**, 574–577 (1995)

Burcar, B.T., Barge, L.M., Trail, D., Watson, E.B., Russell, M.J., McGown, L.B.: RNA oligomerization in laboratory analogues of alkaline hydrothermal vent systems. Astrobiology. **15**, 509–522 (2015)

Callahan, M.P., Smith, K.E., Cleaves, H.J., Ruzicka, J., Stern, J.C., Glavin, D.P., House CH, Dworkin, J.P.: Carbonaceous meteorites contain a wide range of extraterrestrial nucleobases. Proc. Natl. Acad. Sci. **108**, 13995–13998 (2011)

Castañeda, A.D., Li, Z., Joo, T., Benham, K., Burcar, B.T., Krishnamurthy, R., Liotta, C.L., Ng, N.L., Orlando, T.M.: Prebiotic phosphorylation of uridine using diamidophosphate in aerosols. Sci. Rep. **9**, 1–8 (2019)

Catling, D.C., Zahnle, K.J.: The Archean atmosphere. Sci. Adv. **6** (2020). https://doi.org/10.1126/sciadv.aax1420

Champion, D., Smithies, R.H.: Geochemistry of Paleoarchean granites of the East Pilbara Terrane, Pilbara Craton, Western Australia: implications for early Archean crustal growth. In: Van Kranendonk, M.J., Bennett, V.C., Hoffmann, J.E. (eds.) Earth's oldest rocks. Elsevier, Amsterdam (2019)

Charnay, B., Forget, F., Wordsworth, R., Leconte, J., Millour, E., Codron, F., Spiga, A.: Exploring the faint young Sun problem and the possible climates of the Archean Earth with a 3-D GCM. J. Geophys. Res. Atmos. **118** (2013). https://doi.org/10.1002/jgrd.50808

Chi Fru, E., Rodríguez, N.P., Partin, C.A., Lalonde, S.V., Andersson, P., Weiss, D.J., El Albani, A., Rodushkin, I., Konhauser, K.O.: Cu isotopes in marine black shales record the great oxidation event. Proc. Natl. Acad. Sci. **113**, 4941–4946 (2016)

Chyba, C.F.: Energy for microbial life on Europa. Nature. **403**, 381–382 (2000)

Cleaves II, H.J.: The prebiotic geochemistry of formaldehyde. Precambrian Res. **164**, 111–118 (2008)

Cleaves, H.J., Miller, S.L.: Oceanic protection of prebiotic organic compounds from UV radiation. Proc. Natl. Acad. Sci. **95**, 7260–7263 (1998)

Cleaves, H.J., Chalmers, J.H., Lazcano, A., Miller, S.L., Bada, J.L.: A reassessment of prebiotic organic synthesis in neutral planetry atmospheres. Origin Life Evol. Biosph. **38**, 105–115 (2008)

Cody, G.D.: Transition metal sulfides and the origins of metabolism. Annu. Rev. Earth Planet. Sci. **32**, 569–599 (2004)

Da Silva, J.F., Williams, R.J.P.: The Biological Chemistry of the Elements: The Inorganic Chemistry of Life. Oxford University Press, Oxford, UK (2001)

de Wit, M.J., Furnes, H.: 3.5-Ga hydrothermal fields and diamictites in the Barberton Greenstone Belt - Paleoarchean crust in cold environments. Sci. Adv. **2** (2016). https://doi.org/10.1126/sciadv.1500368

Deamer, D.: The role of lipid membranes in life's origin. Life. **7**, 5 (2017)

Dittmar, T., Stubbins, A.: Dissolved organic matter in aquatic systems. In: Treatise on Geochemistry, vol. 12, 2nd edn, pp. 125–156. Elsevier, Oxford (2014)

Dobson, C.M., Ellison, G.B., Tuck, A.F., Vaida, V.: Atmospheric aerosols as prebiotic chemical reactors. Proc. Natl. Acad. Sci. **97**, 11864–11868 (2000)

Dodd, M.S., Papineau, D., Grenne, T., Slack, J.F., Rittner, M., Pirajno, F., O'Neil, J., Little, C.T.: Evidence for early life in Earth's oldest hydrothermal vent precipitates. Nature. **543**, 60–64 (2017)

Ducluzeau, A.-L., van Lis, R., Duval, S., Schoepp-Cothenet, B., Russel, M.J., Nitschke, W.: Was nitric oxide the first deep electron sink? Trends Biochem. Sci. **34**, 9–15 (2008)

Feller, G.: Cryosphere and psychrophiles: insights into a cold origin of life? Life. **7** (2017). https://doi.org/10.3390/life7020025

Ferris, J.P.: Montmorillonite-catalysed formation of RNA oligomers: the possible role of catalysis in the origins of life. Philos. Trans. R. Soc. B Biol. Sci. **361**, 1777–1786 (2006)

Fitzsimmons, J.N., Boyle, E.A., Jenkins, W.J.: Distal transport of dissolved hydrothermal iron in the deep South Pacific Ocean. Proc. Natl. Acad. Sci. **111**, 16654–16661 (2014)

Frost, D.J., McCammon, C.A.: The redox state of Earth's mantle. Annu. Rev. Earth Planet. Sci. **36**, 389–420 (2008)

Furukawa, Y., Chikaraishi, Y., Ohkouchi, N., Ogawa, N.O., Glavin, D.P., Dworkin, J.P., Abe, C., Nakamura, T.: Extraterrestrial ribose and other sugars in primitive meteorites. Proc. Natl. Acad. Sci. **116**, 24440–24445 (2019)

Gardiner, N.J., Hickman, A.H., Kirkland, C.L., Lu, Y.J., Johnson, T.E., Zhao, J.X.: Processes of crust formation in the early earth imaged through Hf isotopes from the east Pilbara terrane. Precambrian Res. **297**, 56–76 (2017)

Gibson, D.G., Glass, J.I., Lartigue, C., Noskov, V.N., Chuang, R.Y., Algire, M.A., Benders, G.A., Montague, M.G., Ma, L., Moodie, M.M., Merryman, C.: Creation of a bacterial cell controlled by a chemically synthesized genome. Science. **329**, 52–56 (2010)

Glein, C.R., Baross, J.A., Waite, J.H.: The pH of Enceladus' ocean. Geochim. Cosmochim. Acta. **162**, 202–219 (2015)

Goldblatt, C., Claire, M.W., Lenton, T.M., Matthews, A.J., Watson, A.J., Zahnle, K.J.: Nitrogen-enhanced greenhouse warming on early Earth. Nat. Geosci. **2**, 891–896 (2009)

Golding, S.D., Duck, L.J., Young, E., Baublys, K.A., Glikson, M., Kamber, B.S.: Earliest seafloor hydrothermal systems on earth: comparison with modern analogues. In: Earliest Life on Earth: Habitats, Environments and Methods of Detection, pp. 15–49. Springer, Dordrecht (2011)

Gough, D.O.: Solar interior structure and luminosity variations. Sol. Phys. **74**, 21–34 (1981)

Grotzinger, J.P., Sumner, D.Y., Kah, L.C., Stack, K., Gupta, S., Edgar, L., Rubin, D., Lewis, K., Schieber, J., Mangold, N., Milliken, R., Conrad, P.G., Des Marais, D.J., Farmer, J., Siebach, K., Calef, F., Hurowitz, J., SM, M.L., Ming, D., Vaniman, D., Crisp, J., Vasavada, A., Edgett, K.S., Malin, M., Blake, D., Gellter, R., Mahaffy, P., Wiens, R.C., Maurice, S., Grant, J.A., Wilson, S., Anderson, R.C., Beegle, L., Arvidson, R., Hallet, B., Sletten, R.S., Rice, M., Bell III, J., Griffes, J., Ehlman, B., Anderson, R.B., Bristow, T.F., Dietrich, W.E., Dromart, G., Eigenbrode, J.L., Fraeman, A., Hardgrove, C., Herkenhoff, K., Jandura, L., Kocurek, G., Lee, S., Leshin, L.A., Leveille, R., Limonadi, D., Maki, J., McCloskey, S., Meyer, M., Minitti, M., Newsom, H., Oehler, D., Okon, A., Palucis, M., Parker, T., Rowland, S., Schmidt, M., Squyres, S., Steele, A., Stolper, E., Summons, R., Treiman, A., RJP, W., Yingst, A., Team, M.S.: A habitable fluvio-lacustrine environment at Yellowknife Bay, Gale Crater. Mar. Sci. **343** (2014). https://doi.org/10.1126/science.1242777

Halevy, I., Bachan, A.: The geologic history of seawater pH. Science. **355**, 1069–1071 (2017)

Haqq-Misra, J., Kasting, J.F., Lee, S.: Availability of O_2 and H_2O_2 on pre-photosynthetic earth. Astrobiology. **11**, 293–302 (2011)

Harris, A.C., White, N.C., McPhie, J., Bull, S.W., Line, M.A., Skrzeczynski, R., Mernagh, T.P., Tosdal, R.M.: Early Archean Hot Springs above epithermal veins, north pole, Western Australia: new insights from fluid inclusion microanalysis. Econ. Geol. **104**, 793–814 (2009)

Harrison, T.M.: The Hadean Crust: evidence from >4 Ga Zircons. Annu. Rev. Earth Planet. Sci. **37**, 479–505 (2009)

Herschy, B., Chang, S.J., Blake, R., Lepland, A., Abbott-Lyon, H., Sampson, J., Atlas, Z., Kee, T.P., Pasek, M.A.: Archean phosphorus liberation induced by iron redox geochemistry. Nat. Commun., 9 (2018). https://doi.org/10.1038/s41467-41018-03835-41463

Herzberg, C., Condie, K., Korenaga, J.: Thermal history of the Earth and its petrological expression. Earth Planet. Sci. Lett. **292**, 79–88 (2010)

Hunter, K.A., Liss, P.S.: Organic sea surface films. Elsevier Oceanogr. Ser. **31**, 259–298 (1981)

Hutson, D.L., Piranjo, F., Morant, P., Cummins, B., Baker, D., Mernagh, T.P.: Paleoarchean mineral deposits of the Pilbara Craton: Genesis, tectonic environment, and comparisons with younger deposits. In: Van Kranendonk, M.J., Bennett, V.C., Hoffmann, J.E. (eds.) Earth's Oldest Rocks, pp. 519–551. Elsevier, Amsterdam (2019)

Isson, T.T., Planavsky, N.J.: Reverse weathering as a long-term stabilizer of marine pH and planetary climate. Nature. **560**, 471–475 (2018)

Johnson, T.E., Brown, M., Kaus, B.J.P., VanTongeren, J.A.: Delamination and recycling of Archaean crust caused by gravitational instabilities. Nat. Geosci. **7**, 47–52 (2014)

Johnston, D.T.: Multiple sulfur isotopes and the evolution of Earth's surface sulfur cycle. Earth Sci. Rev. **106** (2011)

Kelley, D.S., Karson, J.A., Blackman, D.K., Frueh-Green, G.L., Butterfield, D.A., Lilley, M.D., Olson, E.J., Schrenk, M.O., Roe, K.K., Lebon, G.T., Rivizzigno, P.: An off-axis hydrothermal vent field near the mid-Atlantic ridge at 30 N. Nature. **412**, 145–149 (2001)

Kemp, A.I.S., Wilde, S.A., Hawkesworth, C.J., Coath, C.D., Nemchin, A., Pidgeon, R.T., Vervoort, J.D., DuFrane, S.A.: Hadean crustal evolution revisited: new constraints from Pb–Hf isotope systematics of the Jack Hills zircons. Earth Planet. Sci. Lett. **296**, 45–56 (2010)

Knauth, L.P.: Temperature and salinity history of the Precambrian Ocean: implications for the course of microbial evolution. Palaeogeogr. Palaeoclimatol. Palaeoecol. **219**, 53–69 (2005)

Konhauser, K.O., Pecoits, E., Lalonde, S.V., Papineau, D., Nisbet, E.G., Barley, M.E., Arndt, N.T., Zahnle, K.J., Kamber, B.S.: Oceanic nickel depletion and a methanogen famine before the great oxidation event. Nature. **458**, 750–753 (2009)

Konhauser, K.O., Planavsky, N.J., Hardisty, D.S., Robbins, L.J., Warchola, T.J., Haugaard, R., Lalonde, S.V., Partin, C.A., Oonk, P.B.H., Tsikos, H., Lyons, T.W.: Iron formations: a global record of Neoarchaean to Palaeoproterozoic environmental history. Earth Sci. Rev. **172**, 140–177 (2017)

Koschinsky, A., Garbe-Schönberg, D., Sander, S., Schmidt, K., Gennerich, H.H., Strauss, H.: Hydrothermal venting at pressure-temperature conditions above the critical point of seawater, 5 S on the mid-Atlantic ridge. Geology. **36**, 615–618 (2008)

Krissansen-Totton, J., Arney, G.N., Catling, D.C.: Constraining the climate and ocean pH of the early Earth with a geological carbon cycle model. Proc. Natl. Acad. Sci. **115**, 4105–4110 (2018)

Kuhn, W.R., Atreya, S.K.: Ammonia photolysis and the greenhouse effect in the primordial atmosphere of the Earth. Icarus. **37**, 207–213 (1979)

Lammer, H., Sproß, L., Grenfell, J.L., Scherf, M., Fossati, L., Lendl, M., Cubillos, P.E.: The role of N2 as a geo-biosignature for the detection and characterization of Earth-like habitats. Astrobiology. **19**, 927–950 (2019)

Lane, N., Allen, J.F., Martin, W.: How did LUCA make a living? Chemiosmosis in the origin of life. BioEssays. **32**, 271–280 (2010)

Lasaga, A.C., Holland, H.D., Dwyer, M.J.: Primordial oil slick. Science. **174**, 53–55 (1971)

McCollom, T.M., Seewald, J.S.: Abiotic synthesis of organic compounds in deep-sea hydrothermal environments. Chem. Rev. **107**, 382–401 (2007)

McKay, C.P., Anbar, A.D., Porco, C., Tsou, P.: Follow the plume: the habitability of Enceladus. Astrobiology. **14**, 352–355 (2014)

Miller, S.L.: A production of amino acids under possible primitive Earth conditions. Science. **117**, 528–529 (1953)

Miller, S.L., Orgel, L.E.: The Origins of Life on Earth. Prentice Hall, Hoboken, NJ (1974)

Miller, S.L., Urey, H.C.: Organic compound synthesis on the primitive earth. Science. **130**, 245–251 (1959)

Mojzsis, S.J., Arrhenius, G., McKeegan, K.D., Harrison, T.M., Nutman, A.P., Friend, C.R.: Evidence for life on Earth before 3,800 million years ago. Nature. **384**, 55–59 (1996)

Navarro-González, R., Molina, M.J., Molina, L.T.: Nitrogen fixation by volcanic lightning in the early Earth. Geophys. Res. Lett. **25**, 3123–3126 (1998)

Nilson, F.P.R.: Possible impact of a primordial oil slick on atmospheric and chemical evolution. Orig. Life Evol. Biosph. **32**, 247–253 (2002)

Nissenbaum, A.: Scavenging of soluble organic matter from the prebiotic oceans. Orig. Life. **7**, 413–416 (1976)

Nutman, A., McGregor, V.R., Friend, C.L., Bennett, V.C., Kinny, P.D.: The Itsaq gneiss complex of southern West Greenland; the world's most extensive record of early crustal evolution (3900-3600 Ma). Precambrian Res. **78**, 1–39 (1996)

Nutman, A.P., Bennett, V.C., Friend, C.R., van Kranendonk, M.J., Chivas, A.R.: Rapid emergence of life shown by discovery of 3,700-million-year-old microbial structures. Nature. **537**, 535–538 (2016)

Pasek, M.A., Gull, M., Herschy, B.: Phosphorylation on the early earth. Chem. Geol. **475**, 149–170 (2017)

Pizzarello, S., Cronin, J.R.: Non-racemic amino acids in the Murray and Murchison meteorites. Geochim. Cosmochim. Acta. **64**, 329–338 (2000)

Preiner, M., Xavier, J., Sousa, F., Zimorski, V., Neubeck, A., Lang, S., Greenwell, H., Kleinermanns, K., Tüysüz, H., McCollom, T., Holm, N.: Serpentinization: connecting geochemistry, ancient metabolism and industrial hydrogenation. Life. **8** (2018). https://doi.org/10.3390/life8040041

Rasmussen, B., Buick, R.: Redox state of the Archean atmosphere: evidence from detrital heavy minerals in ca. 3250-2750 Ma sandstones from the Pilbara craton, Australia. Geology. **27**, 115–118 (1999)

Reimink, J.R., Davies, J.H.F.L., Chacko, T., Stern, R.A., Heaman, L.M., Sarkar, C., Schaltegger, U., Creaser, R.A., Pearson, D.G.: No evidence for hadean continental crust within Earth's oldest evolved rock unit. Nat. Geosci. **9**, 777–780 (2016)

Ricardo, A., Carrigan, M.A., Olcott, A.N., Benner, S.A.: Borate minerals stabilize ribose. Science. **303**, 196 (2004)

Robbins, L.J., Lalonde, S.V., Planavsky, N.J., Partin, C.A., Reinhard, C.T., Kendall, B., Scott, C., Hardisty, D.S., Gill, B.C., Alessi, D.S., Dupont, C.L.: Trace elements at the intersection of marine biological and geochemical evolution. Earth Sci. Rev. **163**, 323–348 (2016)

Roerdink, D.L., Mason, P.R.D., Farquhar, J., Reimer, T.: Multiple sulfur isotopes in Paleoarchean barites identify an important role for microbial sulfate reduction in the early marine environment. Earth Planet. Sci. Lett. **331-332**, 177–186 (2012)

Rosing, M.T.: ^{13}C-depleted carbon microparticles in > 3700-Ma sea-floor sedimentary rocks from West Greenland. Science. **283**, 674–676 (1999)

Rosing, M.T., Bird, D.K., Sleep, N.H., Bjerrum, C.J.: No climate paradox under the faint early sun. Nature. **464**, 744–747 (2010)

Russell, M.J., Martin, W.: The rocky roots of the acetyl-CoA pathway. Trends Biochem. Sci. **29**, 358–363 (2004)

Russell, M.J., Hall, A.J., Martin, W.: Serpentinization as a source of energy at the origin of life. Geobiology. **8**, 355–371 (2010)

Ryder, G.: Bombardment of the Hadean Earth: wholesome or deleterious? Astrobiology. **3**, 3–6 (2003)

Schidlowski, M.: Carbon isotopes as biogeochemical recorders of life over 3.8 Ga of Earth history: evolution of a concept. Precambrian Res. **106**, 117–134 (2001)

Schoepp-Cothenet, B., Van Lis, R., Atteia, A., Baymann, F., Capowiez, L., Ducluzeau, A.-L., Duval, S., Brink, F.T., Russell, M.J., Nitschke, W.: On the universal core of bioenergetics. Biochim. Biophys. Acta Bioenerg. **1827**, 79–93 (2012)

Shen, Y., Buick, R., Canfield, D.E.: Isotopic evidence for microbial sulphate reduction in the early Archaean era. Nature. **410**, 77–81 (2001)

Smithies, R.H., Champion, D.C., Van Kranendonk, M.J.: Formation of Paleoarchean continental crust through infracrustal melting of enriched basalt. Earth Planet. Sci. Lett. **281**, 298–306 (2009)

Soares, A.R., Taniguchi, M., Chandrashaker, V., Lindsey, J.S.: Primordial oil slick and the formation of hydrophobic tetrapyrrole macrocycles. Astrobiology. **12**, 1055–1068 (2012)

Stüeken, E.E., Anderson, R.E., Bowman, J.S., Brazelton, W.J., Colangelo-Lillis, J., Goldman, A.D., Som, S.M., Baross, J.A.: Did life originate from a global chemical reactor? Geobiology. **11**, 101–126 (2013)

Sugitani, K., Mimura, K., Takeuchi, M., Lepot, K., Ito, S., Javaux, E.J.: Early evolution of large micro-organisms with cytological complexity revealed by microanalyses of 3.4 Ga organic-walled microfossils. Geobiology. **13**, 507–521 (2015)

Swanner, E.D., Planavsky, N.J., Lalonde, S.V., Robbins, L.J., Bekker, A., Rouxel, O.J., Saito, M.A., Kappler, A., Mojzsis, S.J., Konhauser, K.O.: Cobalt and marine redox evolution. Earth Planet. Sci. Lett. **390**, 253–263 (2014)

Tashiro, T., Ishida, A., Hori, M., Igisu, M., Koike, M., Méjean, P., Takahata, N., Sano, Y., Komiya, T.: Early trace of life from 3.95 Ga sedimentary rocks in Labrador, Canada. Nature. **549**, 516–518 (2017)

Thomas, D.N., Dieckmann, G.S.: Antarctic sea ice: a habitat for extremophiles. Science. **295**, 641–644 (2002)

Thomassot, E., O'Neil, J., Francis, D., Cartigny, P., Wing, B.A.: Atmospheric record in the Hadean Eon from multiple sulfur isotope measurements in Nuvvuagittuq Greenstone Belt (Nunavik, Quebec). Proc. Natl. Acad. Sci. **112**, 707–712 (2015)

Tian, F., Kasting, J.F., Zahnle, K.: Revisiting HCN formation in Earth's early atmosphere. Earth Planet. Sci. Lett. **308**, 417–423 (2011)

Tosca, N.J., Guggenheim, S., Pufahl, P.K.: An authigenic origin for Precambrian greenalite: implications for iron formation and the chemistry of ancient seawater. GSA Bull. **128**, 511–530 (2016)

Trail, D., Watson, E.B., Tailby, N.D.: The oxidation state of Hadean magmas and implications for early Earth's atmosphere. Nature. **480**, 79–82 (2011)

Trinks, H., Schröder, W., Biebricher, C.K.: Ice and the origin of life. Origins Life Evol. Biosph. **35**, 429–445 (2005)

Tuck, A.: The role of atmospheric aerosols in the origin of life. Surv. Geophys. **23**, 379–409 (2002)

Ueno, Y., Ono, S., Rumble III, D., Maruyama, S.: Quadruple sulfur isotope analysis of ca. 3.5 Ga dresser formation: new evidence for microbial sulfate reduction in the early Archean. Geochim. Cosmochim. Acta. **72**, 5675–5691 (2008)

Vaida, V.: Prebiotic phosphorylation enabled by microdroplets. Proc. Natl. Acad. Sci. **114**, 12359–12361 (2017)

Van Kranendonk, M.J., Hugh Smithies, R., Hickman, A.H., Champion, D.C.: Review: secular tectonic evolution of Archean continental crust: interplay between horizontal and vertical processes in the formation of the Pilbara Craton, Australia. Terra Nova. **19**, 1–38 (2007)

Van Kranendonk, M.J., Philippot, P., Lepot, K., Bodorkos, S., Pirajno, F.: Geological setting of Earth's oldest fossils in the ca. 3.5 Ga dresser formation, Pilbara Craton, Western Australia. Precambrian Res. **167**, 93–124 (2008)

van Zuilen, M.A., Lepland, A., Arrhenius, G.: Reassessing the evidence for the earliest traces of life. Nature. **418**, 627–630 (2002)

Viehmann, S., Bau, M., Hoffmann, J.E., Münker, C.: Geochemistry of the Krivoy Rog banded Iron formation, Ukraine, and the impact of peak episodes of increased global magmatic activity on the trace element composition of Precambrian seawater. Precambrian Res. **270**, 165–180 (2015)

von Damm, K.L.: Seafloor hydrothermal activity: Black smokers chemistry and chimneys. Annu. Rev. Earth Planet. Sci. **18**, 173–204 (1990)

Walter, M.R., Buick, R., Dunlop, J.S.R.: Stromatolites 3,400–3,500 Myr old from the north pole area, Western Australia. Nature. **284**, 443–445 (1980)

Whitehouse, M.J., Dunkley, D.J., Kusiak, M.A., Wilde, S.A.: On the true antiquity of Eoarchean chemofossils–assessing the claim for Earth's oldest biogenic graphite in the Saglek Block of Labrador. Precambrian Res. **323**, 70–81 (2019)

Wolf, E.T., Toon, O.B.: Hospitable Archean climates simulated by a general circulation model. Astrobiology. **13**, 656–673 (2013)

Yuto, T., Yoshihiro, F., Takamichi, K., Toshimori, S., Naoki, T., Takeshi, K.: Impact-induced amino acid formation on Hadean Earth and Noachian Mars. Sci. Rep., 10 (2020). https://doi.org/10.1038/s41598-41020-66112-41598

Zahnle, K.J., Lupu, R., Catling, D.C., Wogan, N.: Creation and evolution of impact-generated reduced atmospheres of early earth. Planet. Sci. J., 1 (2020). https://doi.org/10.3847/PSJ/ab3847e3842c

Chapter 4
Homochirality: A Prerequisite or Consequence of Life?

Axel Brandenburg

Abstract Many of the building blocks of life such as amino acids and nucleotides are chiral, i.e., different from their mirror image. Contemporary life selects and synthesizes only one of two possible handednesses. In an abiotic environment, however, there are usually equally many left- and right-handed molecules. If homochirality was a prerequisite of life, there must have been physical or chemical circumstances that led to the selection of a certain preference. Conversely, if it was a consequence of life, we must identify possible pathways for accomplishing a transition from a racemic to a homochiral chemistry. After a discussion of the observational evidence, we review ideas where homochirality of any handedness could emerge as a consequence of the first polymerization events of nucleotides in an emerging RNA world. These mechanisms are not limited to nucleotides, but can also occur for peptides, as a precursor to the RNA world. The question of homochirality is, in this sense, intimately tied to the origin of life. Future Mars missions may be able to detect biomolecules of extant or extinct life. We therefore also discuss possible experimental setups for determining the chirality of primitive life forms in situ on Mars.

4.1 Introduction

The occurrence of handedness in biology is not uncommon. The difference between our left and right hands is the most obvious occurrence in the macroscopic world. In ancient Greek, the word $\chi\epsilon\iota\rho$ means hand, which explains the origin of the word chirality. Also some trees exhibit a preference for a left-handed swirl and others for

---

A. Brandenburg (✉)
Nordita, KTH Royal Institute of Technology and Stockholm University, Stockholm, Sweden

Department of Astronomy, AlbaNova University Center, Stockholm, Sweden

McWilliams Center for Cosmology, Carnegie Mellon University, Pittsburgh, PA, USA
e-mail: brandenb@nordita.org

A. Neubeck, S. McMahon (eds.), *Prebiotic Chemistry and the Origin of Life*,
Advances in Astrobiology and Biogeophysics,
https://doi.org/10.1007/978-3-030-81039-9_4

a right-handed swirl. Snails are another such example. In the microscopic world, the biological significance of a preferred handedness was discovered by Pasteur (1853) by analyzing the effect of tartaric acid on polarized light.

Polarization is a property of transversal waves when the wave shows oscillations perpendicular to the direction of propagation. This is different for sound waves that are longitudinal. Unpolarized light consists of a superposition of waves with all the different polarization orientations of the wave plane. Using a polarizer, which is an optical filter with maximum transmission for a particular wave plane, one can determine the orientation of polarization. It turns out that natural sugar in solution has the property of rotating the plane of polarization of polarized light in the right-handed sense, so they are called dextrorotatory, denoted by (+), while many amino acids in solution rotate polarized light in the left-handed sense, denoted by (−). Pasteur (1853) also inspected the shapes of crystals of tartaric acid under the microscope and found upon separating them that the two rotate polarized light in opposite senses.

Handedness of biomolecules is primarily a consequence of the tetrahedral shape of the of carbon compounds; see Fig. 4.1. If each of the four bonds of the carbon atom connect to a different group, its three dimensional structure would be different from that of its mirror image. In the case of complex molecules, there can be several carbon atoms that cause a violation of mirror symmetry. Those carbon atoms are then called chiral centers. In the case of tartaric acid (Fig. 4.2), there are two chiral centers. There is then also the possibility that only one of the two chiral centers is different. That version is called meso-tartaric acid and it is achiral, i.e., it is mirrorsymmetric.

There is no immediate connection between the handedness of molecules (Fig. 4.2) and the handedness hidden in the structure of a crystal (Fig. 4.3). In fact, the association of a given chiral structure with left or right relies on some convention. This also explains that there is nothing strange in having right-handed sugars in our DNA and left-handed amino acids in our proteins. Nevertheless, the

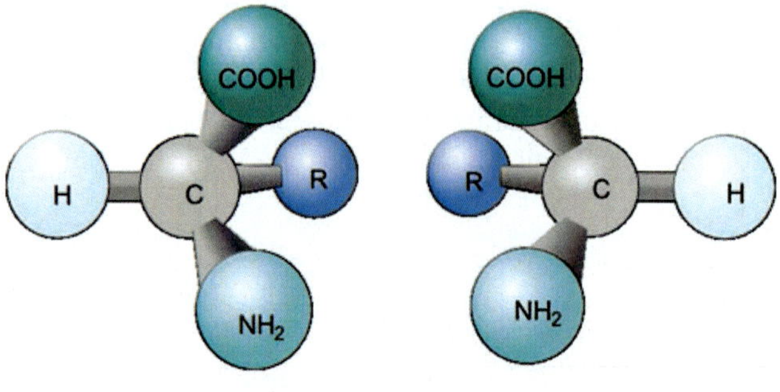

Fig. 4.1 An amino acid that is chiral whenever the residue R is different from H. For example, when R = CH_3, we have alanine, but when R = H, the molecule is glycine, which is the same as its mirror image, i.e., it is achiral. (Source: https://chem.libretexts.org/@api/deki/files/19089/molecule.png?revision=1)

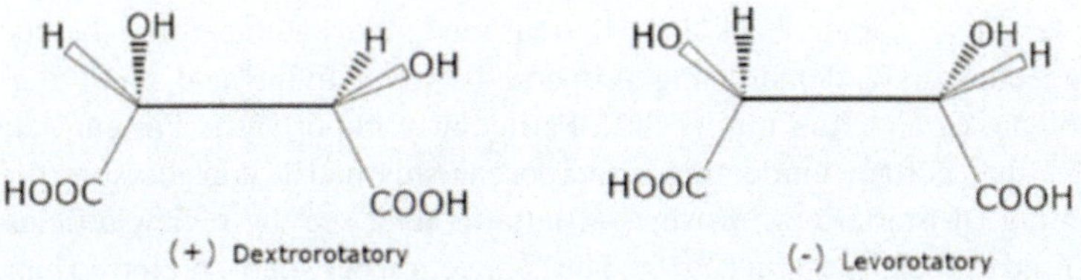

Fig. 4.2 Dextrorotatory (left) and levorotatory (right) tartaric acid. Adapted from Sevin (2015)

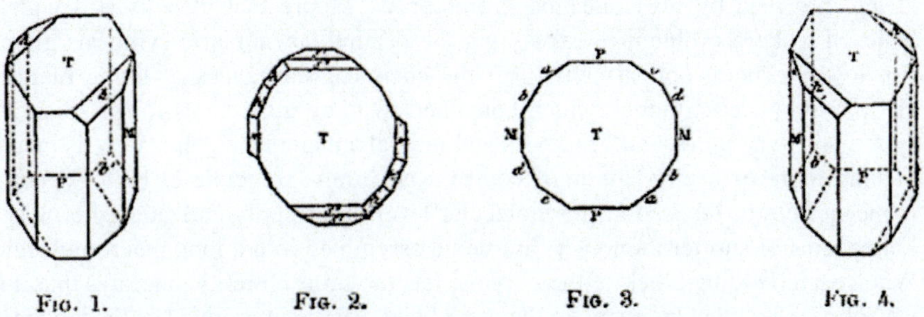

Fig. 4.3 Original drawing from Pasteur's publication (Pasteur 1922) showing dextrorotatory (left, denoted Fig. 4.1) and levorotatory (right, denoted Fig. 4.4) tartaric acid. Adapted from Sevin (2015)

very fact that something can occur in two possible forms that are mirror images of one another is non-trivial and requires some underlying structure that can also be subdivided into two opposite mirror images of each other. It is therefore plausible that one of the two handednesses of the L-tartaric acid molecule crystallizes into macroscopic structures of one form, and the D-tartaric acid into its mirror image (Derewenda 2008). In the molecular context, these two forms are called enantiomers.[1]

[1] We must emphasize that the terminology in terms of levorotatory and dextrorotatory is quite different from that in terms of D and L. Levorotatory/dextrorotatory is the physical property for a compound to induce the rotation of polarized light to the left/right. This property is abbreviated (−)/(+). By contrast, L/D refers to a structural property of a molecule to denote its handedness, that is solely based on conventions. This convention only applies to specific biomolecules, including amino acids and sugars. This convention has been taken in such a way that all biogenic sugars are D, and all biogenic amino acids are L. There is yet another terminology in which R and S refer to a structural property denoting the handedness of a given chiral carbon in a molecule. This is also based on a convention, which applies to any chiral organic compound. This convention has been taken in such a way that any chiral carbon can be assigned uniquely an R or S handedness given a precise set of rules and is thus a drastically different convention from L/D. For example, biogenic L-alanine is dextrorotatory (+) and its chiral carbon is of configuration S; biogenic L-serine is levorotatory (−) and its chiral carbon is of configuration S; biogenic L-cysteine is dextrorotatory (+) and its chiral carbon is of configuration R. Regarding sugars, D-glucose is dextrorotatory and D-fructose is levorotatory. Note also that common table sugar (sucrose, i.e., a D-glucose–D-fructose dimer) is dextrorotatory. If one hydrolyzes it, one obtains a 1:1 mixture of

Interestingly, already back then, Pasteur made the statement that the occurrence of handedness is a demarcating property between living and nonliving matter; see Goldanskii and Kuz'min (1989). Particularly important is Pasteur's discovery of 1857 that certain unidentified microorganisms had a considerable preference consuming (+)-tartaric acid over (−)-tartaric acid; see the review articles by Gal (2008) and especially Sevin (2015). The connection between a preferred handedness of biomolecules and living matter was reinforced in a number of subsequent papers. The first important one was by Frank (1953), who started his paper by saying "I am informed by my colleague Professor W. Moore that there is still widely believed to be a problem of explaining the original asymmetric synthesis giving rise to the general optical activity of the chemical substances of living matter." He then proposed a model, which contained two key ingredients for producing a systematic handedness: autocatalysis and mutual antagonism. Autocatalysis means making more of itself. This is of course a governing principle of biology, but it is meant here to be used at the molecular level during polymerization, i.e., when long chains of shorter monomers are being assembled into a long macromolecule. When each building block of the polymer has the same chirality, one says that it is isotactic. Mutual antagonism, on the other hand, can be interpreted as the tendency for a monomer of the wrong handedness to spoil the polymerization, so that the polymer would no longer be isotactic.

The basic principle discovered by Frank has been governing many of the ideas reflected in subsequent work in the field of homochirality. One such example was the work of Fajszi and Czégé (1981), who also proposed a mathematical model closely related to that of Frank. However, there are various other clues to the question of homochirality on Earth. One is that there is handedness in one of the four basic forces in nature, the weak force. We explain the details below, but this discovery implies that certain properties of a chiral molecule, for example the dissociation energy, can be different for the two enantiomers. The energy difference is usually a *very* small fraction—below 10^{-10} of the energy of the molecule itself; see Bonner (2000) for a review. Because of the smallness, it is not obvious that this alone can be responsible for achieving full homochirality. Thus, it is generally believed that some amplification mechanism is always needed.

An interesting astrobiological connection emerges when considering circularly polarized light from astrophysical sources. This is light where the polarization plane rotates with time or position. Star-forming regions in the Orion constellation have been found to emit circularly polarized light preferentially in only one of two possible senses; see Bailey et al. (1998), Bailey (2001); see Fig. 4.4 for an image of circular polarization in the Orion molecular cloud (OMC). This is interesting because different enantiomers can dissociate or degrade differently under the influence of circularly polarized light. There is further support for this line of

---

D-glucose:L-fructose, corresponding to a mixture that is levorotatory, and thus its common name of "inverted sugar".

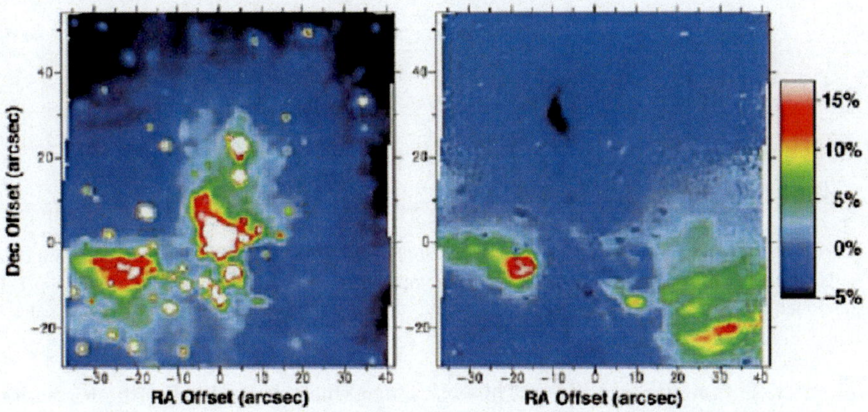

Fig. 4.4 Circular polarization measurements of the star-forming region OMC-1 in the Orion constellation. Note that the circular polarization is predominantly positive in the bulk of the molecular cloud. Courtesy of Bailey et al. (1998)

thought in that the chirality of amino acids in space, for example in meteorites, is found to show a slight preference for the levorotatory ones.

We mentioned already that the connection between chirality and the origin of life goes back to an early suggestion by Pasteur. Another connection to astrobiology arises when considering origins of life on other worlds. We will return to this at the end when we discuss possible ways of assessing the reality of extinct or extant life on Mars.

4.2 Enantiomeric Cross Inhibition: The Need for Homochirality

We mentioned already that the connection between the origin of homochirality and the origin of life has been suspected since the early work of Pasteur. This connection became more concrete with an important discovery of Joyce et al. (1984). He performed experiments with polynucleotide templates, which facilitate polymerization with the complementary monomers of the same handedness.[2]

It was thought that polynucleotide templates of one handedness would direct the pairing with monomers of the same handedness and therefore favor the selection of nucleotides of the same chirality. Joyce et al. (1984) performed experiments with polymers of dextrorotatory (D) cytosine (C) nucleobases, poly(C_D), that are expected to pair with guanosine (G) mono-nucleotides to form short strands,

[2] Instead of polymerization, one sometimes talks about polycondensation to emphasize the fact that polymerization implies the removal of water in the reaction.

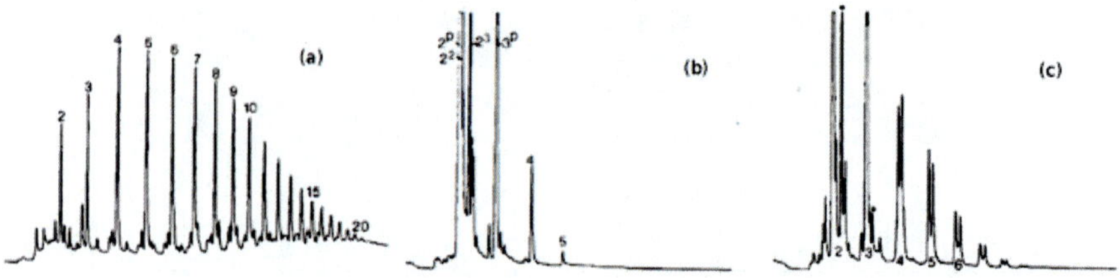

Fig. 4.5 Chromatograms from the work of Joyce et al. (1984) showing template-directed polycondensation of oligo(G_D) on poly(C_D) templates with D mononucleotide (left panel), L mononucleotide (middle panel), and a racemic mixture of D and L mononucleotide (right panel)

oligo(G_D), along the poly(C_D). This was indeed the case and led to the formation of up to 20 base pairs if the solution contained only monomers that are also dextrorotatory; see Fig. 4.5a. By contrast, when the solution contained only levorotatory monomers, no polycondensation occurred; see Fig. 4.5b.

This was also expected, because base pairs with opposite handedness do not fit together. The surprise came when using a racemic mixture of D and L mono-nucleotides. A racemic mixture would indeed be expected under prebiotic conditions. However, in that case there was no significant polycondensation—not even with the D mononucleotides; see Fig. 4.5c. Thus, the idea of using template-directed polycondensation to select only one of two handednesses did not work out. This phenomenon, which is known as enantiomeric cross inhibition, turned therefore out to be a major problem for the RNA world (Gilbert 1986), unless there was a reason to expect that only monomers of one handedness would be around. Joyce et al. (1984) wrote that "this inhibition raises an important problem for many theories of the origin of life". Bonner (1991) credited Gol'danskii and Kuz'min in saying "that a biogenic scenario for the origin of chiral purity was not viable even in principle, since without preexisting chiral purity the selfreplication characteristic of living matter could not occur." This is where the discovery of the weak force comes into play. It provides a reason why one particular handedness might be preferred. This will be discussed next.

4.3 The Weak Force: Non-mirrorsymmetry in Nature

At the atomic level, there is the strong and the weak force. They are two of the four fundamental forces in nature: gravity, the electromagnetic force, the weak force, and the strong force; see the early review by Ulbricht (1975) in the astrobiological context. The weak force is still rather strong compared with gravity (10^{24} times stronger than gravity), but weak compared with the electromagnetic force (10^{11} times weaker). The weak force is responsible for the decay of free neutrons, whose half-time is only about 10 m. The neutron (n) decays then into a proton (p) and an electron (e). This, as well as the reverse process (electron capture), occur also

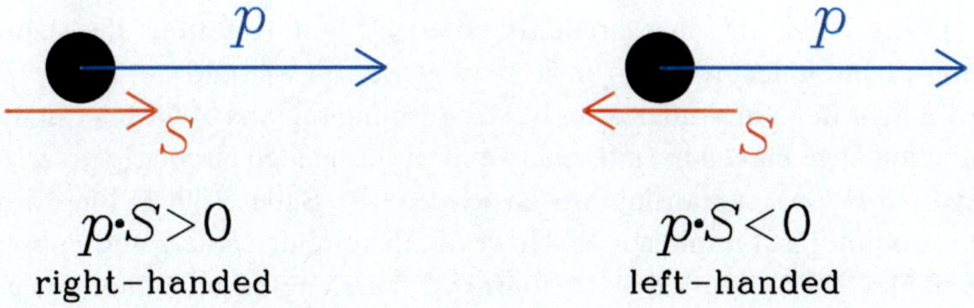

Fig. 4.6 Illustration of lepton helicity. The momentum $\boldsymbol{p}$ is a polar vector, while the spin $\boldsymbol{S}$ is an axial vector, so their dot product is a pseudoscalar, so it changes its sign when inspected in a mirror. The electron from the beta decay of a neutron has $\boldsymbol{p} \cdot \boldsymbol{S} < 0$ and is referred to as *left-handed*

in the nuclei of atoms, for example in the decay of radioactive potassium-40 into calcium-40 and argon-40, where the half-time is 1.25 Gyr. While there is significant astrobiological significance in this, for example for dating rocks,[3] we are here concerned with the fact that the electrons from the decay of neutrons are always left-handed. This means that the spin of the electron is anti-aligned with its momentum; see Fig. 4.6. At low energies, however, the spin can flip relative to the momentum, so the handedness of electrons is predominantly a high energy phenomenon.

The fact that electrons produced by β decay are chiral is remarkable, because it means that our physical world is, at least in some respects, different from its mirror image. This goes back to a remarkable discovery by Lee and Yang (1956), which earned them the Nobel Prize in Physics of 1957 "for their penetrating investigation of the so-called parity laws which has led to important discoveries regarding the elementary particles."

The connection between the chirality of electrons and that of biomolecules is not immediately evident. There are two different ways of establishing a connection between the handedness imposed by the electroweak force and the handedness in the biomolecules. One is through the fact that the bremsstrahlung emission from chiral electrons rotating around magnetic field lines is circularly polarized with a sense of polarization that depends on the chirality of the electrons. This implies that the sense of polarization from bremsstrahlung is always negative and that this radiation destroys preferentially right-handed amino acids through photolysis. This was found by Goldhaber et al. (1957) and McVoy (1957) in back-to-back papers in the Physical Review almost immediately after the influential paper by Lee and

[3] Measuring the argon inclusions in solidified rocks is the basis for determining the age of rocks. The potassium-40 isotope constitutes only 0.01% of naturally occurring potassium. Its half-time is 1.25 Gyr, making it ideal for geochronology. The two decay reactions are

$$ {}^{40}_{19}\mathrm{K} \longrightarrow {}^{40}_{20}\mathrm{Ca}: \quad \mathrm{n} \rightarrow \mathrm{p} + \mathrm{e} + \bar{\nu}_\mathrm{e} \quad (\beta \text{ decay}), \tag{4.1} $$

$$ {}^{40}_{19}\mathrm{K} \longrightarrow {}^{40}_{18}\mathrm{Ar}: \quad \mathrm{p} + \mathrm{e} \rightarrow \mathrm{n} + \nu_\mathrm{e} \quad (\text{electron capture}). \tag{4.2} $$

The latter reaction is responsible for the argon in the atmospheres of Earth and Mars.

Yang (1956). If the idea that circularly polarized light can affect the stability and selection of biomolecules is to make any sense, one should be able to discover polarized light in nature. Interestingly, star-forming regions of OMC-1 in the Orion constellation have indeed been found to emit right-handed circularly polarized light (Bailey et al. 1998), supporting this basic idea; see Bailey (2001) for a discussion of the astrobiological implications. However, the circular polarization observed by Bailey et al. (1998) occurred at near-infrared wavelengths and is not related to the mechanism of Goldhaber et al. (1957) and McVoy (1957), who considered circularly polarized bremsstrahlung. Bailey et al. (1998) argued that the observed circular polarization is caused by Mie scattering of unpolarized light, but this mechanism is unrelated to the weak force. It is therefore conceivable that also left-handed circularly polarized light could have been produced in the opposite direction.

Instead of relying on starlight, there is yet another possibility. Muons, like electrons, belong to the group of fermions that tend to have a certain handedness. Muons are about 200 times more massive than electrons and can therefore be more effective in producing strongly circularly polarized radiation. Muons occur in the cosmic radiation on Earth. They are only produced when an energetic cosmic particle hits the Earth's atmosphere and produces a muon shower. For this reason, the muons in the cosmic radiation can play a significant role in affecting the chirality of biomolecules (Globus and Blandford 2020). Unlike the observed circular polarization from the OMC-1 in the Orion constellation, the sense of circular polarization from this mechanism is connected with the weak force and therefore, just like in the case of bremsstrahlung, only one of the two senses are possible, giving rise to the preferential destruction of right-handed amino acids.

There is another completely different connection between biomolecules and the weak force. Quantum-mechanical calculations have shown that the dissociation energies for D and L molecules are slightly different (Hegstrom 1984; Hegstrom et al. 1980; Mason and Tranter 1984). Therefore, the D and L amino acids in a racemic mixture will degrade at different rates, which leads to an excess of L amino acids.

4.4 Chiral Amino Acids in Meteorites

Amino acids have been found in some meteorites (Engel and Macko 1997). Two particular meteorites are often discussed in connection with the enantiomeric excess of amino acids: the Murray and the Murchison meteorites (Pizzarello and Cronin 2000). Those are carbonaceous chondrites, which means that they are carbon-rich. They are also rich in organics, as was superficially evidenced by the smell reported by initial eyewitnesses of the Murchison meteorite. Interestingly, Table 1.5 of Rothery et al. (2008) lists 18 different amino acids that have been found not only in the Murchison meteorite, but also in the Miller–Urey experiment (Miller 1953). Twelve of them are not found in proteins on Earth. This is interesting, because it suggests that those amino acids were indeed originally present in the meteorite and could not have come from contamination by life after the meteorite landed on Earth. Those amino acids that are found on Earth include glycine, alanine, valine, proline, aspartic acid, and glutamic acid.

The sense of the enantiomeric excess is the same in the two meteorites, corresponding to levorotatory amino acids, but the amount is different (Pizzarello and Cronin 2000). In addition, there is the possibility that the enantiomeric excess may be caused by terrestrial contamination (Bada 1995). But, as emphasized above, this would only apply to the six amino acids that are also found on Earth. In particular, those amino acids that have the clearest enantiomeric excess are also those that are most vulnerable to contamination; see Ehrenfreund et al. (2001) for a discussion of terrestrial contaminants in connection with the carbonaceous chondrites Orgueil and Ivuna. They are of the type CI (I for Ivuna) and are extremely fragile and therefore susceptible to terrestrial weathering. In Orgueil, alanine was found to be racemic and was argued to be abiotic in origin (Ehrenfreund et al. 2001). They could not, however, support the suggestion of terrestrial contamination with corresponding soil samples. Incidently, the Orgueil meteorite is also known for a famous contamination hoax; see Anders et al. (1964), who discusses the paper by Cloez (1864) claiming the existence of life on the meteoritic parent body a few weeks after Pasteur's famous lecture to the French Academy on the spontaneous generation of life.

Among the possible causes for the enantiomeric excess of meteoritic amino acids, there is the aforementioned effect of circularly polarized starlight. Circularly polarized ultraviolet light could have preferentially destroyed one of the two chiralities through photolysis (Zeldovich et al. 1977). The experiments of Bonner et al. (1981) with a D L mixture of leucine showed that right-handed circular polarized light leads to a preferential destruction of D leucine, while left-handed circular polarized leads to a preferential destruction of L leucine; see also (Meierhenrich and Thiemann 2004) for recent experiments. To explain the systematic L excess of amino acids on Earth, one would need the protosolar nebula to be irradiated by right-handed polarized light. Indeed, the star-forming region OMC-1 has been found to emit right-handed circularly polarized light, supporting this basic idea (Bailey 2001; Bailey et al. 1998; Boyd et al. 2018). However, as discussed above, also left-handed circularly polarized light could have been produced in the opposite direction. Therefore, any systematic L excess of amino acids caused by this mechanism would have been by chance.

The enantiomeric excess found in some amino acids is at most around 1–2%. This would be too small to avoid the problem reported by Joyce et al. (1984). So, even if there is an external effect producing a systematic enantiomeric excess, we always need an amplification mechanism. Therefore, we discuss next the Frank mechanism and move then to some variants of it that avoid either autocatalysis or enantiomeric cross inhibition. We begin by explaining first the basic idea.

4.5 The Basic Idea Behind the Frank Mechanism

The essence of the mechanism of Frank (1953) is the *combination* of two ingredients operating in a substrate: catalysis of molecules for their own production and "anticatalysis" that corresponds to some antagonism or deleterious effect. He even talks about "poisoning" one of the two enantiomers out of existence. In fact, he called his simple mathematical model a "life model", suggesting already back then that he was thinking of them as being processes acting at the moment when the first life emerged.

The essence of Frank's model is perhaps best explained graphically. For this purpose, it is most instructive to begin with the deleterious effect by assuming that an equal amount of D and L enantiomers eliminate each other in each reaction step. This is illustrated in Fig. 4.7, where we indicate the amount of D enantiomers with blue bars and the amount of L enantiomers with red bars.

We see that, in the end, only D is left (see the blue bars), but the amount is very small, namely just as big as the initial difference by which one of the two enantiomers exceeded the other. This is why we also need autocatalysis. Autocatalysis is a process that is not enantioselective, i.e., it works the same way for the D and L enantiomers. This is demonstrated by stretching out the columns by a factor such that the highest column always retains the original height; see Fig. 4.8. It is instructive to quantify here the enantiomeric excess (e.e.) as the ratio of the difference to the sum of the concentrations of right- and left-handed compounds, i.e.,

$$\text{e.e.} = \frac{[D]-[L]}{[D]+[L]} \tag{4.3}$$

At each each, the value of e.e. in Figs. 4.7 and 4.8 is the same: $1/(5+4) = 1/9$ initially, then $1/(4+3) = 1/7$, $1/(3+2) = 1/5$, $1/(2+1) = 1/3$, and finally $1/1 = 1$.

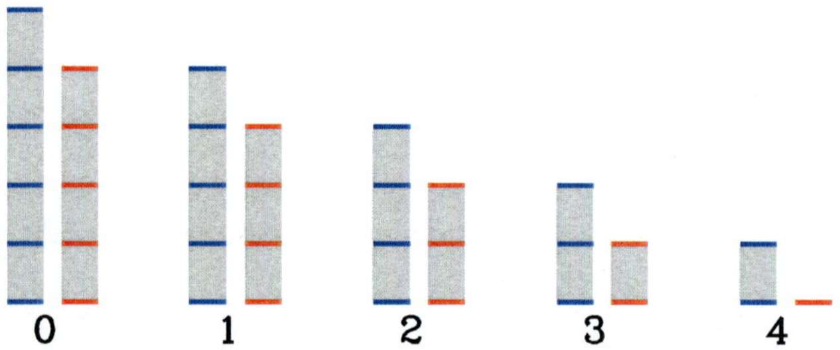

Fig. 4.7 Sketch showing the effect of enantiomeric cross inhibition only. Red and blue bars indicate opposite enantiomers, with the blue one being initially in the majority by one "unit", the separation between subsequent bars. The gray columns indicate the total amounts, which is 5 units for the column with blue bars and 4 units for the column with red bars. In the end, in step 4, only one unit of the enantiomer marked with blue bars survives

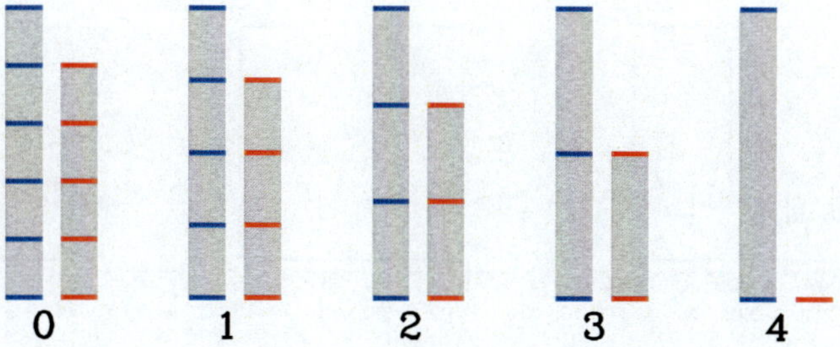

Fig. 4.8 Similar to Fig. 4.7, but in each reaction step, the separations between subsequent bars has been stretched by a certain factor such that the column with the blue bars retains the same height. The stretching emulates the effect of autocatalysis. In the end, again only the enantiomers marked with blue bars survive, but now, because of the stretching, the amount is no longer small

Frank's paper was mathematical, which may have been a reason why it was not widely recognized in the biology community at the time. In fact, it was not quoted in the work of Joyce et al. (1984), who discovered enantiomeric cross inhibition in the context of nucleotides. It was only with the paper of Sandars (2003) that Frank's mutual antagonism was identified with enantiomeric cross inhibition. It was clear from Frank's work that, as long as both reactions, autocatalysis and antagonism, remain active, the racemic state is unstable and there will be a bifurcation into a chiral state with an excess of either D or L enantiomers; see also Sandars (2005), where enantiomeric cross inhibition was no longer regarded as a problem, but as an essential ingredient in achieving full homochirality.

In the Frank mechanism, there must be at least a very small initial imbalance which will then be amplified. However, this is not a problem because, even if we tried to construct a purely racemic mixture in the laboratory, there will always remain a tiny imbalance. This is just for the same reasons that in a cup of blueberries we will hardly ever have exactly the same number twice.[4]

Given that the racemic state is unstable, the enantiomeric excess, as defined in Eq. (4.3), will grow exponentially in time and it does therefore not matter how small the initial imbalance in the concentrations of D and L was. To demonstrate this more clearly, we use here a figure of Brandenburg et al. (2007), who considered the model of Plasson et al. (2004), which we discuss later in more detail in Sect. 4.10. This model also has the property that the racemic state is unstable and that the system evolves toward one of the two homochiral states.

In the following, we discuss an ensemble of solutions of the model of Plasson et al. (2004) with different realizations or initial states, which consisted of a racemic

[4] If you take a cup of blueberries, for example, the exact number varies between 65 and 70 (https://www.howmuchisin.com/produce_converters/blueberries), so we must always expect there to be a small imbalance in the number if we say we have an equal *amount* of D and L enantiomers. Mathematically, this imbalance grows with the square root of the number of molecules (or blueberries) and would be about $\pm 10^{12}$ for one mole with $\mathcal{N} = 6 \times 10^{23}$ molecules (or ± 8 for 65 blueberries); the fractional imbalance is $1/\sqrt{\mathcal{N}} = 10^{-12}$ in one mole (or 12% for 65 blueberries).

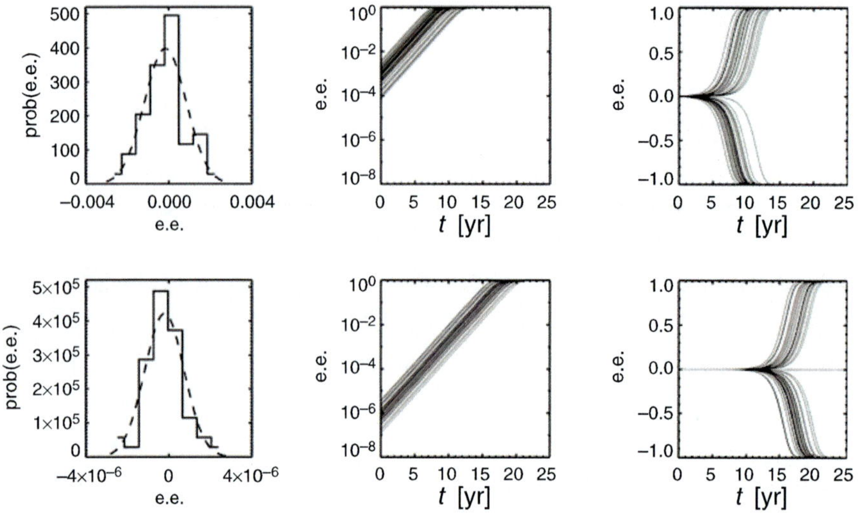

Fig. 4.9 Probability distribution of the initial enantiomeric excess (e.e.) for racemic mixtures with 10^6 and 10^{12} molecules together with the resulting evolution of e.e., both in logarithmic and linear representations. The dashed lines give a gaussian fit to the distribution function. Adapted from Brandenburg et al. (2007)

mixture of equally many D and L enantiomers. Figure 4.9 shows that one always obtains a fully homochiral state, but in about 50% of the cases (or in 50% of the realizations of the same experiment), one obtains eventually a state with either just D enantiomers, and in the other 50% of the cases or realizations, one with only L enantiomers. When we talk about different cases or realizations, we must realize that the genesis of life on Earth is just one such realization. Another one may have occurred on Mars, or in the atmosphere of Venus, or elsewhere in the Galaxy. Of course, there is also the possibility of multiple geneses on Earth alone, with certain lifeforms being either completely or partially wiped out (Davies and Lineweaver 2005). The latter case may be particularly interesting in models where we allow for chemical evolution in models with spatial extent, which will also be discussed later in Sect. 4.9.

4.6 Evidence for Autocatalysis

Unlike the process of enantiomeric cross inhibition, where we have referred to the experiments of Joyce et al. (1984), the actual evidence for autocatalysis is poor. In fact, there is only the classical reaction of Soai et al. (1995) that exhibits autocatalysis and can lead to a finite enantiomeric excess; see Gehring et al. (2010) and Athavale et al. (2020) for more recent work clarifying the implications of the

Soai reaction. However, the basic idea of autocatalysis remains plausible, especially since the discovery by Guerrier-Takada and Altman (1984) and Cech (1986) that RNA molecules can exhibit autocatalytic functionality. This was a very important discovery that earned Sidney Altman and Thomas R. Cech the Nobel Prize in Chemistry in 1989 "for their discovery of catalytic properties of RNA." It is this mechanism that is at the heart of the idea of an RNA world (Gilbert 1986).

It is important to realize that the existing evidence for autocatalysis is irrelevant from an astrobiological viewpoint. This is particularly clear in view of the fact that the Soai reaction requires zinc alkoxides as an additional crucial catalyst. Those compounds are not generally believed to play a role on the early Earth.

Autocatalysis in the sense of making more of itself is obviously a basic principle of life, but this is already at a rather complex and not at the level of individual molecules. It is therefore possible that autocatalysis does not play a significant role and that it is rather the process of network catalysis (Plasson 2015), i.e., the combined action of different molecules that lead to the desired appearance of what is in the end equivalent to autocatalysis. We will return to this in Sect. 4.10, when we discuss a particular sequence of reactions that, in the end, have the effect of autocatalysis, even though autocatalysis is not present in any individual reaction.

4.7 The Effect of an External Chiral Influence

In the beginning of this review, we have discussed extensively the possibility of a systematic bias resulting eventually from the fact that the weak force introduces a preference of one of two handednesses through one or several possible effects. Those would always favor L amino acids and D sugars. On the other hand, we have now seen that the Frank mechanism can result in full homochirality of either chirality. Does this mean that the bias introduced by the weak force is unimportant? Maybe not quite. It depends on how strong the external influence is in comparison with the speed of autocatalysis, which determines the rate of the instability. This was first discussed in the work of Kondepudi and Nelson (1983, 1985) in papers that appeared just at the time as that of Joyce et al. (1984), but, at the time, neither of those authors mentioned the work of Frank.

The paper by Kondepudi and Nelson (1983, 1985) was in principle quite general and therefore applicable to other symmetry breaking instabilities. In essence, the effect of the bias is that it makes the bifurcation asymmetric. A symmetric bifurcation is one where the enantiomeric excess (positive or negative) departs away from strictly zero as some bifurcation parameter increases. Sandars (2003) identified this bifurcation parameter with the fidelity of the autocatalytic process, which measures the probability with which the catalytic process does indeed facilitate the polymerization with monomers of the same chirality instead of the opposite one. The fidelity f is unity (zero) when the autocatalytic process always (never) produces polymerization with the same handedness.

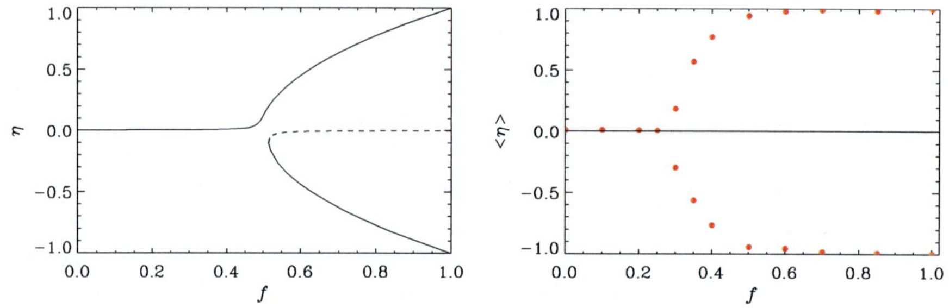

Fig. 4.10 Bifurcation diagrams showing a slight preference for positive enantiomeric excess (e.e., here denoted by η). The left panel has been adapted from Brandenburg et al. (2005), where the bifurcation begins for a fidelity f that is clearly below the otherwise critical value of $f = 0.5$. (The dashed line denotes the unstable solution.) The right panel has been adapted from Brandenburg (2019), who considered a stochastic model where $f = 0.2$ was assumed

In Fig. 4.10 we show a bifurcation diagram from the work of Brandenburg et al. (2005), where we see that for all values of the fidelity f, the solution with positive enantiomeric excess (η) is stable. For $\eta \gtrsim 0.5$, the solution with negative η is also stable, but to reach this solution, the initial fluctuations must be large enough. The complete bifurcation diagram also contains an unstable solution, which corresponds to the watershed between the two stable branches. In the left hand plot of Fig. 4.10, it is shown as a dashed line. Similar diagrams have also appeared in the works of Kondepudi and Nelson (1983) and later in the review of Avetisov et al. (1991).

4.8 Polymerization Model of Sandars (2003)

Looking at the chromatographs of Joyce et al. (1984), we see that the ultimate goal is to assemble long polymers. For this reason, Sandars (2003) developed a polymerization model for D and L nucleotides, where he also allowed for enantiomeric cross inhibition. In his model, monomers of the D and L forms are being produced at rates, Q_D and Q_L, respectively, that are proportional to same reaction rate k_C and the concentration of some substrate $[S]$, i.e.,

$$Q_D = k_C[S]\left\{\tfrac{1}{2}(1+f)C_D + \tfrac{1}{2}(1-f)C_L + C_{0D}\right\}, \tag{4.4}$$

$$Q_L = k_C[S]\left\{\tfrac{1}{2}(1+f)C_L + \tfrac{1}{2}(1-f)C_D + C_{0L}\right\}, \tag{4.5}$$

where $0 \le f \le 1$ is the fidelity, C_L and C_D are parameters describing the global handedness of the system (the concentrations of the longest possible chains of left- and right-handed polymers for Sandars 2003 and quantities proportional to the masses of all polymers of the D and L forms for Brandenburg et al. 2005). These parameters are introduced in such a way that for $f > 0$ in Eqs. (4.4) and (4.5), Q_D increases with C_D, and Q_L increases with C_L. The parameters C_{0D} and C_{0L} allow for the possibility of non-catalytic production of left- and right-handed monomers. They can be different from zero when there is an external bias or external influence. When $C_{0D} = C_{0L} = 0$, Eqs. (4.4) and (4.5) show that $Q_D = k_C[S]C_D$ and $Q_L = k_C[S]C_L$ when $f = 1$, while $Q_D = Q_L = k_C[S](C_D + C_L)/2$ when $f = 0$.

Sandars (2003) assumed that the catalytic effect depends on the concentrations of the longest possible chains of left and right handed polymers. Brandenburg et al. (2005) adopted a similar model, but assumed that C_D and C_L to be proportional to the masses of all polymers of the D and L forms, respectively. This allowed them to extend the model to much longer polymers without needing to wait for the longest one to appear before autocatalysis became possible at all.

The full set of reactions included in the model of Sandars (2003) is (for $n \geq 2$)

$$L_n + L_1 \xrightarrow{2k_S} L_{n+1}, \tag{4.6}$$

$$L_n + D_1 \xrightarrow{2k_I} L_n D_1, \tag{4.7}$$

$$L_1 + L_n D_1 \xrightarrow{k_S} L_{n+1} D_1, \tag{4.8}$$

$$D_1 + L_n D_1 \xrightarrow{k_I} D_1 L_n D_1, \tag{4.9}$$

where k_S and k_I are suitably chosen reaction rates for symmetric autocatalysis and (non-symmetric) inhibition, respectively. For all four equations we have the complementary reactions obtained by exchanging $L \rightleftarrows D$. The polymerization starts from a large but limited set of monomers that all begin to develop longer polymers. Because the number of monomers was limited, the theoretically obtained chromatograms show a characteristic wave-like motion with increasing time; see Fig. 4.11.

Particularly important is of course the case where monomers of both chiralities exist. In that case, the result depends on the fidelity of the autocatalytic reactions; see Fig. 4.12 for such a result. We see that longer polymers can only be produced when the fidelity is relatively high. The lack of sufficient fidelity therefore explains the limited length of polymers found in the work of Joyce et al. (1984).

4.9 Spatiotemporal Chirality Dynamics

In all the chemical reactions discussed so far, the assumption was made that the system is well mixed. This means that the concentrations $[D_n]$, $[D_n L]$, etc, are the same everywhere. On larger length scales, this assumption must eventually break down. Even on the scale of alkaline hydrothermal vents, where many scientists place the origin of life (Russell 2006) the relevant chemical reactions would take place

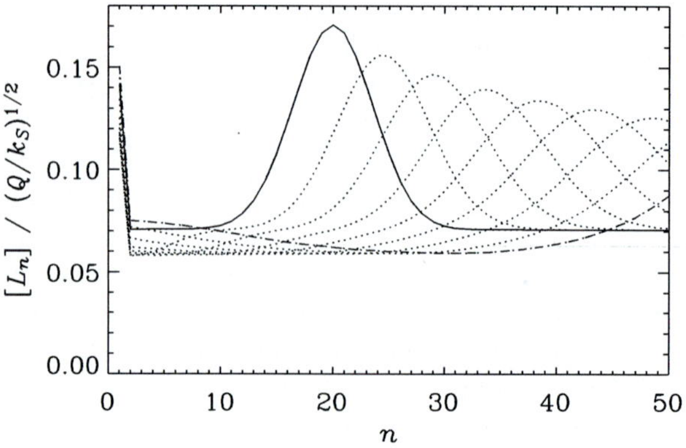

Fig. 4.11 Normalized concentration $[L_n]$ versus n showing a wave-like evolution of an initial Gaussian profile (solid line). All later times are shown as dashed lines, except for the last time, which is shown as a long-dashed line. Adapted from Brandenburg et al. (2005)

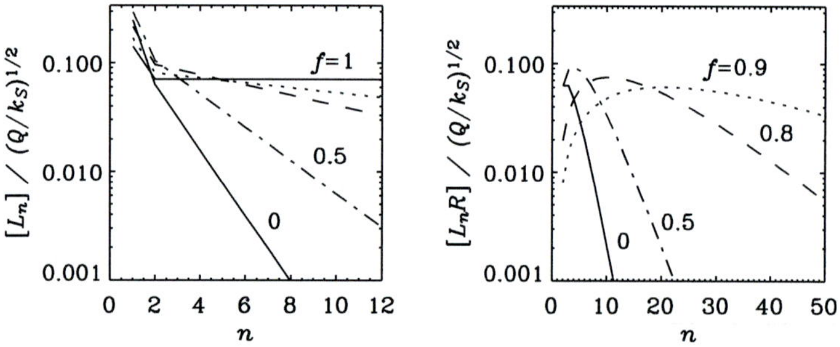

Fig. 4.12 $[L_n]$ (left) and $[L_n D]$ (right) of equilibrium solutions for different values of f. For $f = 1$ we have $[L_n D] = 0$, which cannot be seen in the logarithmic representation. Adapted from Brandenburg et al. (2005)

within small semiporous cells. It is then conceivable that similar reactions take place in neighboring compartments that would be formed by the sulfurous precipitants from these vents. Russell (2006) draws here an analysis to the chemical gardens that would allow for a growing arrangement of new compartments, which could act as primitive cells and would, in principle, allow for Darwinian evolution as these chemical reactions propagate from one layer of compartments to the next; see also Russell et al. (2014) and Barge et al. (2017, 2019) for more recent developments. In each of these compartments, strong spatial gradients and 10^8-fold concentration enhancements can be achieved through thermal convective flows when the aspect ratio of the compartment is sufficiently large (Baaske et al. 2007). This setup can also lead to oscillations, which can locally lead to exponential replication of nuclei acids, analogous to the polymerase chain reaction (Braun and Libchaber 2002).

In the scenario described above, we can no longer talk about a well mixed system. Therefore, the concentrations must be regarded as function not only of time, but also of space. Because the chemistry in neighboring compartments is loosely coupled by diffusion terms, there would be spatio-temporal evolution. In that case, the chemical reaction equations attain a spatial diffusion term. The resulting system of equations is usually referred to as reaction–diffusion equations. Such models, but for only one instead of several species, have frequently been employed in modeling the dynamics of diseases such as the black death (Noble 1974) or rabies (Källén et al. 1985; Murray et al. 1986). It has also been used to model the spreading of the novel coronavirus, where the total number of cases was found to follow a quadratic or piecewise quadratic growth behavior (Brandenburg 2020).

To address the question of homochirality in an extended system, Brandenburg and Multamäki (2004) employed a similar approach, but with two or multiple species. Multiple species occur when we invoke polymers of different length and composition of different species for the D and L forms. They found that a given species tends to spread through front propagation. It turned out that, once two populations of opposite chirality meet, the front can no longer propagate and the evolution comes to a halt. This result was first obtained in a one-dimensional model, where the concentrations of D and L depend on just one spatial coordinate x and on time t. The result is shown in Fig. 4.13, for the evolution of short polymers. These are all regarded as separate species. The initial condition consists of a small number of monomers of the L form at one position (at $x/\lambda = -0.1$ in Fig. 4.13, where λ is the length of the domain) and a three times larger number of monomers of the D form at another position (at $x/\lambda = +0.1$). The theoretical chromatograms are stacked next to each other for each x position. We see that in both positions, longer polymers are produced, indicated by the yellow-reddish colors. The initially threefold larger number of D monomers is insignificant, because the growth is exponentially fast soon saturates at the same level as that for the L monomers, when the polymers have reached their maximum size. At the same time, polymers of the same handedness can still be produced by diffusion to the neighboring positions. This leads to a propagation front. However, when polymers of opposite chirality emerge at neighboring positions, the front stops (here at $x \approx 0$), while on the other two sides, the reaction fronts still diffuse further outward.

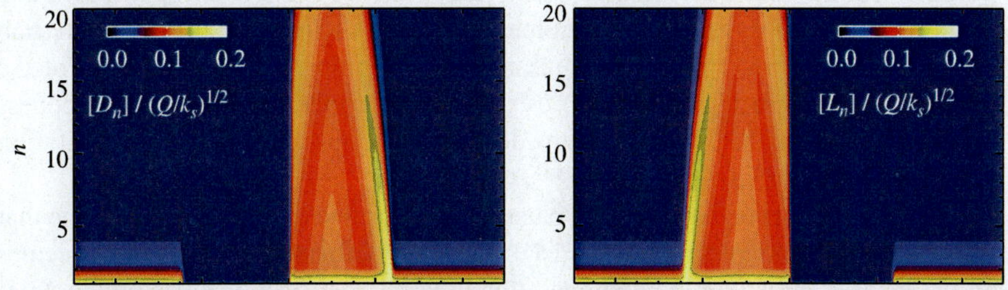

Fig. 4.13 Color scale plots of $[D_n]$ and $[L_n]$ after 0.8 diffusion times as a function of position x and polymer length n. In $0 < x/\lambda < 0.25$, only D polymers exist (left) and no L polymers at all (right), while in $-0.25 < x/\lambda < 0$ it is the other way around. Adapted from Brandenburg and Multamäki (2004)

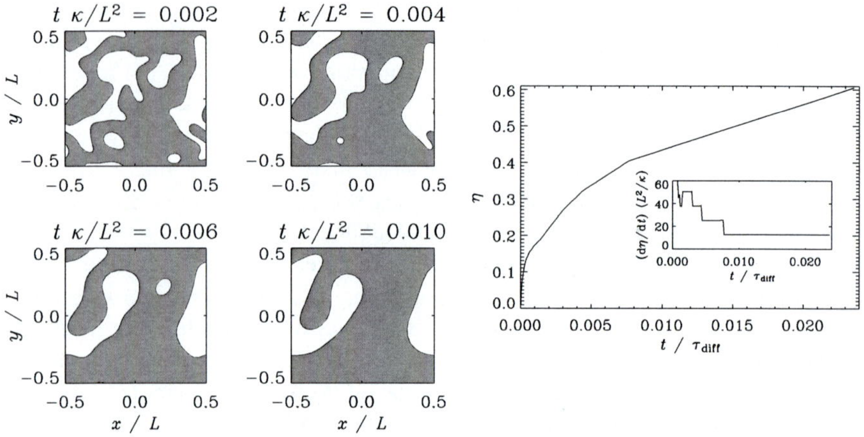

Fig. 4.14 *Left*: Fractional concentration of one chirality versus position (x and y) at four different times t, normalized by the diffusivity κ per total surface area λ^2, so $t\kappa/\lambda^2$ is nondimensional. In this numerical simulation, $\kappa/(\lambda^2\lambda_0) = 2 \times 10^{-4}$, and the resolution was 1024^2 mesh points. The number of disconnected regions decreases from 4 in the first plot to 3, 2, and 1. *Right*: Evolution of enantiomeric excess η for the model shown in the left. The inset shows the normalized slope. Note the four distinct regimes with progressively decreasing slope. Adapted from Brandenburg and Multamäki (2004)

More interesting dynamics is possible when the system is two-dimensional, corresponding to different locations on the Earth's surface. In that case, the fronts between regions with monomers and polymers of opposite handedness can be curved. It turned out that then the fronts are never precisely straight and can therefore still propagate. Interestingly, the propagation is always in the direction of maximum curvature. This result has also been obtained by Gleiser and Walker (2012). This then implies that a closed circular front will always shrink and never expand; see left Fig. 4.14. The speed of shrinking depends just on the number of individual closed fronts. Each time a closed front merges with another one to form a single one, the speed decreases; see the right panel of Fig. 4.14. In particular, this means that, even if, say, the L enantiomers were initially in the majority, but in such a way that they would be enclosed in an island surrounded by enantiomers of the opposite handedness, the enantiomeric excess would develop toward the handedness that was present on the periphery of the domain; see Fig. 4.15 for such an illustration with the model of Plasson et al. (2004).

The reason for this particular propagation direction is quite simple. Imagine that we place D and L molecules around a circular front, then the number of molecules on the inner front is would be one less than the number of molecules on the outer front; see Fig. 4.16 for an illustration.[5]

[5] To understand why the difference in the number of molecules between the outer and inner circles is always just three, let us imagine the molecules being represented by little discs of radius r on

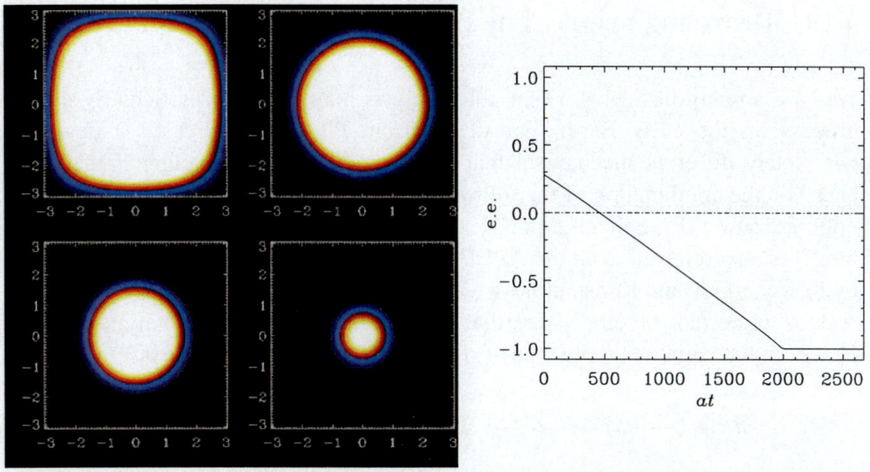

Fig. 4.15 *Left*: Shrinking of an initially large patch of molecules of the D form surrounded by molecules of the L form. *Right*: The resulting enantiomeric excess (e.e.) versus time t, scaled with the activation rate a, so at is nondimensional. Note that it was initially positive, but reaches later complete homochirality with $\eta = -1$

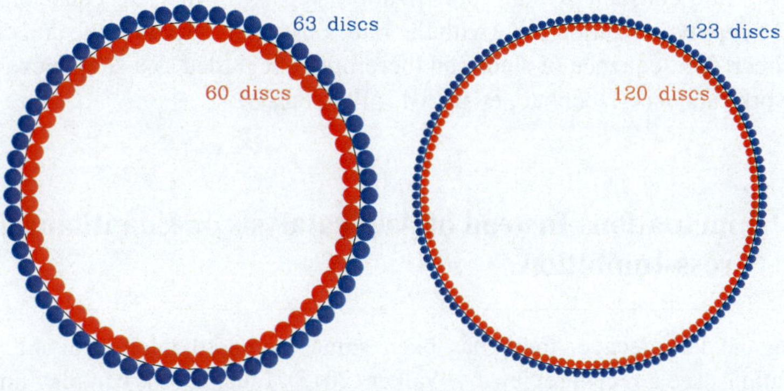

Fig. 4.16 Sketch illustrating that densely packed discs inside the periphery of a circle differ in their number from those outside the periphery by just 3. This result is independent of the total number; compare the left and right illustrations with 60 and 120 discs, respectively. As time goes on, pairs of red and blue discs get eliminated and the circle shrinks, because the number of discs inside the periphery is slightly smaller (by three) than the number of discs outside the periphery. The smallness of the difference in the numbers on the inner and outer peripheries is the reason for the shrinking of the circle slowing down

---

the periphery of a circle of radius R. The circumferences of the outer and inner peripheries are $2\pi(R \pm r) = 2\pi R \pm \pi r$ for the upper and lower signs, respectively. The difference is therefore $2\pi r \approx 3d$, where $d = 2r$ is the diameter of each disc. The difference in the number of discs is there for three.

4.10 Recycling Frank: The Peptide Model of Plasson et al.

We said already in Sect. 4.6 that autocatalysis may not be a particularly evident process on the early Earth. For that reason, Plasson et al. (2004) devised a completely different mechanism that they advertised as "recycling Frank". It is based on the combination of the following four important reactions: activation (A), polymerization (P), epimerization (E), and depolymerization (D). So the resulting model is also referred to as the APED model. A variant of this model was studied by Konstantinov and Konstantinova (2018).

It is important to emphasize that there is no explicit autocatalytic reaction. However, the combined sequence of reactions (Brandenburg et al. 2007)

$$D + L \xrightarrow{a} D^* + L \xrightarrow{p} DL \xrightarrow{e} LL \xrightarrow{h} L + L, \tag{4.10}$$

$$L + D \xrightarrow{a} L^* + D \xrightarrow{p} LD \xrightarrow{e} DD \xrightarrow{h} D + D. \tag{4.11}$$

does effectively result in an autocatalytic reaction, but it is not a direct one. First of all, it requires an activation step, indicated by asterisk, a polymerization step (with the rate constant p), an epimerization step (with the rate constant e), and finally a depolymerization step (with the rate constant h). Because the autocatalysis in indirect, this sequence of steps can therefore be regarded as a simple example of a network catalysis (Hochberg et al. 2017; Plasson 2015).

4.11 Fluctuations Instead of Autocatalysis or Enantiomeric Cross-Inhibition

During the last decade, there has been some increased interest in the role of fluctuations; see a recent review by Walker (2017). Fluctuations can play important roles in diluted systems, in which the number of molecules is small. In such case, rate equations no longer provide a suitable description of the relevant kinetics when the system is dilute and reactions are rare (Gillespie 1977; Toxvaerd 2014). In that case, a stochastic approach must be adopted. This may be relevant to the work of Toxvaerd (2013), where homochirality has been found without apparent autocatalysis or enantiomeric cross-inhibition. Instead of solving rate equations, as discussed in the previous sections, one solves stochastic equations. This means that at each reaction step, the state of the system changes, but with a reaction that is taken to depend on chance with a certain probability. The system is then described by vector $\boldsymbol{q} = (n_A, n_D, n_L)$, where n_A denotes the numbers of achiral molecules and n_D and n_L denotes the number of molecules of the D and L forms, respectively. In the model of Brandenburg (2019), seven different reactions were considered, each with a certain probability. Not all those seven reactions need to be possible in a certain experiment, so the probability for some reactions can be zero. Applying a single reaction step with enantiomeric cross inhibition implies

$$(n_A, n_D, n_L) \xrightarrow{k_\times} (n_A + 2, n_D - 1, n_L - 1), \tag{4.12}$$

i.e., the numbers of D and L get reduced by one, and that of A increases by two. We can also include spontaneous deracemization reactions, i.e.,

$$(n_A, n_D, n_L) \xrightarrow{k_+} (n_A - 1, n_D + 1, n_L), \tag{4.13}$$

$$(n_A, n_D, n_L) \xrightarrow{k_+} (n_A - 1, n_D, n_L + 1). \tag{4.14}$$

To model different reaction rates, the different reactions must happen with different probabilities. This is done by taking at each reaction step a random number between zero and one. Suppose we want to model enantiomeric cross inhibition together with spontaneous deracemization, then the probability that the first reaction happens is proportional to $k_\times$, and the probability that one of the other two reactions in Eqs. (4.14) and (4.14) occurs is proportional to $k_+/2$. If we also allow for the possibility that nothing happens (probability proportional to k_0), then our scheme with $\boldsymbol{q} \to \boldsymbol{q} + \Delta\boldsymbol{q}$ is as follows:

$$\text{if} \quad 0 \le r < r_1 \equiv k_\times \qquad \text{then } \Delta\boldsymbol{q} = (2, -1, -1), \tag{4.15}$$

$$\text{if} \quad r_1 \le r < r_2 \equiv r_1 + k_+/2 \qquad \text{then } \Delta\boldsymbol{q} = (-1, 1, 0), \tag{4.16}$$

$$\text{if} \quad r_2 \le r < r_3 \equiv r_2 + k_+/2 \qquad \text{then } \Delta\boldsymbol{q} = (-1, 0, 1), \tag{4.17}$$

$$\text{if} \quad r_3 \le r < 1 \qquad \text{then } \Delta\boldsymbol{q} = \boldsymbol{0} (\text{no reaction}). \tag{4.18}$$

Note that $k_\times + k_+ + k_0 = 1$ is here assumed. This particular experiment was referred to as experiment III in Brandenburg (2019), where $k_0 = 0$ was assumed. As he varied $k_\times$, k_+ was assumed to vary correspondingly such that $k_+ = 1 - k_\times$. The results of this experiment are similar to those with spontaneous deracemization replaced by autocatalysis, which is referred to as experiment I in Fig. 4.17. Here, the autocatalysis rate is varied such that $k_C = 1 - k_\times$. This is the standard Frank model, but for a diluted system, while model III is close to that of Sugimori et al. (2008, 2009), who were the first to find a transition to full homochirality even without autocatalysis. Next, in experiment II, there is autocatalysis, but no enantiomeric cross inhibition and just spontaneous racemization instead. This type of model was first considered by Jafarpour et al. (2015, 2017). The transition to full homochirality was originally though impossible in such a model (Stich et al. 2016).

In a comparative study, all these processes were studied within a single unified model. In Fig. 4.17 we show the results of the four different experiments. We mentioned already experiments I–III. In experiment IV, by comparison, there is just racemization and deracemization, but neither autocatalysis nor enantiomeric cross inhibition. In that case, the average of the modulus of the enantiomeric excess, $\langle|\eta|\rangle$,

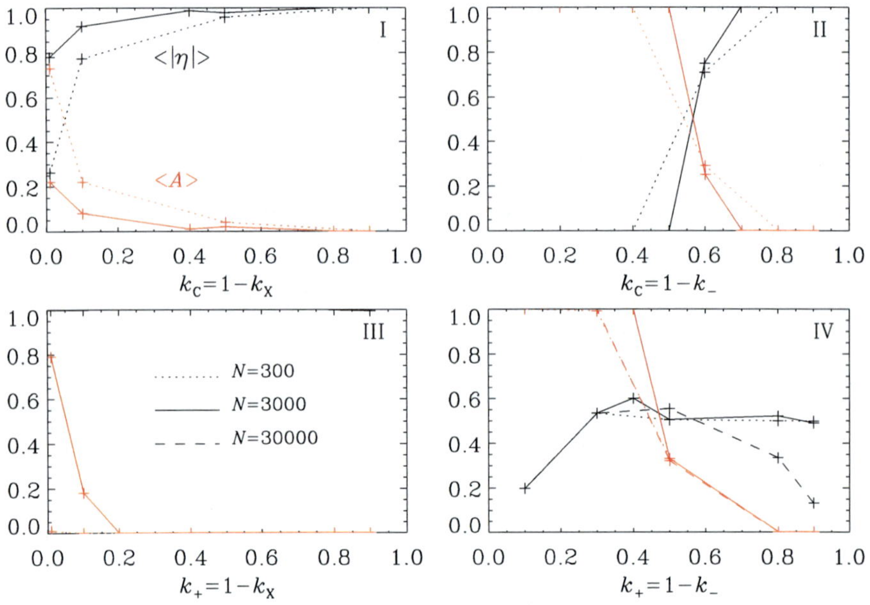

Fig. 4.17 Bifurcation diagrams of $\langle|\eta|\rangle$ (black) and $\langle A\rangle$ (red) for $N = 3000$ (solid lines) and $N = 300$ (dotted lines) as a function of parameters for models I, II, III, and IV. Adapted from Brandenburg (2019)

no longer reaches unity, but levels off at about 0.5 when $k_+ \gtrsim 0.4$. We also see from the red lines that the achiral compounds (A) get depleted in favor of producing chiral ones either of the D or the L form.

4.12 Chirality from a Martian Labeled Release Experiment

Back in 1976, when the Viking I and II landers visited the Chryse Planitia and Utopia Planitia regions, respectively, many of the things we now know about Mars were still unclear. In particular, the existence of water on Mars was still very much an open question. Nevertheless, one was relatively optimistic at the time. Both landers came with advanced experiments on board to look for life. One of the experiments, the Labeled Release (LR) experiment, was actually successful (Levin and Straat 1976, 1977), but another experiment never detected any organics, which was decisive enough to conclude that no life was detected after all (Klein et al. 1976).

The idea behind the LR experiment is simple: take Martian soil, mix it with water and organics as nutrients, and see whether a metabolic reaction occurs that decomposes the nutrients and produces a gaseous waste product, for example methane or carbon dioxide; see the recent account by Levin and Straat (2016), where detailed tests with various terrestrial soils were presented. The carbon atoms of the nutrients were labelled with carbon-14 isotopes, a technique commonly used in medicine, which allows one to trace those labeled carbon atoms by their radioactivity. To identify the gaseous waste product, one simply measured the level of radioactivity. Control experiments with sterilized soil showed that only fresh Martian soil produced a reaction. The Viking laboratories were flexible enough to perform additional experiments with lower sterilization temperatures. The critical temperature below which no sterilization occurred was found to be around 50°. Those temperatures would appear reasonable for Martian cryophiles, but are generally too low for sterilization on Earth. The experiment was tested in various deserts on Earth and it was able to detect metabolism at measurable levels.

It is only since 2012 that organics were detected on the Martian surface by the Curiosity rover; see Voosen (2018) for a popular account. We also know that organics get quickly destroyed by perchlorates, in particular $KClO_4$, which were discovered on the Martian surface by the Phoenix lander in 2008 (Hecht et al. 2009). Such processes could potentially result in reactions found with the LR experiment (Quinn et al. 2005), but it remains puzzling why a critical sterilization temperature of 50° was found, and not much higher, for example. Thus, while an explanation in terms of abiotic processes has not been fully conclusive (Valdivia-Silva 2012), the explanation that life was actually detected might seem more straightforward (Levin and Straat 2016). However, as already noted by Carl Sagan, "the more extraordinary the claim, the more extraordinarily well-tested the evidence must be. The person making the extraordinary claim has the burden of proving to the experts at large that his or her belief has more validity than the one almost everyone else accepts." In any case, it seems justified to repeat this experiment to clarify the phenomenon that the Viking landers discovered back in 1976; see also the recent paper by Carrier et al. (2020).

Given that the experiment is relatively simple and can detect life under harsh conditions on Earth, it would be interesting to repeat some variants of it in the future. One such variant would be to allow for the detection of handedness (Levin 2009). This would constitute a more conclusive signature of life than just the discovery of some metabolism. Such an experiment can be done by using chiral nutrients, which goes back to the old findings of Pasteur of 1857 that certain microorganisms had a preference for consuming (+)-tartaric acid over (−)-tartaric acid; see the reviews by Gal (2008) and Sevin (2015).

In Fig. 4.18 we show the result of recent experiments by Sun et al. (2009) using different types of eukarya, bacteria, and archaea, which were given either D sugars or L sugars. In most of the cases there was a clear preference in the microbes taking up the naturally occurring D sugars compared with the synthetically produced L sugars. Subsequent work showed that the specificity for some microbes is low and that some of those can use sugars of the opposite chirality also (Moazeni et al. 2010). Although the dependence on the type of nutrients has not yet been studied in detail, it may be important to allow for a broad range of different ones in an attempt to account for such ambiguities.

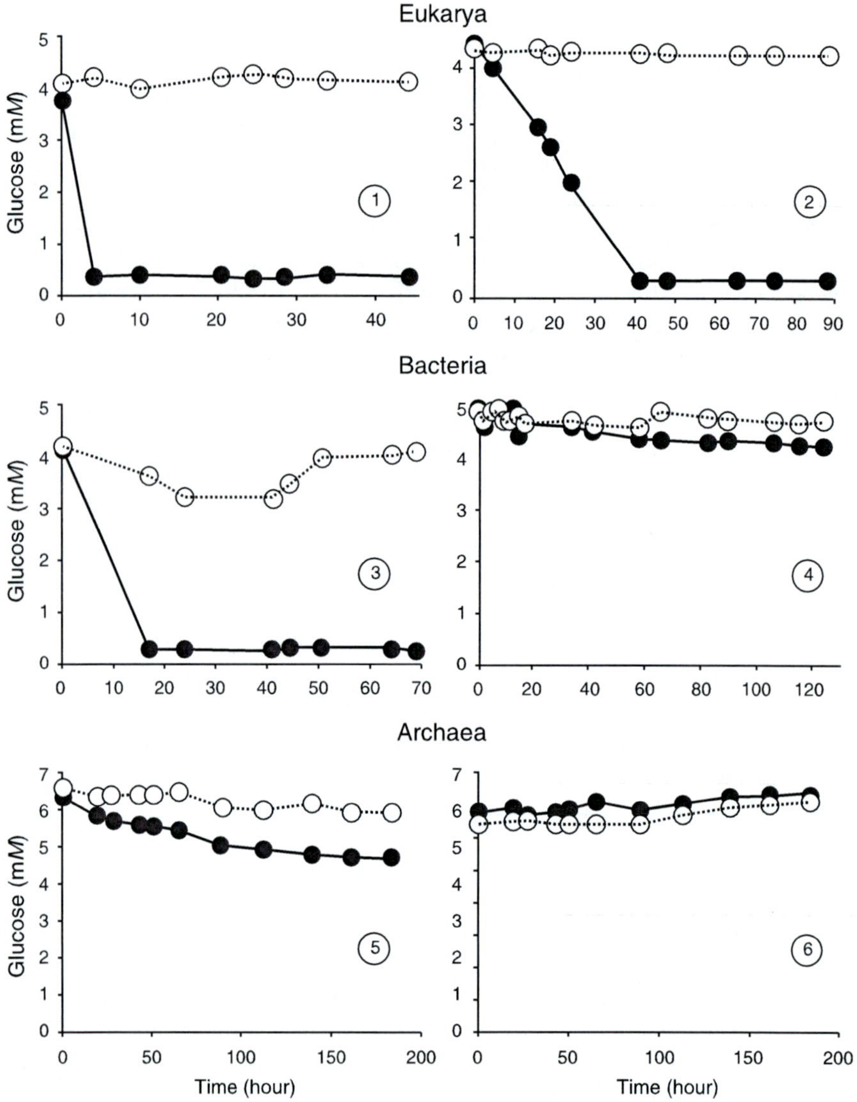

Fig. 4.18 Metabolic consumption of D -glucose (filled symbols) and L -glucose (open symbols) and by (1) *Saccharomyces cerevisiae*, (2) *Penicilium expansum*, (3) *E. coli*, (4) *Micrococcus luteus*, (5) *Natronobacterium sp.*, and (6) *Halostagnicola sp.* Adapted from Sun et al. (2009)

4.13 Conclusions

Louis Pasteur was well ahead of his time when he identified the biological role of chirality in living matter. Particular remarkable is his realization that, during fermentation, the metabolic uptake of nutrients of opposite chiralities is different.

To understand why this experiment was not put in the context of extraterrestrial life detection, we have to realize that in those years, it was not uncommon to think of extraterrestrial life on Mars. When the astronomer Herschel (1784) discovered seasons on Mars, he wrote in his paper in the Philosophical Transactions of the Royal Society that this "planet has a considerable but moderate atmosphere so that its inhabitants probably enjoy a situation in many respects similar to ours." So, not just the existence of life, but the existence of *intelligent* of life on Mars was commonly expected. This only changed in 1964, when Mariner 4 returned the first flyby pictures of Mars, which suggested that any life there would probably only be of microbial nature. But that Vikings 1 and 2 would not even find any organics on Mars was such a shock to many that the search for life in the Universe appeared fairly hopeless, and Mars exploration was put on hold for the next two decades. This all changed since the turn of the century with the discovery of extremophiles on Earth and the realization that terrestrial life has existed since the time that stable continents existed. Gradually, with the conclusive detection of water on Mars, the search for extinct or extant life on Mars restarted, and Pasteur's discovery of different metabolic uptakes of D and L nutrients may finally turn into an actual Martian experiment. As shown by Sun et al. (2009), this property can be used to detect the presence of homochirality through in situ experiments. Homochirality can also be detected through remote sensing by looking for circular polarization. This approach has been persued by Patty et al. (2019), who found negative or left-handed circularly polarized light emitted from terrestrial plant life at about 680 nm; see also Avnir (2021) for a recent review. To what extent this technique can be used as a biomarker still needs to be seen, but it is amazing to see once more how Pasteur's early discoveries have shaped some important aspects of astrobiology.

While homochirality remains a property that is strongly associated with life—or at least some chemical process that keeps the system far from equilibrium—it is not clear whether we should expect it to have the same or the opposite handedness as an Earth (Bada 1996). To answer this question, one would need to have more realistic models with meaningful estimates for the concentrations of suitable chemicals in some protocells. This would allow for an estimate of the level of fluctuations in relation to the strengths of the small but systematic effects resulting from the weak force. It would be the only way of guaranteeing that each genesis of life always produces the same chirality. But for now, we should be satisfied if one could find (and understand) any type of an extraterrestrial metabolic process that works differently for nutrients of D and L forms. Thus, the discussed possibilities would need to be put on a quantitatively meaningfully basis. And if it is not life, it certainly is interesting enough to deserve serious attention!

Acknowledgments I am grateful to the two referees for many useful comments and suggestions that have led to improvements of the manuscript. In particular, I wish to acknowledge Raphaël's clarifications on the differences in the terminology in terms of levorotatory and dextrorotatory on the one hand and that in terms of D and L on the other. This work was supported in part through the Swedish Research Council, grant 2019-04234. I acknowledge the allocation of computing resources provided by the Swedish National Allocations Committee at the Center for Parallel Computers at the Royal Institute of Technology in Stockholm.

References

Anders, E., DuFresne, E.R., Hayatsu, R., Cavaille, A., DuFresne, A., Fitch, F.W.: Contaminated meteorite. Science **146**, 1157–1161 (1964)

Athavale, S.V., Simon, A., Houk, K.N., Denmark, S.E.: Demystifying the asymmetry-amplifying, autocatalytic behaviour of the Soai reaction through structural, mechanistic and computational studies. Nat. Chem. **12**, 412–423 (2020)

Avetisov, V.A., Goldanskii, V.I., Kuz'min, V.V.: Handedness, origin of life and evolution. Phys. Today **44**, 33–41 (1991)

Avnir, D.: Critical review of chirality indicators of extraterrestrial life. New Astron. Rev. **92**, 101596 (2021)

Baaske, P., Weinert, F.M., Duhr, S., Lemke, K.H., Russell, M.J., Braun D.: Extreme accumulation of nucleotides in simulated hydrothermal pore systems. Proc. Nat. Acad. Soc. **104**, 9346–9351 (2007)

Bada, J.L.: Origins of homochirality. Nature **374**, 594–595 (1995)

Bada, J. L.: Amino acid homochirality on Earth and Mars. Orig. Life Evol. Biosph. **26**, 518–519 (1996)

Bailey, J.: Astronomical sources of circularly polarized light and the origin of homochirality. Orig. Life Evol. Biosph. **31**, 167–183 (2001)

Bailey, J., Chrysostomou, A., Hough, J.H., Gledhill, T.M., McCall, A., Clark, S., Ménard, F., Tamura, M.: Circular polarization in star forming regions: implications for biomolecular homochirality. Science **281**, 672–674 (1998)

Barge, L.M., Branscomb, E., Brucato, J.R., Cardoso, S.S.S., Cartwright, J.H.E., Danielache, S.O., Galante, D., Kee, T.P., Miguel, Y., Mojzsis, S., Robinson, K.J., Russell, M.J., Simoncini, E., Sobron, P.: Thermodynamics, disequilibrium, evolution: Far-from-equilibrium geological and chemical considerations for origin-of-life research. Orig. Life Evol. Biosph. **47**, 39–56 (2017)

Barge, L.M., Flores, E., Baum, M.M., VanderVelde, D.G., Russell, M.J.: Redox and pH gradients drive amino acid synthesis in iron oxyhydroxide mineral systems. Proc. Nat. Acad. Soc. **116**, 4828–4833 (2019)

Bonner, W.A., Blair, N.E., Dirbas, F.M.: Experiments on the abiotic amplification of optical activity. Orig. Life **11**, 119–134 (1981)

Bonner, W.A.: The origin and amplification of biomolecular chirality. Orig. Life Evol. Biosph. **21**, 59–111 (1991)

Bonner, W.A.: Parity violation and the evolution of biomolecular homochirality. Chirality **12**, 114–126 (2000)

Boyd, R.N., Famiano, M.A., Onaka, T., Kajino, T.: Sites that can produce left-handed amino acids in the supernova neutrino amino acid processing model. Astrophys. J. **856**, 26 (2018)

Brandenburg, A.: The limited roles of autocatalysis and enantiomeric cross-inhibition in achieving homochirality in dilute systems. Orig. Life Evol. Biosph. **49**, 49–60 (2019)

Brandenburg, A.: Piecewise quadratic growth during the 2019 novel coronavirus epidemic. Infect. Dis. Model. **5**, 681–690 (2020)

Brandenburg, A., Andersen, A.C., Höfner, S., Nilsson, M.: Homochiral growth through enantiomeric cross-inhibition. Orig. Life Evol. Biosph. **35**, 225–241 (2005)

Brandenburg, A., Lehto, H.J., Lehto, K.M.: Homochirality in an early peptide world. Astrobiol. **7**, 725–732 (2007)

Brandenburg, A., Multamäki, T.: How long can left and right handed life forms coexist? Int. J. Astrobiol. **3**, 209–219 (2004)

Braun, D., Libchaber, A.: Trapping of DNA by thermophoretic depletion and convection. Phys. Rev. Lett. **89**, 188103 (2002)

Carrier, B.L., Beaty, D.W., Meyer, M.A., Blank, J.G., Chou, L., DasSarma, S., Des Marais, D.J., Eigenbrode, J.L., Grefenstette, N., Lanza, N.L., Schuerger, A.C., Schwendner, P., Smith, H.D., Stoker, C.R., Tarnas, J.D., Webster, K.D., Bakermans, C., Baxter, B.K., Bell, M.S., Benner, S.A., et al.: Mars extant life: what's next? Conference report. Astrobiology **20**, 785–814 (2020)

Cech, T.R.: A model for the RNA-catalyzed replication of RNA. Proc. Nat. Acad. Soc. **83**, 4360–4363 (1986)

Cloez, S.: Note sur la composition chimique de la pierre. Compt. Rend. Acad. Sci. Paris **58**, 986 (1864)

Davies, P.C.W., Lineweaver, C.H.: Finding a second sample of life on Earth. Astrobiology **5**, 154–163 (2005)

Derewenda, Z.S.: On wine, chirality and crystallography. Acta Cryst. **A64**, 246–258 (2008)

Ehrenfreund, P., Glavin, D.P., Botta, O., Cooper, G., Bada, J.L.: Extraterrestrial amino acids in Orgueil and Ivuna: tracing the parent body of CI type carbonaceous chondrites. Proc. Nat. Acad. Soc. **98**, 2138–2141 (2001)

Engel, M.H., Macko, S.A.: Isotopic evidence for extraterrestrial non-racemic amino acids in the Murchison meteorite. Nature **389**, 265–268 (1997)

Fajszi, Cs., Czégé, J.: Critical evaluation of mathematical models for the amplification of chirality. Orig. Life **11**, 143–162 (1981)

Frank, F.C.: On spontaneous asymmetric synthesis. Biochim. Biophys. Acta **11**, 459–464 (1953)

Gal, J.: The discovery of biological enantioselectivity: Louis Pasteur and the fermentation of tartaric acid, 1857—a review and analysis 150 yr later. Chirality **20**, 5–19 (2008)

Gehring, T., Busch, M., Schlageter, M., Weingand, D.: A concise summary of experimental facts about the Soai reaction. Chirality **22**, E173–E182 (2010)

Gilbert, W.: Origin of life—the RNA world. Nature **319**, 618–618 (1986)

Gillespie, D.T.: Exact stochastic simulation of coupled chemical reactions. J. Phys. Chem. **81**, 2340–2361 (1977)

Gleiser, M., Walker, S.I.: Life's chirality from prebiotic environments. Int. J. Astrobiol. **11**, 287–296 (2012)

Globus, N., Blandford, R.D.: The chiral puzzle of life. Astrophys. J. Lett. **895**, L11 (2020)

Goldanskii, V.I., Kuz'min, V.V.: Spontaneous breaking of mirror symmetry in nature and the origin of life. Sov. Phys. Usp. **32**, 1–29 (1989)

Goldhaber, M., Grodzins, L., Sunyar, A.W.: Evidence for circular polarization of bremsstrahlung produced by beta rays. Phys. Rev. **106**, 826–828 (1957)

Guerrier-Takada, C., Altman, S.: Catalytic activity of an RNA molecule prepared by transcription in vitro. Science **223**, 285–286 (1984)

Hecht, M.H., Kounaves, S.P., Quinn, R.C., West, S.J., Young, S.M.M., Ming, D.W., Catling, D.C., Clark, B.C., Boynton, W.V., Hoffman, J., DeFlores, L.P., Gospodinova, K., Kapit, J., Smith, P.H.: Detection of Perchlorate and the Soluble Chemistry of Martian Soil at the Phoenix Lander Site. Science **325**, 64–67 (2009)

Hegstrom, R.A.: Parity nonconservation and the origin of biological chirality—theoretical calculations. Orig. Life **14**, 405–414 (1984)

Hegstrom, R.A., Rein, D.W., Sandars, P.G.H.: Calculation of the parity nonconserving energy difference between mirror-image molecules. J. Chem. Phys. **73**, 2329–2341 (1980)

Herschel, W.: On the remarkable appearances at the polar regions of the planet Mars, the inclination of its axis, the position of its poles, and its spheroidical figure; with a few hints relating to its real diameter and atmosphere. Philos. Trans. R. Soc. Lond. **74**, 233–273 (1784)

Hochberg, D., Bourdon G., Rubén D., Ágreda B, Jesús A., Ribó, J. M.: Stoichiometric network analysis of spontaneous mirror symmetry breaking in chemical reactions. Phys. Chem. Chem. Phys. **19**, 17618–17636 (2017)

Jafarpour, F., Biancalani, T., Goldenfeld, N.: Noise-induced mechanism for biological homochirality of early life self-replicators. Phys. Rev. Lett. **115**, 158101 (2015)

Jafarpour, F., Biancalani, T., Goldenfeld, N.: Noise-induced symmetry breaking far from equilibrium and the emergence of biological homochirality. Phys. Rev. E **95**, 032407 (2017)

Joyce, G.F., Visser, G.M., van Boeckel, C.A.A., van Boom, J.H., Orgel, L.E., Westrenen, J.: Chiral selection in poly(C)-directed synthesis of oligo(G). Nature **310**, 602–603 (1984)

Källén, A., Arcuri, P., Murray, J.D.: A simple model for the spatial spread and control of rabies. J. Theor. Biol. **116**, 377–393 (1985)

Klein, H.P., Horowitz, N.H., Levin, G.V., Oyama, V.I., Lederberg, J., Rich, A., Hubbard, J.S., Hobby, G.L., Straat, P.A., Berdahl, B.J., Carle, G.C., Brown, F.S., Johnson, R.D.: The Viking biological investigation: preliminary results. Science **194**, 99–105 (1976)
Kondepudi, D.K., Nelson, G.W.: Chiral symmetry breaking in nonequilibrium systems. Phys. Rev. Lett. **50**, 1023–1026 (1983)
Kondepudi, D.K., Nelson, G.W.: Weak neutral currents and the origin of biomolecular chirality. Nature **314**, 438–441 (1985)
Konstantinov, K.K., Konstantinova, A.F.: Chiral symmetry breaking in peptide systems during formation of life on Earth. Orig. Life Evol. Biosph. **48**, 93–122 (2018)
Lee, T.D., Yang, C.N.: Question of parity conservation in weak interactions. Phys. Rev. **104**, 254–258 (1956)
Levin, G.V.: Comment on "Stereo-specific glucose consumption may be used to distinguish between chemical and biological reactivity on Mars: a preliminary test on Earth". Astrobiol. **9**, 503 (2009)
Levin, G.V., Straat, P.A.: Labeled release—experiment in radiorespirometry. Orig. Life Evol. Biosph. **7**, 293–311 (1976)
Levin, G.V., Straat, P.A.: Recent results from the Viking Labeled Release Experiment on Mars. J. Geophys. Res. **82**, 4663–4667 (1977)
Levin, G.V., Straat, P.A.: The case for extant life on Mars and its possible detection by the Viking Labeled Release Experiment. Astrobiol. **16**, 798–810 (2016)
Mason, S.F., Tranter, G.E.: The parity-violating energy difference between enantiomeric molecules. Mol. Phys. **53**, 1091–1111 (1984)
McVoy, K.W.: Circular polarization of bremsstrahlung from polarized electrons in Born approximation. Phys. Rev. **106**, 828–829 (1957)
Meierhenrich, U.J., Thiemann, W.H.-P.: Photochemical concepts on the origin of biomolecular asymmetry. Orig. Life Evol. Biosph. **34**, 111–121 (2004)
Miller, S.L.: A production of amino acids under possible primitive Earth conditions. Science **117**, 528–529 (1953)
Moazeni, F., Zhang, G., Sun, H.J.: Imperfect asymmetry of life: Earth microbial communities prefer D-lactate but can use L-lactate also. Astrobiol. **10**, 397–402 (2010)
Murray, J.D., Stanley, E.A., Brown, D.L.: On the spatial spread of rabies among foxes. Proc. Roy. Soc. Lond. Ser. B **229**, 111–150 (1986)
Noble, J.V.: Geographic and temporal development of plagues. Nature **250**, 726–728 (1974)
Pasteur, L.: Umwandlung der Weinsäure in Traubensäure. Entdeckung von unwirksamer Weinsäure. Neue Methode der Zerlegung von Traubensäure in Rechts- und in Linksweinsäure. Ann. Phys. **166**, 504–509 (1853)
Pasteur, L.: Recherches sur les propriétés spécifiques des deux acides qui composent l'acide racémique. In: Annales de chimie et physique (ed. Paris: Masson), pp. 56–99, 3rd edn, XXVIII (1922)
Patty, C.H.L., ten Kate, I.L., Buma, W.J., van Spanning, R.J.M., Steinbach, G., Ariese, F., Snik, F.: Circular spectropolarimetric sensing of vegetation in the field: possibilities for the remote detection of extraterrestrial life. Astrobiol. **19**, 1221–1229 (2019)
Pizzarello, S., Cronin, J.R.: Non-racemic amino acids in the Murray and Murchison meteorites. Geochim. Cosmochim. Acta **64**, 329–338 (2000)
Plasson, R., Bersini, H., Commeyras, A.: Recycling Frank: spontaneous emergence of homochirality in noncatalytic systems. Proc. Nat. Acad. Soc. **101**, 16733–16738 (2004)
Plasson, R.: Chemical reaction network. In: Gargaud, M., et al. (eds.) Encyclopedia of Astrobiology, Id. 483. Springer, Berlin Heidelberg (2015)
Quinn, R.C., Zent, A.P., Grunthaner, F.J., Ehrenfreund, P., Taylor, C.L., Garry, J.R.C.: Detection and characterization of oxidizing acids in the Atacama Desert using the Mars oxidation instrument. Planet. Space Sci. **53**, 1376–1388 (2005)
Rothery, D.A., Gilmour, I., Sephton, M.A.: An Introduction to Astrobiology, Cambridge University Press, Cambridge (2008)
Russell, M.: First life. Am. Sci. **94**, 32–39 (2006)

Russell, M.J., Barge, L.M., Bhartia, R., Bocanegra, D., Bracher, P.J., Branscomb, E., Kidd, R., McGlynn, S., Meier, D.H., Nitschke, W., Shibuya, T., Vance, S., White, L., Kanik, I.: The drive to life on wet and icy worlds. Astrobiol. **14**, 308–343 (2014)

Sandars, P.G.H.: A toy model for the generation of homochirality during polymerization. Orig. Life Evol. Biosph. **33**, 575–587 (2003)

Sandars, P.G.H.: Chirality in the RNA world and beyond. Int. J. Astrobiol. **4**, 49–61 (2005)

Sevin, A.: Pasteur and molecular chirality. Bibnum [Online], Chimie **459**, 1–10 (2015)

Soai, K., Shibata, T., Morioka, H., Choji, K.: Asymmetric autocatalysis and amplification of enantiomeric excess of a chiral molecule. Nature **378**, 767–768 (1995)

Stich, M., Ribó, Josep M., Blackmond, D.G., Hochberg, D.: Necessary conditions for the emergence of homochirality via autocatalytic self-replication. J. Chem. Phys. **145**, 074111 (2016)

Sugimori, T., Hyuga, H., Saito, Y.: Fluctuation induced homochirality. J. Phys. Soc. Jpn. **77**, 064606 (2008)

Sugimori, T., Hyuga, H., Saito, Y.: Fluctuation induced homochirality in an open system. Phys. Soc. Jpn. **78**, 034003 (2009)

Sun, H.J., Saccomanno, V., Hedlund, B., McKay, C.P.: Stereo-specific glucose consumption may be used to distinguish between chemical and biological reactivity on Mars: a preliminary test on Earth. Astrobiol. **9**, 443–446 (2009)

Toxvaerd, S.: The role of carbohydrates at the origin of homochirality in biosystems. Orig. Life Evol. Biosph. **43**, 391–409 (2013)

Toxvaerd, S.: Discrete dynamics versus analytic dynamics. J. Chem. Phys. **140**, 044102 (2014)

Ulbricht, T.L.V.: The origin of optical asymmetry on Earth. Orig. Life **6**, 303–315 (1975)

Valdivia-Silva, J.E., Navarro-González, R., de la Rosa, J. McKay, C.P.: Decomposition of sodium formate and L- and D-alanine in the Pampas de La Joya soils: implications as a new geochemical analogue to Martian regolith. Adv. Spa. Res. **49**, 821–833 (2012)

Voosen, P.: NASA Curiosity rover hits organic pay dirt on Mars: Carbon molecules in rocks from ancient lakebed resemble kerogen, a "goopy" fossil fuel building block on Earth. Science **360**, 1055–1055 (2018)

Walker, S.I.: Origins of life: a problem for physics, a key issues review. Rep. Prog. Phys. **80**, 092601 (2017)

Zeldovich, B. Ya., Saakyan, D.B., Sobelman, I.I.: Energy difference between right-hand and left-hand molecules, due to parity nonconservation in weak interactions of electrons with nuclei. JETP Lett. **25**, 94–96 (1977)

Chapter 5
Origin of Nucleic Acids

Frank Trixler

Abstract Nucleic acids are more than classical chemistry, but not quite biology. They are a manifestation of the transition from self-assembly (highly ordered) to self-organization (functional complexity) in the course of prebiotic chemical evolution. We might still be far away from conceiving the big picture of this epochal transition during the prebiotic era, but a closer look at the puzzle pieces we got so far from different research disciplines can already shape the view on nucleic acids, their origin, and the emergence of life fundamentally. This chapter sheds light on some of these puzzle elements and their relations in order to give an insight into the multi-faceted character of nucleic acids emergence and the interdisciplinary approach required to find insightful answers and new good questions.

5.1 Introduction

What are nucleic acids? The answer is easy to give and difficult at the same time. This is because it depends from which point of view we look at nucleic acids and ask that question.

From the perspective of biochemistry, we watch living systems and focus on how they work at the nanoscale. The answer to the initial question can thus easily be given by referring to the current understanding of what nucleic acids are with respect to their structure, function, and role in the intracellular network of functional biomolecules (Neidle 2008; Blackburn et al. 2006; Doonan 2004). In structural terms nucleic acids (RNA, DNA) are the largest organized molecules known so far. They are natural polymers. Four different species of a certain type of molecules (nucleotides) act as structural building blocks for RNA and DNA

F. Trixler (✉)
Center for NanoScience (CeNS), Ludwig-Maximilians-Universität München, Munich, Germany

Department of Earth and Environmental Sciences, Ludwig-Maximilians-Universität München, Munich, Germany
e-mail: trixler@lrz.uni-muenchen.de

A. Neubeck, S. McMahon (eds.), *Prebiotic Chemistry and the Origin of Life*, Advances in Astrobiology and Biogeophysics,
https://doi.org/10.1007/978-3-030-81039-9_5

respectively. When these monomeric units polymerize into a chain of nucleotides they form polynucleotides. Nucleic acids are examples of such polynucleotides. The nucleotide sequence in nucleic acids is of organized complexity. That means that its complexity is based on encoded information. The information that defines the sequence is transferred via replication—a capability of nucleic acids which is outstanding among all other known polymers. In the context of a living system the information coded, organized complexity is functional: it enables nucleic acid sequences to provide all instructions that encode the assembly of biomolecular nanomachines and the structures of cells, organs, and organisms. In sum, nucleic acids can be defined as biopolymers with the characteristics of organized functional complexity that has the ability to replicate. This unique combination of properties makes nucleic acids to one of the most intriguing molecules and places them within the fuzzy zone between chemistry and biology.

When we change our point of view on nucleic acids and try to understand them from the perspective of prebiotic chemistry things become weird along all central aspects of their existence. Intriguing answers given by molecular biology and biochemistry are replaced by absorbing questions asked in astrobiology (Table 5.1). First, let us take a look at the building blocks: we see four nucleic acid nucleotides in RNA and DNA respectively. Decades of progress towards the abiotic synthesis of canonical and noncanonical building blocks and their subunits provided deep insights into possible prebiotic synthesis pathways and conditions (Fialho et al. 2020; Yadav et al. 2020) but many uncertainties remain with respect to prebiotic plausibility and there are numerous open issues still unsolved. Among the most puzzling aspects are the questions: why are there exactly four nucleotide species for a given type of nucleic acids? And why is adenosine monophosphate—a subunit of one of these five building blocks—also found in many other functions and molecular structures all over the biochemistry of cellular life? Coming to the aspect of polymerization of nucleotides in water we see that it is immersed in a nagging paradox—the so called "water problem". The functional complexity of nucleic acid sequences also appears strange from the perspective of prebiotic chemistry because it is still mostly in the dark what gave birth to it initially. Biochemistry revealed that the sequence guides the assembly of living systems, but what makes a sequence functional, and what is that encoded sequence about, basically? The aspect

Table 5.1 Aspects of nucleic acids from the viewpoints of biochemistry and prebiotic chemistry

	Biochemistry	Prebiotic chemistry
Building blocks	Four nucleotide species for a given type of nucleic acids	Why just four? Why is AMP so common?
Polymerization	Via polymerase enzymes	How to solve the water paradox?
Sequence	For protein synthesis, directing the functions of living systems	What makes a sequences functional? Encoding for what, basically?
Copying	Replication of genetic information	How to solve the DNA-protein paradox? Why copying at all?

of copying poses another paradox: similar to the question of what came first—chicken or egg—the rise of the close cooperation between nucleic acids (crucial in the copying and assembly of proteins) and proteins (crucial in the copying and assembly of nucleic acids) it is still an unsolved problem in prebiotic chemistry. But copying is more than just making more complex molecules and living systems—it is the expansion of life deep into the dimension of time. Now, what does that mean for the definition and evolution of life?

The scope of this chapter is to dive deeper into these questions. It aims to provide a clearer understanding of why it is so difficult to define what nucleic acids are when looking at them from the viewpoint of prebiotic chemistry. But exactly the understanding of this difficulty can be the starting point for scientists from various disciplines who take up the challenge and try to assemble the big picture of nucleic acids prebiotic origin and the emergence of life.

5.2 The Strange Ubiquity of AMP

If we look at biochemistry from a broader perspective and avoid getting lost in too many details something remarkable becomes apparent: in many very different biochemical processes which are not directly related to each other, a specific molecule appears as a crucial element: adenosine monophosphate (AMP) (Jauker et al. 2015). This nucleotide consists of the nucleobase adenine, a pentose, and one phosphate group (Fig. 5.1). But what is so special about AMP? The question is still open so far, but we can survey its multi-faceted role in biochemistry and try to find properties that at least correlate with its outstanding role.

When we shed light on nucleic acids in our survey we see that AMP is one of its canonical building blocks. But there is more: AMP is also highly relevant for the process of gene expression in the form of terminating poly(A) tails. Such tails consist of multiple AMPs. Comparative genomics suggests that polyadenylation (the addition of poly(A) tails) is very common in all known life, so the last universal common ancestor (LUCA) probably already had such a polyadenylation system (Anantharaman et al. 2002). Beyond these tasks for nucleic acids AMP has many more jobs in the molecular factory of a cell. It is involved in the form of AMP residues in various biomolecules which are of central importance for metabolism and signal transduction. Such biomolecules are, for example, adenosine triphosphate (ATP: energy current of the cell, also involved in intra- and extracellular signalling and many other biochemical processes), cofactors such as nicotinamide adenine dinucleotide (NAD), flavin adenine dinucleotide (FAD) and acetyl conenzyme A (Acetyl-CoA) or molecules for intracellular signal transduction and other functions such as cyclic adenosine monophosphate (cAMP) and diadenosine tetraphosphate (Ap4A) (Fig. 5.1).

Why is AMP such an all-rounder in the biochemistry of all known life? The multi-functional role of AMP suggests that it has been selected very early in the prebiotic chemical evolution towards the origin of cellular life. Now, what property

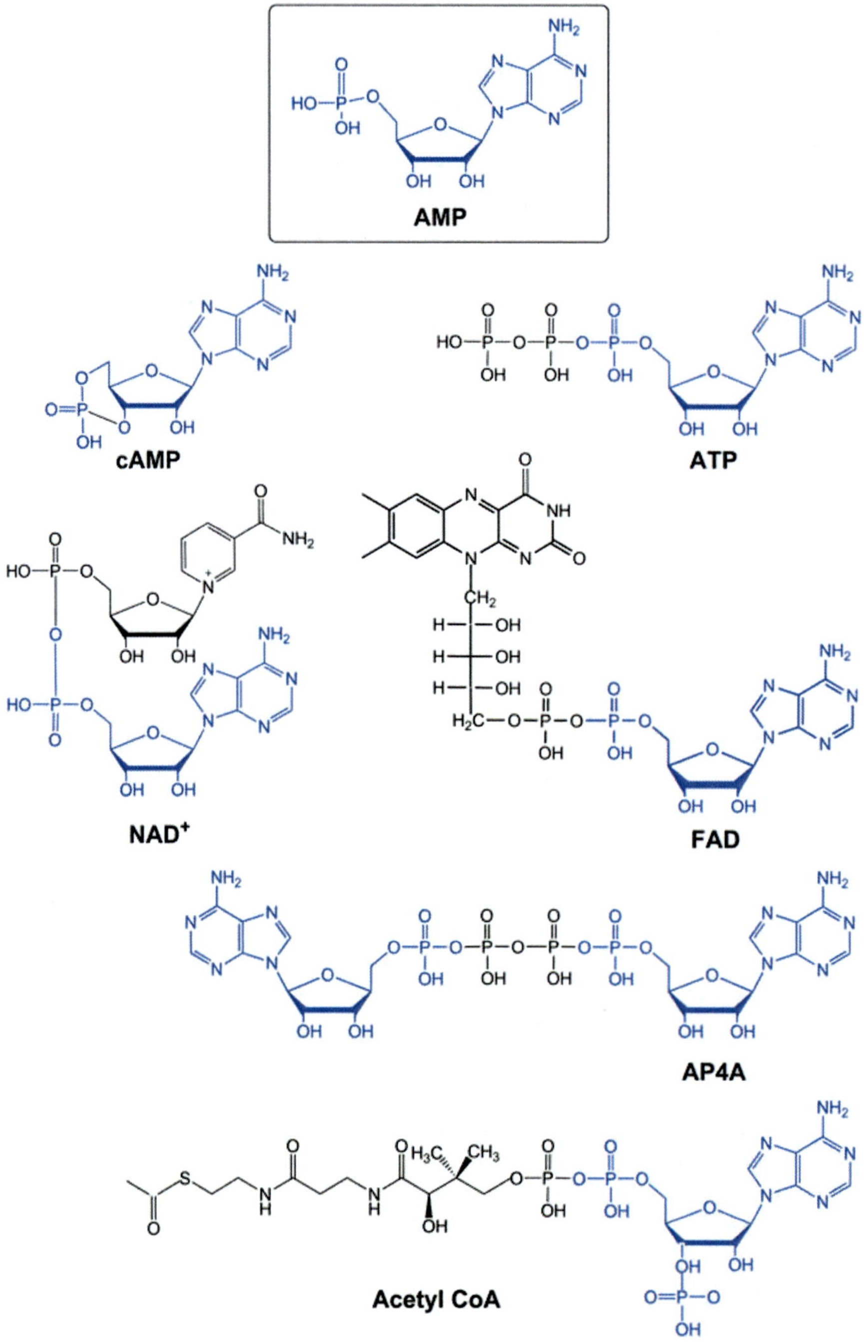

Fig. 5.1 AMP residues (blue) in various biomolecules which are crucial for metabolism and signal transduction

makes AMP so outstanding during this era? There is no definitive answer so far, but we have some indications which wait to further be explored.

One indication arises in experiments in which peptidyl RNAs are synthesized under conditions established for genetic copying (Jauker et al. 2015). These conditions include a mixture of a heterocyclic catalyst and a condensing agent. When RNAs and amino acids are added to this mixture in the absence of chemical preactivators, mineral surfaces, and enzymes a spontaneous formation of peptidyl RNA is reported. What is relevant for this discussion is the observation that AMP turned out to be of outstanding reactivity in these synthesis processes. In addition, AMP reacts under the same experimental conditions with added NMN+, FMN and the tetrasodium salt of inorganic pyrophosphate to NAD^+, FAD, and ATP, respectively. All these results might indicate that AMP has favorable reactive properties for the generation of a variety of biomolecules which have a central role in cellular processes.

Another indication for an outstanding property of AMP comes from experiments that investigated a possible interplay between nucleotides and nanofluidic phenomena emerging in aqueous suspensions of particles where temporal nanoconfinements of water are formed. These experiments initially aimed to increase the efficiency of a certain interfacial effect (organic solid/solid wetting) for nanotechnological applications (Eberle et al. 2018). In the context of the experiments added phosphates turned out to be efficient catalysts for the solid/solid wetting behavior of molecular assemblies. However, when turning to biomolecular phosphates, AMP surprised with having an outstanding catalytic effect. The catalytic activity is of very high statistical significance not only in comparison to all tested inorganic phosphates but also with respect to other nucleotides and phosphate-residue containing biomolecules (Greiner and Trixler 2016). As the dense aqueous suspension of particles in these experiments can be regarded as an inorganic version of the crowded intracellular environment where dispersed macromolecules with nano-scale separation also create nanofluidic effects on water between them these findings have prebiotic relevance: the results suggest that catalytic reactions in temporal nanoconfinements of water may have played an important part in the prebiotic chemical evolution with AMP as a crucial actor. But which property can be responsible for the described outstanding catalytic role of AMP? An interesting correlation appears when comparing the stacking equilibrium constants of nucleoside monophosphates with the corresponding organic solid/solid wetting results. It turns out that AMP not only increases the extent of nanofluidic phenomena in these experiments more than any other used nucleoside monophosphates but also has the highest self-stacking constant among them (Norberg and Nilsson 1995) and the ability to form indefinite stacks (Tribolet and Sigel 1987).

These examples indicate that the strange ubiquity of AMP in biochemistry could be a message from the era of prebiotic chemistry transmitted by the conservative nature of evolution. Further research on that issue could thus help us to better develop and select possible scenarios about the origin of nucleic acids.

5.3 Nucleotides and the Creative Power of Standardization

Why are there exactly four nucleotide building blocks life on Earth uses to build RNA or DNA respectively? Couldn't there be more? There is a definite "yes", but also an obvious "no".

Let us a first focus on the "yes": according to results from synthetic biology there are indeed a lot more building blocks possible: the diversity of nucleic acid analogues that have been synthesized so far in order to explore the variety of structural and chemical parameters for information storage, heredity, and in vitro evolution reveals that the chemical space for synthetic genetic polymers is even tremendous (Cleaves et al. 2019; Nie et al. 2020). The progress in the field of xenobiology has shown that nucleic acid analogues can successfully be created by replacing all three moieties of nucleotides with non-natural counterparts. Such modifications have been reported with respect to nucleobases, the sugar moiety (xeno-nucleic acids, termed XNAs (Chaput and Herdewijn 2019)) or the phosphodiester linkage. Even combined modifications are possible, as e.g. in the form of peptide nucleic acids (PNAs) (Nielsen and Appella 2014) where a synthetic peptide backbone replaces the natural sugar phosphate backbone (Fig. 5.2). Apart from replacing structural elements of nucleic acids approaches to extend their

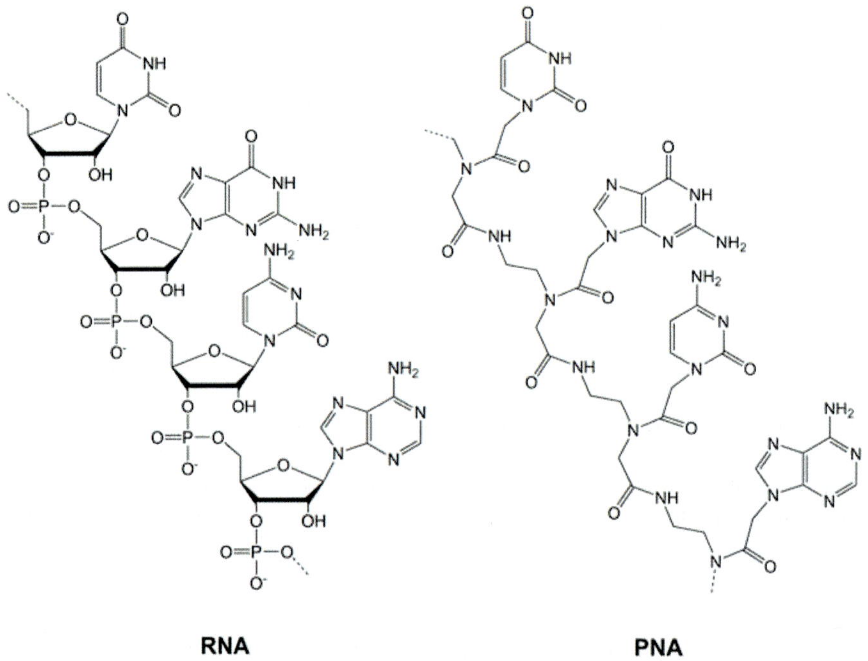

Fig. 5.2 Structural differences between RNA and PNA (Peptide Nucleic Acid). In PNA the ribosephosphate-backbone of RNA is replaced by a synthetic peptide backbone

number in a genetic system have also successfully been demonstrated. For example, hachimoji RNA and DNA represent a synthetic genetic system based on eight different nucleotide species each instead of four (Nie et al. 2020). All these results demonstrate that the structural scope of functional nucleic acids is huge. But for some reason this scope is not tapped by life as we know it.

This leads us to the "no" as the answer to the initial question. The huge structural scope of genetic polymers makes it very likely that a complex mixture of both non-canonical and canonical nucleic acids may have existed during the prebiotic era. As a consequence, some selection processes must have occurred which resulted in natural "standardization" and led to the few canonical versions of nucleic acids that are now used by all known life on Earth. What could possibly have led to such a selection before LUCA went on stage?

We can first take a look at this question from the perspective of biochemistry. Several possible processes and mechanisms have been discussed so far (Kim et al. 2020) which may have played a role in the transition from a heterogeneous mixture of genetic polymers to the homogeneous set used by known life. For example, by shedding light on the effect of UV photolysis a selective impact on nucleotides becomes apparent: experiments indicate that canonical nucleobases have a higher stability against degrading photochemical reactions in comparison to non-canonical ones. Selective effects can also occur with respect to the sugar moiety and the phosphodiester linkage. The effects are based on steric constraints and differences in chemical reactivity which may have had an impact on the composition of prebiotic nucleic acids. There are also indications that selective pressure for superior function might have played a role in the process of natural "standardization" during the prebiotic era. This was derived from observations which revealed that mixed RNA/DNA oligonucleotides are of lower functionality compared to the known homogeneous ones. Another selective effect has experimentally been demonstrated with regard to non-enzymatic template directed primer extension reactions. Such reactions turned out to be much less efficient when running with a non-canonical nucleotide. Though being less efficient, the incorporation of a non-canonical nucleotide was reported to be possible, but, however, it inhibits continued primer extension reactions. Investigations also revealed that chimeric oligonucleotides which were formed by both canonical and non-canonical nucleotides turned out to be effective templates for RNA synthesis. These results suggest that during the prebiotic era untemplated polymerization initially generated chimeric oligonucleotides but that the formation of RNA was then favored by template copying chemistry (Kim et al. 2020).

But biochemistry is not the only discipline we can ask what possibly led to the few canonical versions of nucleotides before the wake of LUCA. Complexity science has something to add. Let us start with the genetic system: it consists of multiple, diverse elements that are highly interconnected in a coordinative and cooperative way with having the autonomy to adapt. The sum of all these properties classifies this system to be a complex one. It is a system with functional complexity. Nucleic acids, as being elements of the genetic system, are by themself complex systems. They are a manifestation of a special category of complexity termed "Functional Sequence Complexity". We will see later in this chapter that functional sequence complexity can only exist in a narrow section of the complexity spectrum and that this section is closer to randomness than to order. What is relevant for this discussion is the relation between novelty and standards within this narrow section. If a complex, evolutionary system is either too sensitive or too resistant to variations and mutations its adaptability is corrupted. This is where standardization becomes relevant (Higginbotham 2017). From the viewpoint of standardization, the selection of just a few versions of nucleotides from their broad chemical space may have been provided an essential contribution to poise the complexity of nucleic acids at the functional and adaptable part of the complexity spectrum, thus ensuring that the genetic system is neither too resistant nor too sensitive to mutations and thus well prepared for coping with genetic innovations. A comparable situation can be found in some economic systems where technical innovations with high market relevance can lead to competing technical standards defined by different suppliers. This can result in so called "format wars" (Anscombe 2008). They usually end when one format catches on and becomes a general standard. This leads to a much better cooperative and integrative interconnection of the new technological elements within the complex socio-economical system, thus catalysing innovations enabled by the defined technical standards.

In sum, recent results from different disciplines such as xenobiology, biochemistry, and systems theory support the view on the canonical set of nucleotides as being just a small fraction of a much broader chemical space available for this class of molecules. The existence of this restriction can be interpreted as information about conditions during the prebiotic era. This message, if properly deciphered and better understood, could reveal aspects of the origin of nucleic acids that were of crucial importance for prebiotic chemistry before LUCAs descendants took the helm in evolution. The few, very basic insights we achieved so far are already stunning, but there is much work ahead to get an idea of the big picture about this facet of nucleic acids origin.

5.4 A Paradox Falls into Water

Earth is a very dry planetary body. Although is seems to be a water world, appearances are deceptive: the oceans are an extremely thin varnish with respect to Earth's diameter. A census of all water both on its surface and within the lithosphere

(Pearson et al. 2014) reveals that it is just about 0.04% of Earth's total mass (Williams 2014). But despite water is a very rare compound on Earth, it is of central importance for the origin of terrestrial life: it has a highly destructive potential for prebiotic chemistry as it impedes the formation and weakens the stability of dissolved biopolymers such as nucleic acids. Nevertheless, life is spread all over the Earth. And it is just water-based.

Being attracted by paradoxes helps a lot when trying to understand life's origin. Among the amazing zoo of paradoxes (Benner 2014) in the field of abiogenesis is the so-called "water problem" (Benner et al. 2012). It is one of the most nagging but also fascinating ones.

The problem with water comes from the fact that it hampers the formation and stability of biopolymers such as nucleic acids (Lang et al. 2019). If we take a look at phosphodiester bonds in nucleic acids we see that these bonds are formed in the process of a condensation-dehydration reaction. Such a reaction releases water. However, a product of a reaction is difficult to form when being at the same time the solvent. Thus, when occurring within an aqueous solution, a condensation-dehydration reaction is highly unfavorable. As a consequence, a spontaneous, water releasing formation of nucleic acids is prohibited by water. Even if RNA has been formed under some conditions it is spontaneously corroded by the effect of hydrolysis when being dissolved in water. RNA hydrolysis (Voet et al. 2019) means that the phosphodiester bonds of the polymer get spontaneously ruptured by water molecules.

We thus have an intriguing paradox: water inhibits the synthesis of essential biopolymers such as RNA and corrode them via hydrolysis, but, at the same time, is essential for running the biochemistry of all life as we know it.

Several methods have been proposed so far on how to overcome the "water problem" for the abiotic formation of oligonucleotides. The focus of these studies is mostly set by the scientific community on RNA. This focus is motivated by the influence of the popular RNA-World Hypothesis which is based on the assumption that before the wake of LUCA RNA was the main actor, playing a dual role not only as an information carrier but also as a catalyst for critical biochemical reactions (Le Vay and Mutschler 2019). Approaches that have successfully produced RNA-like oligonucleotides are playing with parameters such as temperature, activating dehydration or reducing water activity. Such methods are based on just heating dissolved nucleotides at 160 °C in the presence of phosphates, adding condensing agents such as cyanamide to support dehydration reactions in water, tapping the eutectic phase between water ice crystals, using multi-component solvents such as a mixture of ammonium formate, urea, and water in order to achieve a solvent with low water activity, or even using water-free solvents such as formamide (Yadav et al. 2020; Lang et al. 2019; Lazcano 2018). Another approach which is widely discussed is the application of wet-dry-cycles (Higgs 2016). This method is also based on reducing water activity in the dry phase in which polymerization occurs. Although hydrolysis breaks some of the bonds during the wet phase, repositioning of the molecules enable the formation of new bonds in the subsequent dry phase, so each cycle increases the length of nucleotide polymers.

However, in view of these achievements which give valuable insights into the spectrum of possible pathways to circumvent the water problem, some weak spots appear: high temperatures far outside the known tolerance spectrum of terrestrial life, drying events, non-aqueous solvents, or the presence of condensing agents which are toxic even for primitive lifeforms are of limited prebiotic plausibility when taking evolutionary conservatism into account—the principle that evolution builds on existing solutions. In addition, as terrestrial life is happy with just using water for its biochemistry the water problem is obviously soluble in watery solutions.

A possible approach to achieve condensation reactions under aqueous, mild conditions and without needing condensing agents is based on diamidophosphate (DAP) as a crucial agent. DAP is very efficient in phosphorylating various species of prebiotic building blocks including nucleotides but the same reaction conditions turned out to be also suitable to generate oligonucleotides (Gibard et al. 2018). DAP is water-soluble and a possibly prebiotic relevant pathway to its synthesis is known which involves iron phosphate minerals (Gibard et al. 2019). In the next section, we will see that minerals have much more to say in finding prebiotically plausible solutions for the water paradox.

5.5 The Crystalline Womb

The huge diversity of life is accompanied by a huge diversity of minerals on Earth. As far as we know it, no other celestial body in the solar system presents such a tremendous variety of mineral species and morphologies (Hazen et al. 2008; Grosch and Hazen 2015). Are the presence of life and the highly diverse mineralogy on Earth just a coincidence or is there more?

We have seen that water destabilizes biopolymers and hampers condensation reactions. Life seems not only to cope with it but even use water as an essential element for running its complex biochemical system. Such an aqueous interplay between destruction and construction can also be observed with respect to geology: when astrochemistry and orbital dynamics work together and cast liquid water to the geology of a planetary body we see a massive impact to the geology in causing vast geological transformations. Water transformed the geology of Earth into a highly dynamic geo-*system* with multiple feedback loops. At the same time, water transformed the geochemistry on early Earth partly into *systems chemistry* and finally into biology—a system with organized complexity. Both geology and chemistry were told by water to become increasingly complex, thus gaining new possibilities of creation from destruction.

When focusing on mineral surfaces, we note that they adsorb much attention from the prebiotic chemistry community due to their considerable set of interactions with molecular building blocks of life. Mineral surfaces have been shown to be able to select and concentrate relevant organic molecules, increase their thermal stabilities and enable chiral symmetry breaking (Hazen and Sverjensky 2010).

Within our context of nucleotide polymerization, the clay mineral montmorillonite deserves special attention. According to results from numerous experiments montmorillonite shows catalytic effects on the polymerization of activated nucleotides (Kitadai and Maruyama 2018). In some experiments oligomers of up to 40 nt have been detected. The catalytic effects can be traced back to the adsorption of nucleotides on the clay layers, mediated by Van-der-Waals interactions between the nucleobase groups of the nucleotides and the montmorillonite layers. After the adsorption nucleotides are oriented in such a way that the formation of 3′,5′-phosphodiester bonds becomes favorable. The catalysed polymerization reactions seem to occur only at specific active sites within the clay interlayers. But montmorillonite can do more: it is able to accelerate the formation of lipid vesicles. Such formed vesicles have often been observed to incorporate the clay particles during the process. This observation points to a possible prebiotic scenario in which nucleotide oligomers are not only formed catalytically via clay particles but also subsequently get encapsulated within a lipid vesicle.

All these approaches are focusing on the catalytic *surfaces* of minerals. But there has recently been described a completely different way to polymerize nucleotides into nucleic acids with the help of mineral particles. This approach uses the *interspace* between particles suspended in water (Greiner and Trixler 2019). Due to nanofluidic phenomena emerging within temporal, nanoscopic interspaces between suspended particles the activity of water gets reduced. This is relevant for nucleotide polymerization due to thermodynamic reasons: in the previous section we have seen that the formation of phosphodiester bonds during a condensation reaction releases water, thus making nucleotide polymerization in water highly unfavorable. However, a significant reduction of water activity, caused by an interplay of various nanofluidic phenomena emerging within the nanoscopic particle interspaces can significantly reduce the thermodynamic barrier for the condensation reaction, thus making nucleotide polymerization favorable. This possibility to polymerize nucleotides is not restricted to a specific mineral species as comparative experiments with different, but highly common minerals such as graphite, magnetite, and quartz have shown. Aqueous suspensions of such minerals are geologically ubiquitous.

The interplay between minerals and organic molecules seems to be of significant help in trying to solve the water paradox in prebiotic chemistry. This help from minerals might be an indication that the presence of life and the highly diverse mineralogy on Earth is probably not just a coincidence.

5.6 Informed Molecules Chain Order with Chaos

Where should we search for life in the Universe? An obvious answer would focus on celestial bodies where indications for liquid water exist. The selection of targets for exobiology research such as Mars, dwarf planets, or icy moons in the outer solar systems can be seen against the background of a common strategy: "Follow the water" (Abrevaya et al. 2016). The conservative nature of this strategy is a good starting point, but we should be aware of a selection bias as we would only find water-based life. If a herd of dragons would lurk in the vast methane seas of Saturn's moon Titan (Hofgartner et al. 2014) we would simply miss them. But there is another strategy which may be termed as "follow the waste": from the viewpoint of physics life can be regarded as a dissipative chemical system of organized complexity which is maintained by a constant flux of energy. The intake of energy is needed to locally reduce entropy, but the reduction is at the cost of an entropy increase of the surroundings or, in other words, the production of "waste". For example, oxygen in Earth's atmosphere is waste from this viewpoint as it is a byproduct of oxygenic photosynthesis. Due to its high reactivity, it would not be stable in our atmosphere without a constant resupply via lifeforms running oxygenic photosynthesis. As a consequence, oxygen on Earth can be considered as being a biomarker. We may have a similar situation on Mars with respect to methane. Thus, robotic missions that are in search for atmospheric methane on Mars can be regarded as missions that "follow the waste". The advantage of this strategy: it is not restricted to a specific chemical species but is based on a criterion of a more general nature. Another example of a more general strategy is based on thermodynamic issues and is termed "follow the energy" (Abrevaya et al. 2016). In the overall view, the purpose of all these strategies is to scout possible extraterrestrial locations which meet our assumptions about habitability.

But there is also a non-geographical location where habitable conditions exist. When we cast a light on the theory of complex systems in the search for life we find it thriving within a narrow section of the transition zone between order and disorder (Hidalgo et al. 2014). This brings us back to nucleic acids and their origin.

The outstanding character of nucleic acids among all other known molecules is the fact that they contain messages with meanings. The messages are addressed to the gene expression and regulation system. They are based on information which is encoded within the sequence of nucleic acids in a linear, segregated way. If we look at the genetic code from the viewpoint of information theory we see a unit of digital information: the two nucleotide triplets in a codon-anticodon pairing represent a six-bit byte of digital information (Yokey 2005). So we have information stored in nucleic acids that is linear, segregated, and digitally encoded. But what does that mean for the overall nucleotide sequence? To answer this question we should have a look at complexity.

The complexity of a nucleotide sequence can be described in relation to aspects such as order, algorithmic compressibility and algorithmic function (Abel and Trevors 2005). A highly ordered sequence with a simple, repeating motive such as a poly(A) tail of mRNA has the lowest possible complexity. When complexity gradually increases it reaches the maximum at randomness where no order exists at all. Thus, the degree of complexity of a given sequence can be described in relation to two extrema: order (minimal complexity) and randomness (maximum complexity) (Durston et al. 2007).

In addition to order/disorder, the aspect of algorithmic compressibility can be related to sequence complexity. What does algorithmic compressibility mean within this context? Let us take the example of a poly(A) tail of mRNA with the sequence

"AAAAA". An algorithmic compression program for this ordered simple repetitive sequence can be very short by just reading "5A". This tells us that a sequence with low complexity (high degree of order) has high algorithmic compressibility. When the sequence complexity is increased its algorithmic compressibility decreases.

But there is even more about sequence complexity. It can also be related to a third aspect: algorithmic function. It is an aspect that refers to what a sequence can do itself (and not what can be done to it as in the case of algorithmic compression) and thus indicates the *functionality* of a sequence. Functionality means in this context the ability of a sequence to be instructional. When including the dimension of functionality in the description of sequence complexity something very remarkable becomes apparent: the ability of a sequence to be instructional (and thus functional) turns out to only be possible within a rather narrow part of the complexity spectrum. This part appears much closer to randomness than to the ordered end of complexity and opens up a new category of sequence complexity. This category is termed "*Functional Sequence Complexity* (Abel and Trevors 2005). The relevant aspect of this discussion is that it enables algorithmic instructions. Outside the functional part of the complexity spectrum sequences of various degrees of complexity exist of course. But they cannot be instructive due to a very low level of algorithmic function (Fig. 5.3).

In sum, the theory of complex systems enables us to describe nucleic acids as polymers with a unique type of complexity—the functional sequence complexity. Its characteristic degree of aperiodicity opens up a window to algorithmic programming of biological organization. Thus, functional sequence complexity is of central

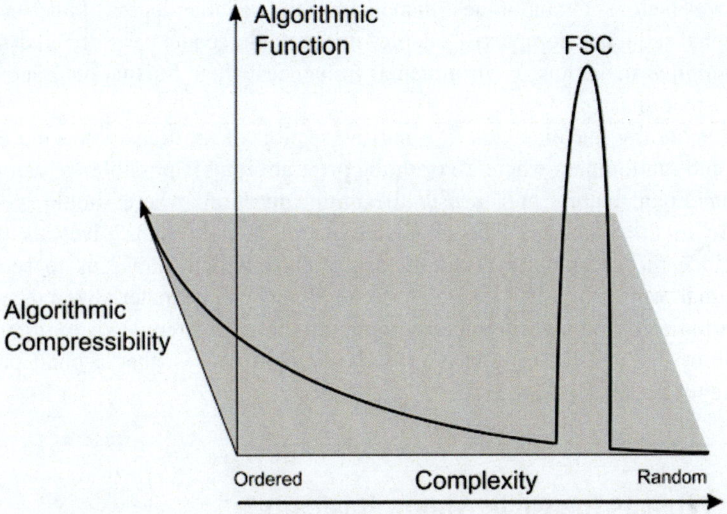

Fig. 5.3 Illustration of the relationship between complexity, algorithmic compressibility, and algorithmic function. FSC: Functional Sequence Complexity. Adapted from (Hofgartner et al. 2014)

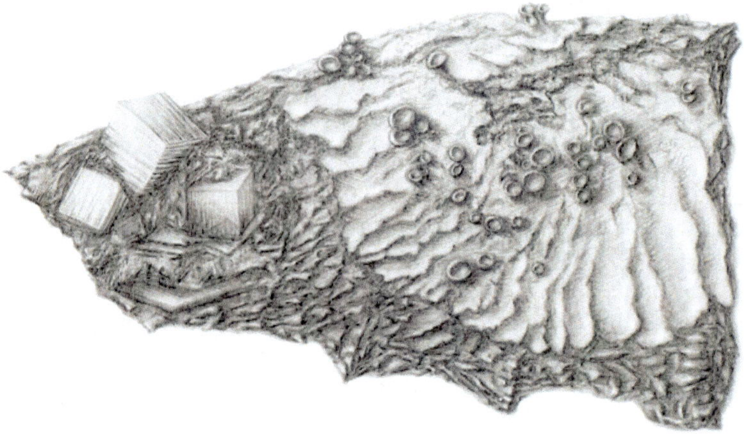

Fig. 5.4 Illustration of different complexity categories. A lichen (upper right part) representing *functional complexity* grows on a rock representing nearly *random complexity* and is accompanied by pyrite cubic crystals (left part of the rock) which represent *ordered complexity*. The functional complexity of the lichen is located between order and randomness, but closer to the random side. Its functional complexity is algorithmically programmed by nucleic acids which itself represent *functional sequence complexity*. Credit: H. Turhyt

importance for understanding the origin, function, and evolution of nucleic acids and can be regarded as the outcome of the nonrandom selection of prescriptive sequences during chemical and biological evolution (Abel and Trevors 2005). As being much closer to randomness than to order the peculiar degree of aperiodicity of functional sequence complexity enables nucleic acid sequences to be *instructional* and *adaptive* to various environmental influences, their fluctuations, and overall transformations.

After finding nucleic acids in a narrow section of the transition zone between order and randomness where algorithmic programming is possible we can pick up this functional feature and turn to the initial question: where should we search for life in the universe? The characteristic of nucleic acids gives us the hint to include the complexity vector in our toolbox which allows us to search for functional complexity. This category of complexity can be materialized in any form of aperiodicity with algorithmic programming that ultimately leads to organisation instead of just self-ordering (e.g. crystals) or randomness (the "asphalt problem" (Benner et al. 2012)) (Fig. 5.4).

5.7 What Is the Code About, Basically?

That question is easy to answer—until someone asks. Of course, we know from biochemistry that the genetic code of nucleic acids provides all instructions that encode the assembly and regulation of subcellular entities and the structures of cells, organs, and organisms. But why taking the trouble to run an encoded, organized assembly and regulation system? What motivated prebiotic chemistry to gradually become biology?

An analogy may be of some help to approach an answer to the initial question. We can ask: what is the construction plan of a house about, basically? Of course, to build a house. But that answer is only halfway to the core: what is the purpose of a house? It is to give a stable, protective environment for its inhabitants. This is usually needed because the builders want to settle in an environment that differs from those in which their distant ancestors initially have been evolved. So, the construction of a house, designed in adaption to its environment, is basically to allow humans to exist in locations where they haven't been adapted physically but want to stay for work by tapping the local resources for maintaining their existence. The house has to be designed in such a way that it is an open system, having a constant throughput of energy (electricity, heat, waste) and an intake of water in order to remain stable. Thus, when looking at a construction plan from this perspective, aspects known from evolutionary biology and thermodynamics become apparent. Against the background of this analogy, we now can approach the initial question via a combination of evolutionary biology and thermodynamics (Black 1978; Wicken 1998).

A genetic code contains all instructions to assemble a living system. From the viewpoint of thermodynamics, a living system is a dissipative structure with functional complexity. A dissipative structure is a structure far from thermodynamic equilibrium. This means that it requires a net flow of energy and matter to maintain its dynamically ordered structure via degrading free energy. The energy degrading kinetic pathways in living systems are genetically instructed. An exchange of energy and matter with the environment is essential as it reduces the entropy of the system so a dynamically ordered structure can exist. Due to the essential exchange of energy and matter with the environment living systems are ecological entities that are dependent on their environment. Selection favors those living systems which manage either to increase the energy throughput or to use a given one more efficiently for both maintaining their internal organization (avoiding decay) and for replication.

From the perspectives of thermodynamics and evolutionary biology it seems that the genetic code is for instructing a dissipative, complex chemical system not to passively wait for resupply of free energy (as e.g. a candle flame—a non-complex dissipative structure—does) but to use mutation and selection for computing the local environment. This is done to actively explore and tap new thermodynamic disequilibria of any kind so the system can expand its dissipative structure via autopoiesis and replication deep into the dimension of time over billions of years.

5.8 Who Holds the Copyright? The Case Nucleic Acids Versus Amino Acids

Where to start? Answering that question is problematic for prebiotic chemists, but obviously not for prebiotic chemical evolution when it comes to the DNA-protein paradox.

The research on the emergence of life runs again into a severe paradox when facing the strong interdependence between nucleic acids and proteins. The design of possible scenarios for the origin of this close linkage causes considerable headaches as the interdependence is of strict circular nature: DNA contains all instructions for algorithmically programming the organisation of the complex polypeptide sequence that defines the function of proteins while, on the other side, the replication, processing, and repair of DNA are almost exclusively made by proteins. In other words: nucleic acids are essential for the synthesis of proteins, and proteins are essential for the synthesis of nucleic acids. One side cannot come into being without the other side.

It is this DNA-protein paradox that gave rise to the "RNA-World Hypothesis" (Le Vay and Mutschler 2019). The hypothesis is mainly based on the assumption that a primitive ribosome could have been existed in the prebiotic era solely based on RNA. Or, in other words: RNA came first by having a self-replicating capability. This idea was boosted when ribozymes have been discovered with some catalytic function e.g. in terms of self-splicing (Kruger et al. 1982). Further support comes from the fact that the energy currency of the cell—ATP (Fig. 5.1)—is a building block of RNA. Experimental observations also reveal that the core of ribosomes consists solely of RNA and that the activity of protein enzymes often requires RNA-like cofactors (Hayden et al. 2019). However, the hypothesis generates its own numerous paradoxes (Robertson and Joyce 2012) and has extensively been challenged so far (Hayden et al. 2019). In sum, it appears that persisting on an "RNA first" scenario didn't really help to solve the circular RNA-protein paradox in a plausible way and that many unsolved questions remain regarding the origin of an RNA-World.

Thus, models which try to circumvent a decision on what came first become increasingly attractive. Such models are based on the assumption that a prebiotic chemical evolution of polypeptides and nucleic acids was closely connected right from the start. We can define this start as the emergence of prebiotic synthesis pathways resulting in the simultaneous generation of nucleotides and amino acids. Such pathways have been discovered and described in recent years (Frenkel-Pinter et al. 2020). Against this background an experimental finding (Greiner and Trixler 2019) could give a hint to how the initial connection between nucleotides and amino acids may have been started: water within nanoscopic confinements shows various nanofluidic phenomena which alter its properties compared to the bulk. Experimental results suggest that such altered thermodynamic properties of water allow not only to circumvent the water paradox for the polymerization of nucleotides into RNA. Moreover, when an amino acid was added to the aqueous solution of nucleotides it significantly catalyses nucleotide polymerization under nanofluidic conditions. This may be indicative of a very early interplay between the building blocks of nucleic acids and proteins.

When it comes to the complexity level of biopolymers some basic interactions between RNA and peptides have been revealed in which one partner benefits in terms of aspects such as functional structure, reactivity, or stability (Frenkel-Pinter et al. 2020). However, in living systems, we see mutual benefits between both biopolymer species. In this light, the relevance of recent studies becomes apparent that provide indications for a very early co-evolution of polypeptides and polynucleotides starting with mutual stabilizations of RNA and proto-peptides (Frenkel-Pinter et al. 2020).

Despite such substantial achievements in prebiotic chemistry research that uncover some basic functional interactions between both types of biopolymers the big picture needed to solve the DNA-protein paradox remains unclear. There is still a wide-open field for further research for all who take up the challenge and try to solve this paradox from a broader perspective.

5.9 A New Form of Stability Arises

While the oceanic crust on Earth is constantly recycled, supercontinents form and break up again and mountains rise and fall over hundreds of millions of years, the presence of nucleic acids as one of the most vulnerable molecular species persists all over the aeons despite being constantly attacked by hydrolysis, radiation, high temperature and many other destructive influences from the environment. But it is exactly the extreme fragility of nucleic acids that makes their presence on Earth stable over billions of years. To comprehend this seemingly paradoxical issue and to understand how it has been solved conceptually, we have to pay a visit to one of the most influential research institutions of ancient times.

The great library of Alexandria stored thousands of written documents, mainly in the form of rolls of papyrus and parchment. Environmental conditions in that geographical area such as high humidity at the borderline of the Nile delta have led to a comparatively vast decomposition of the rolls. To cope with it, a huge effort was required to constantly replicate the library stocks. This needed a permanent source of funding. Although the Museion, including its great library, was attacked several times in the course of its history, it seems that the very end came when the funding was cut.

Today we can look at this situation from the viewpoint of physics and chemistry: instructional information was stored via fragile biopolymers (the main constituent of papyrus and parchment) as the main constituent of a metastable medium (rolls). The fragility of the medium was compensated by replication (copyists of the Museion). Replication needed a constant flux of energy (salary) by tapping thermodynamic disequilibria (funding). So the stability of the ancient data storage device was not given due to low chemical reactivity but rather due to another kind of stability raised by replication. What survived as information from the ancient documents usually comes from a chain of copies of copies. By far the most amount of rolls from ancient times didn't survive over time, but their content did when having been repeatedly replicated.

This brings us back to nucleic acids: most probably not a single genetic polymer snippet survived the billions of years since LUCA explored the early archean world, but for life, this is not the point. Living systems compensate for the fragility of their components and even their whole system via replication. The emergence of replicative chemistry changed everything in the prebiotic world and can be regarded as being a "phase transition" in prebiotic chemical evolution. This is because a replication reaction is dissimilar to every other chemical reaction. It seems that replicative chemistry started with the emergence of nucleic acid species integrated into a prebiotic chemical system. The rise of replicative chemistry opened up a whole new world beyond chemical stability: it enables a molecular species such as nucleic acids to persist over time despite a constant degradation of individual entities. We thus have a new kind of stability which is termed "Dynamic Kinetic Stability (DKS)" (Pross 2013).

What is the characteristic of DKS? A system based on DKS remains in a steady state as long as its replication rate is equal to or higher than the decay rate. It is opposite to thermodynamic (or "chemical") stability but includes it at the same time. A chemically (meta)stable system has reached a (local) minimum of Gibbs free energy. This can occur spontaneously via a downhill reaction (release of energy). Such a system remains stable as long as there is no intake of energy. In contrast, a system with DKS needs to be continuously reactive via an intake of energy (metabolism) in order to maintain replication. Its elements have to be of low chemical stability in order to overall maintain dynamic kinetic stability.

DKS has many fundamental implications for prebiotic chemistry and biology (Pross 2012). What is important for our discussion is the consequence of DKS for nucleic acids: it makes the presence of this molecular species with replication capability extremely stable over billions of years despite—or even because—of extremely low chemical stability.

5.10 Summary and Outlook

In this chapter, we took a spot on nucleic acids and discussed what happens when we change the perspective on these biopolymers from biochemistry to prebiotic chemistry: the clear biochemical definition of what nucleic acids are became fuzzy from this perspective. What we have seen were paradoxes and messages from the prebiotic era waiting to be fully deciphered. We can view these aspects as puzzle pieces which, in sum, give us a deeper insight into the multi-faceted character of nucleic acids and their origin.

But these instructional polymers tell us more: we can stitch the collected puzzle pieces together and see if something takes shape from a broader perspective.

The ubiquity of AMP in biochemistry comes along with a uniquely high self-association tendency among other nucleoside monophosphates. This points to strong supramolecular interactions which can form assemblies with high thermodynamic (chemical) stability. This kind of stability also becomes relevant within the context of "standardization" of nucleic acids building blocks: canonical nucleobases are more stable against photolysis than non-canonical ones. We also see an important contribution of thermodynamic stability within the context of the "water problem": DNA is used by living systems for long-term genetic information storage due to its higher chemical stability against hydrolysis compared to RNA (Voet et al. 2019). A certain kind of stability takes also shape on the puzzle piece that refers to the aspect of standardization of nucleotides: selection during prebiotic chemical evolution poised the complexity of nucleic acids at the functional and adaptable part of the complexity spectrum. Thus, standardization of nucleic acids building blocks created and still maintains functional stability as well as the stability of adaptability. Furthermore, the replication ability of nucleic acids causes another kind of stability that enables these polymers to be persistent despite their fragility: Dynamic Kinetic Stability.

In sum, when viewed from a broader perspective, nucleic acids seem to be optimized by prebiotic chemical and biological evolution to work with different types of stability at the same time: we see thermodynamic stability with respect to their building blocks and the long-term genetic information storage. There is also functional stability and the stability of being adaptable—both maintained by functional sequence complexity. And we see and dynamic kinetic stability maintained by the ability to replicate. It is easy to run into crystallization (order) or asphaltization (randomness), but how prebiotic chemical evolution managed to reach and combine these different kinds of stability in one molecular entity (nucleic acids) and balance them at the narrow peak of functional complexity could only be revealed by achieving an even bigger picture. But this needs more puzzle elements, waiting for being discovered and collected by those who take the challenge.

References

Abel, D.L., Trevors, J.T.: Three subsets of sequence complexity and their relevance to biopolymeric information. Theor. Biol. Med. Model. **2**(1), 29 (2005)

Abrevaya, X.C., Anderson, R., Arney, G., Atri, D., Azúa-Bustos, A., Bowman, J.S., Brazelton, W.J., Brennecka, G.A., Carns, R., Chopra, A., Colangelo-Lillis, J., Crockett, C.J., DeMarines, J., Frank, E.A., Frantz, C., de la Fuente, E., Galante, D., Glass, J., Gleeson, D., Glein, C.R., Glein, C.R., Goldblatt, C., Horak, R., Horodyskyj, L., Kaçar, B., Kereszturi, A., Knowles, E., Mayeur, P., McGlynn, S., Miguel, Y., Montgomery, M., Neish, C., Noack, L., Rugheimer, S., Stüeken, E.E., Tamez-Hidalgo, P., Walker, S.I., Wong, T.: The astrobiology Primer v2.0. Astrobiology. **16**(8), 561 (2016). https://doi.org/10.1089/ast.2015.1460

Anantharaman, V., Koonin, E.V., Aravind, L.: Nucleic Acids Res. **30**(7), 1427 (2002)

Anscombe, N.: Nature Photon. **2**, 412 (2008)

Benner, S.A.: Orig. Life Evol. Biosph. **44**, 339 (2014)

Benner, S.A., Kim, H.J., Carrigan, M.A.: Acc. Chem. Res. **45**(12), 2025 (2012)

Black, S.: Persp. Biol. Med. **21**(3), 348 (1978)
Blackburn, G.M., Gait, M.J., Loakes, D., Williams, D.M.: Nucleic Acids in Chemistry and Biology. Royal Society of Chemistry, Cambridge (2006)
Chaput, J.C., Herdewijn, P.: Angew. Chem. Int. Ed. **58**(34), 11570 (2019)
Cleaves II, H.J., Butch, C., Burger, P.B., Goodwin, J., Meringer, M.: J. Chem. Inf. Model. **59**, 4266 (2019)
Doonan, S.: Nucleic acids. Royal Society of Chemistry, Cambridge (2004)
Durston, K.K., Chiu, D.K., Abel, D.L., Trevors, J.T.: Theor. Biol. Med. Model. **4**(1), 47 (2007)
Eberle, A., Markert, T., Trixler, F.: J. Am. Chem. Soc. **140**(4), 1327 (2018)
Fialho, D.M., Roche, T.P., Hud, N.V.: Chem. Rev. **120**, 4806–4830 (2020)
Frenkel-Pinter, M., Haynes, J.W., Mohyeldin, A.M., Martin, C., Sargon, A.B., Petrov, A.S., Krishnamurthy, R., Hud, N.V., Williams, L.D., Leman, L.J.: Nat. Commun. **11**(1), 1–14 (2020)
Gibard, C., Bhowmik, S., Karki, M., Kim, E.-K., Krishnamurthy, R.: Nat. Chem. **10**(2), 212–217 (2018)
Gibard, C., Gorrell, I.B., Jiménez, E.I., Kee, T.P., Pasek, M.A., Krishnamurthy, R.: Angew. Chem. **131**(24), 8235–8239 (2019)
Greiner, A., Trixler, F.: Adenosine monophosphate in temporal nanoconfined water catalyzes molecular self-assembly. In: Astrobiology – Life in the Context of Cosmic Evolution, Conference Abstracts, p. 46. Springer, New York (2016)
Greiner, A., Trixler, F.: Prebiotic reaction vessels – RNA formation in nanoconfinements of water. In: 19th EANA Astrobiology Conference. Orléans, France (2019)
Grosch, E.G., Hazen, R.M.: Astrobiology. **15**(10), 922 (2015)
Hayden, E.J., Lehman, N., Unrau, P.J.: RNA and ribozymes in the development of life, Chap. 6.1. In: Kolb, V.M. (ed.) Handbook of Astrobiology. CRC Press, Boca Raton, FL (2019)
Hazen, R.M., Sverjensky, D.A.: Cold Spring Harbor Perspect. Biol. **2**(5), a002162 (2010)
Hazen, R.M., Papineau, D., Bleeker, W., Downs, R.T., Ferry, J.M., McCoy, T.J., Sverjensky, D.A., Yang, H.: Am. Mineral. **93**, 1693 (2008)
Hidalgo, J., Grilli, J., Suweis, S., Muñoz, M.A., Banavar, J.R., Maritan, A.: PNAS. **111**(28), 10095 (2014)
Higginbotham, B.D.: The standardization of standardization: the search for order in complex systems. PhD Thesis, George Mason University (2017)
Higgs, P.G.: Life. **6**(2), 24 (2016)
Hofgartner, J.D., Hayes, A.G., Lunine, J.I., Zebker, H., Stiles, B.W., Sotin, C., Barnes, J.W., Turtle, E.P., Baines, K.H., Brown, R.H., Buratti, B.J., Clark, R.N., Encrenaz, P., Kirk, R.D., Le Gall, A., Lopes, R.M., Lorenz, R.D., Malaska, M.J., Mitchell, K.L., Nicholson, P.D., Paillou, P., Radebaugh, J., Wall, S.D., Wood, C.: Nat. Geosci. **7**(7), 493 (2014)
Jauker, M., Griesser, H., Richert, C.: Angew. Chem. Int. Ed. **54**(48), 14564 (2015)
Kim, S.C., Zhou, L., Zhang, W., O'Flaherty, D.K., Rondo-Brovetto, V., Szostak, J.W.: J. Am. Chem. Soc. **142**(5), 2317 (2020)
Kitadai, N., Maruyama, S.: Geosci. Front. **9**, 1117 (2018)
Kruger, K., Grabowski, P.J., Zaug, A.J., Sands, J., Gottschling, D.E., Cech, T.R.: Cell. **31**(1), 147 (1982)
Lang, C., Lago, J., Pasek, M.A.: Phosphorylation on the early earth: the role of phosphorous in biochemistry and its bioavailability. In: Kolb, V.M. (ed.) Handbook of Astrobiology, Chap. 5.8. CRC Press, Boca Raton, FL (2019)
Lazcano, A.: ACS Nano. **12**(10), 9643 (2018)
Le Vay, K., Mutschler, H.: Emerg. Top. Life. Sci. **3**(5), 469 (2019)
Neidle, S.: Principles of Nucleic Acid Structure. Elsevier, New York (2008)
Nie, P., Bai, Y., Mei, H.: Molecules. **25**(15), 3483 (2020)
Nielsen, P.E., Appella, D.H. (eds.): Peptide Nucleic Acids: Methods and Protocols. Springer, New York (2014)
Norberg, J., Nilsson, L.: J. Am. Chem. Soc. **117**(44), 10832 (1995)
Pearson, D., Brenker, F.E., Nestola, F., McNeill, J., Nasdala, L., Hutchison, M.T., Matveev, S., Mather, K., Silversmit, G., Schmitz, S., Vekemans, B., Vincze, L.: Nature. **507**, 221 (2014)

Pross, A.: What Is Life? How Chemistry Becomes Biology. Oxford University Press, Oxford (2012)

Pross, A.: Curr. Org. Chem. **17**(16), 1702 (2013)

Robertson, M.P., Joyce, G.F.: Cold Spring Harbor Perspect. Biol. **4**(5), a003608 (2012)

Tribolet, R., Sigel, H.: Biophys. Chem. **27**(2), 119 (1987)

Voet, D., Voet, J.G., Pratt, C.W.: Fundamentals of Biochemistry – Life at the Molecular Level, p. 854. Wiley, New York (2019)

Wicken, J.S.: Syst. Res. **15**(5), 365 (1998)

Williams, M.: Universe Today, 12/1. https://www.universetoday.com/65588/whatpercent-of-earth-is-water/ (2014)

Yadav, M., Kumar, R., Krishnamurthy, R.: Chem. Rev. **120**, 4766–4805 (2020)

Yokey, H.P.: Information Theory, Evolution, and the Origin of Life. Cambridge University Press, New York (2005)

Chapter 6
Abiotic Synthesis and Role of Amphiphiles in the Encapsulation Process in Life's Origin

Augustin Lopez, Carolina Chieffo, and Michele Fiore

In memory of Océane

Abstract In a prebiotic world, the first form of cells, the protocells, were able to incorporate functional molecules such as polymers with self-replicative properties thanks to primitive forms of membranes. Lipid boundaries represent the chemico-physical barrier between the inner and the outer part of a modern cellular environment. However, one could expect that other types of boundaries were present in primitive compartments. In this chapter we present a prebiotic chemistry perspective that includes the synthesis of amphiphiles under prebiotic conditions, their role in protocells boundary formation and how synthetic protocells were assembled to simulate plausible.

6.1 Introduction

Encapsulate and its related noun, "capsule," derive from "*capsula*," a diminutive form of the Latin noun *capsa*, meaning box. The original sense of encapsulate "to enclose something in a capsule" fits with the definition of cell. It is roughly a box where the complex lipid membrane represents the perimeter of the capsule encapsulating a content consisting of the systemic memory (genetic code) and the metabolic machinery (proteins) (Albertsen et al. 2014). Cell membranes are three-dimensional molecular films composed of biosurfactants, which define the

Carolina Chieffo and Augustin Lopez contributed equally.

A. Lopez · C. Chieffo · M. Fiore (✉)
Institut de Chimie et Biochimie Moléculaires et Supramoléculaires (ICBMS, UMR 5246), Bâtiment Lederer, Claude Bernard Lyon 1, Université de Lyon, Villeurbanne Cedex, France
e-mail: michele.fiore@univ-lyon1.fr

A. Neubeck, S. McMahon (eds.), *Prebiotic Chemistry and the Origin of Life*, Advances in Astrobiology and Biogeophysics,
https://doi.org/10.1007/978-3-030-81039-9_6

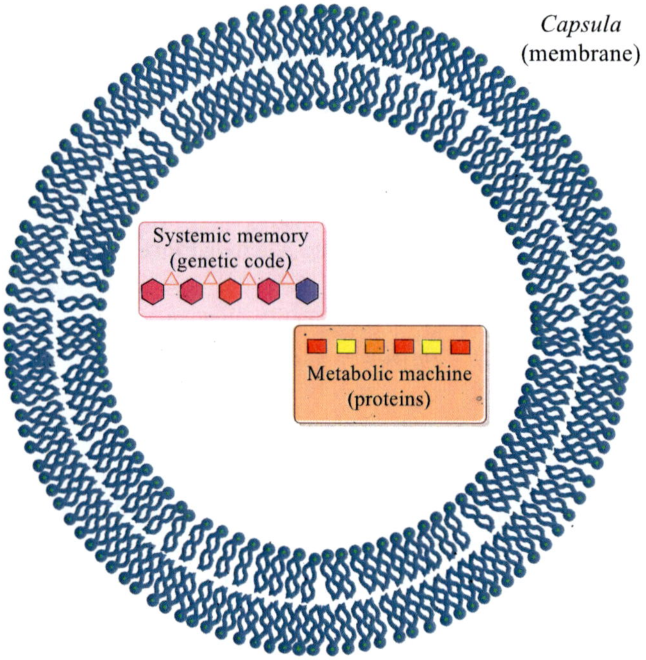

Fig. 6.1 A schematic representation of the cell: the systemic memory (genetic code) and the metabolic machinery (proteins) are encapsulated into a lipidic membrane that is the *capsula*. Cell membranes can be defined as two-dimensional fluid molecular films composed of biosurfactants, which define the boundaries and the compartmented substructures (Deamer 2017)

boundaries and its substructures (Deamer 2017). The role of these boundaries is to spatially separate and co-locate microenvironments while protecting and keeping them at defined concentrations. It enables a multitude of often competing and interfering biochemical reactions to occur simultaneously (Fig. 6.1) (Alberts et al. 2014; Deamer 2017; Monnard and Walde 2015).

Spontaneous appearance of closed membranes composed of bilayers of self-assembling amphiphiles was likely a prerequisite for molecular evolution of protocells (Ruiz-Mirazo et al. 2014). Such compartments would have been dynamic in their composition, sufficiently stable to harbor macromolecules and semi-permeable to allow reactive small molecules to cross the membrane. Although life, as we know it, uses lipid bilayer-based compartments, the diversity of prebiotic chemistry may have enabled primitive living systems to start from other types of boundary systems (Monnard and Walde 2015).

6.2 Why Is the Emergence of Life Related to the Encapsulation Phenomenon?

Defining life has always been a major concern in the origins-of-life studies. In spite of numerous attempts, this achievement seems to be unreachable due to conceptual divergences (Mariscal et al. 2019). A consensus concerning the properties of any living organism has emerged though: this chemical system self-sustains and is capable of Darwinian evolution (Vitas and Dobovišek 2019). Compartmentalization and encapsulation which happened at some point during the origins of life, are part of the process of selection and evolution of life as we know today (Shirt-Ediss et al. 2017).

6.2.1 Encapsulation Improves the Efficiency of a Self-Sustaining Chemical System

A living chemical system represents an organized network which is kept far from equilibrium thanks to energy supply fueling chemical reactions (Danger et al. 2020). It becomes clear now that such an entity would be composed by a multitude of molecules. These compounds could be oligonucleotides, peptides, lipids, carbohydrates and many others. Encapsulation would group all these different molecules forming a real system. One can expect that cooperations, emerging properties and eventually life could come from such system (Kee and Monnard 2017).

Various chemical reactions would take place in this self-sustaining network such as the synthesis and elongation of polymers. Encapsulating the different compounds of this network entails their gathering within a limited space and then the reactivity is strongly impacted. Several experiment, carried out using a large variety of model systems and conditions have proven that chemical reactions under confinement are accelerated due to higher concentrations of reagents and catalysts, increasing the efficiency of such system (Grommet et al. 2020). On the contrary, the same or similar set or reactions carried out in non confined environments would be slowed down as dilution would reduce the rate of reactions (Ruiz-Mirazo et al. 2014). Besides, the chemical reactions selectivity is modified due to the confinement whereas it also stabilizes various chemical species such as complexes (Grommet et al. 2020). Thus, encapsulation allows a reactivity which is very different from the one observed in solution and enables the formation of potentially relevant chemical species. Remarkably, chemical reactions may also occur within the boundaries increasing the catalytic potential of such network (Murillo-Sánchez et al. 2016). We also want to stress that compartmentalization does not mean isolation because the system is permeable as it needs to be fueled with sources of energy and matter but also to remove the waste. Both are guaranteed by the presence of semi-permeable boundaries (Mansy 2010).

6.2.2 Encapsulation Is a Key Step Towards Darwinian Evolution

The information of a supramolecular and living system would be contained in sequenced polymers, the most plausible ones being oligonucleotides (Szostak et al. 2001). This information would define the "identity" of the whole system. One of the advantage of compartmentalization is that it permits to keep the information-encoding polymers encapsulated and then to conserve the identity of the network (Segre et al. 2000).

Besides, the information-encoding polymers of the system can be inherited with few modifications through self-reproductions. Then, the different variants would be in competition and the ones with the highest fitness would be selected: evolution would occur (Ruiz-Mirazo et al. 2017). As higlighted previously, the identity of the network depends on its encapsulation. Besides, only a compartmentalized system could self-reproduce as a whole (the confined network with the compartment itself) and eventually evolve (Lopez and Fiore 2019).

6.3 Which Compartments for the First Encapsulations?

Considering the interesting properties offered by the confinement, it is generally admitted that the encapsulation happened very early forming abiotic systems (Monnard and Walde 2015). These networks at the limit between chemistry and biology which are called protocells would have join non-living with living matter through evolution ultimately leading to LCA (Last Common Ancestor) (Ma and Feng 2015). However, deciphering the nature of the first compartments remains problematic because geology (Javaux 2019) and phylogeny (Betts et al. 2018) evidences cannot reach these protocells. Various candidates have been proposed as suitable prebiotic compartments.

At least three different types of prebiotic plausible compartments can be quoted: the ones made of lipidic molecules (such as fatty acids or phospholipids) which are able to self-assemble forming boundaries; the ones made of non-lipidic molecules (such as α-hydroxy acids (αHAs)) able to form microdroplets encapsulating biomolecules and those that can not be called "compartments" such as minerals that can adsorb guest molecules providing crowding without isolation from the medium.

6.4 Why Lipids as the First Membranogenic Components for Protocells?

One essential characteristic of living species is their ability to create compartmentalization of bioactive molecules. Closed membranes exert confinement and protection of an internalised chemical network, and also support reactions on their hydrophobic region (Szostak 2009). In life, as we know it, these compartments are mainly made of phospholipids along with other compounds (proteins, carbohydrates, other lipids...) (Alberts et al. 2014). The natural enantiomer of all diacyl–glycerol phospholipids in *eucaria* and in most of *bacteria* is D (Fischer convention), 1,2-diacyl-*sn*-glycerol-3-phosphoalcoholine (*sn*-glycerol nomenclature) or 2R—in the Cahn-Ingold-Prelog formalism). The opposite configuration of L-diacylglycerol, 2,3-diacyl-*sn*-glycerol-phosphoalcohol or (2S) is the natural lipidic isoprenoid glycerol ethers of *archaea* membranes (Fig. 6.2) (IUPAC-IUB Commission on Biochemical Nomenclature 1978; Lombard et al. 2012).

Lipids are amphiphiles bearing an hydrophilic head and a hydrophobic tail (cf. Figs. 6.2 and 6.3). The lipids tend to aggregate near a critical aggregation

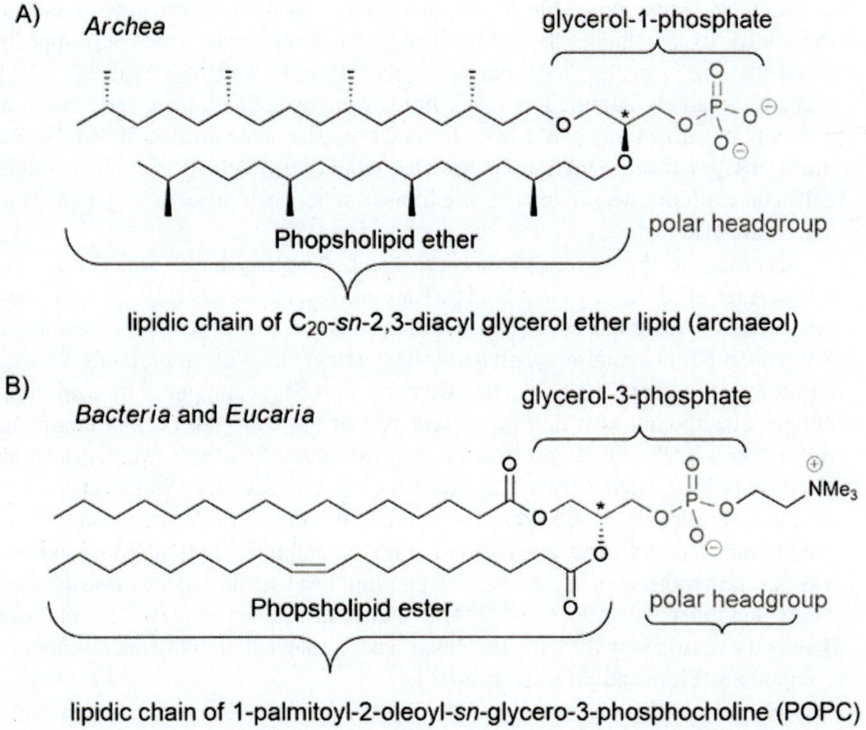

Fig. 6.2 Phospholipid ethers and esters are naturally occurring molecules that form lipid boundaries in *Archea* (**a**) *and Eucaria* and *Bacteria* (**b**)

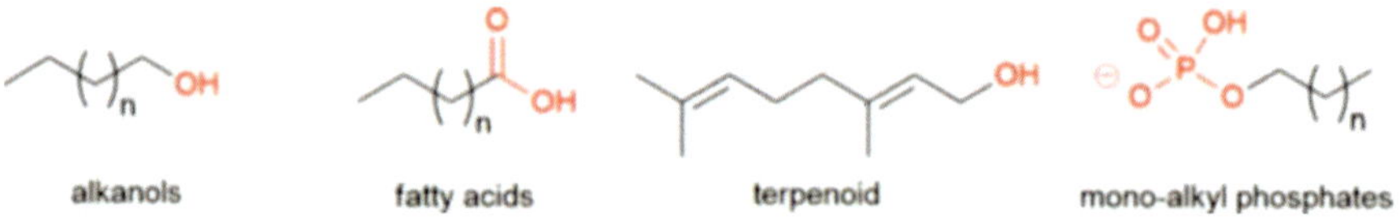

Fig. 6.3 The structures of common lipid-based biosurfactants able to produce vesicles that can encapsulate biomolecules in prebiotic systems chemistry experiments. Among others, fatty alcohols, acids, terpenoids or mono-alkyl phosphates are highlighted as the first membranogeic molecules in lipid-based protocells

concentration (cac). At a certain concentration, the critical micelle concentration (cmc), lipids can form micelles. Micelles are spheroidal aggregates of lipids having their hydrophilic heads on the edges and their the hydrophobic chains inside. At a higher concentration, the critical vesicle concentration (cvc), lipids can form vesicles. Vesicles are compartments with a polar lumen enclosed by a bilayer of amphiphiles: internal leaflet hydrophophilic heads are oriented toward the lumen whereas external leaflet hydrophilic heads are oriented towards the medium (cf. Fig. 6.1).

These self-assembled structures are able to entrap polar compounds in the lumen during their formations. Due to the similarity with modern cell membranes and the ability to spontaneously encapsulate polar compounds, lipids are appealing candidates to form the first compartments (Hanczyc and Monnard 2017). This scenario is largely admitted as many lipids were available on the early Earth and a variety of surfactants could have formed compartments (Walde 2006). Besides, protocells based on membranous vesicles would be relatively similar to modern cells and evolution would explain the transition between these two systems (Fiore and Strazewski 2016).

According to the current view, early membranes may have formed from alcohols, (Apel et al. 2002) fatty acids, (Hargreaves and Deamer 1978) mono-alkyl phosphates, (Albertsen et al. 2014; Fiore et al. 2017; Fiore 2018; Powner and Sutherland 2011), mono-alkyl glycerols, (Maurer et al. 2009) isoprenoids (Ourisson and Nakatani 1994) (Fig. 6.3). They were most probably composed by a mixture of components though as it more representative of the wide diversity of compounds present in a prebiotic soup (Hanczyc and Monnard 2017). However, only pure fatty acids have been very well studied in the last decades (Walde et al. 1994; Zhu and Szostak 2009). Besides, the problem of homochiral/heterochiral fatty acid membranes has not been investigated, with a notable exception (Morigaki et al. 1997). The synthesis of more complex amphiphiles such as racemic phospholipids esters and ethers (Peretó et al. 2004) is harder to achieve in prebiotic conditions. This is the reason why they are less mentioned as relevant prebiotic membranogenic compounds (Fiore and Strazewski 2016).

6.4.1 Which Lipids on the Early Earth?

Even though phospholipids are considered as too complex to be part of the first lipidic membrane, fatty acids are widely acknowledged as prebiotic compounds (Kitadai and Maruyama 2018). Around 4 billions years ago, interplanetary dust particles may have been delivering about 10^8 kg of organic carbon per year (Chyba and Sagan 1992). Among these particles, carbonaceous meteorites (the Murchinson meteorite for instance) contained a wide diversity of compounds. They would have delivered kerogen-like macromolecules but also amino acids and membranogenic compounds such as fatty acids (Pizzarello and Shock 2010). Short-chain monocarboxylic acids are actually the most abundant soluble organic compounds in some meteorites (Huang et al. 2005). Besides, the non-polar molecules extracted from carbonaceous meteorites are able to form boundary structures at alkaline pH (Deamer 1985).

A few experiments were reported in the literature concerning the abiotic formation of single-chain amphiphiles: *n*-alkanols, *n*-alkanoic acids, *n*-alkenes, *n*-alkanes and alkanones. UV irradiation of *n*-alkanes followed by the oxidation and reduction through radical intermediates generated *n*-alkanols (Klein and Pilpel 1973). Mixtures of cold condensed gases (CH_4, NH_3 and H_2O) produced aldo- and keto-compounds of medium length (Simionescu et al. 1981). Otherwise, endogenous production of amphiphilic species could have happened in hydrothermal environments. Fischer-Tropsch type reactions from the disproportion of formic (see **1** in Fig. 6.4) or oxalic acid (**2**) under simulated hydrothermal conditions (100–400 °C) producing C2 to C35 *n*-alkanes derivatives such as *n*-alkanoic acids and long-alkyl chain amphiphilic materials. These compounds such as alcohols, aldehydes and acids (**3–5**) can form, once hydrated, liposome-like objects (Mccollom et al. 1999; Mccollom and Simoneit 1999; Simoneit 2004). Alcohols (**3**), which are able to form closed membranes as well, (Apel et al. 2002) are also considered to be the prebiotic precursors of mono-alkyl phosphates (Albertsen et al. 2014; Fiore et al. 2017) (**6**) and phospholipid ethers (**12**). Indeed, alkanoles can be phosphorylated upon heating at 80 °C with a condensing agent (such as urea or cyanamide) and a phosphate donor (such as ammonium dihydrogen phosphate $NH_4H_2PO_4$ or 2-aminoethyl phosphate) as shown in Fig. 6.4 (Albertsen et al. 2014; Fiore et al. 2017; Fiore 2018; Powner and Sutherland 2011).

6.4.2 Prebiotic Precursors of Lipids

Formamide (**7**), is known as being one of the most simple and plausible source of carbons for the prebiotic synthesis of biomolecules (Saladino et al. 2012). As demonstrated by several scholars, formamide is the natural precursor of several molecules and among others of the glycerol backbone (**9**)—through glyceraldehyde (**8**)—of phospholipids esters and ethers (Fig. 6.6) (Paltauf and Hermetter 1994).

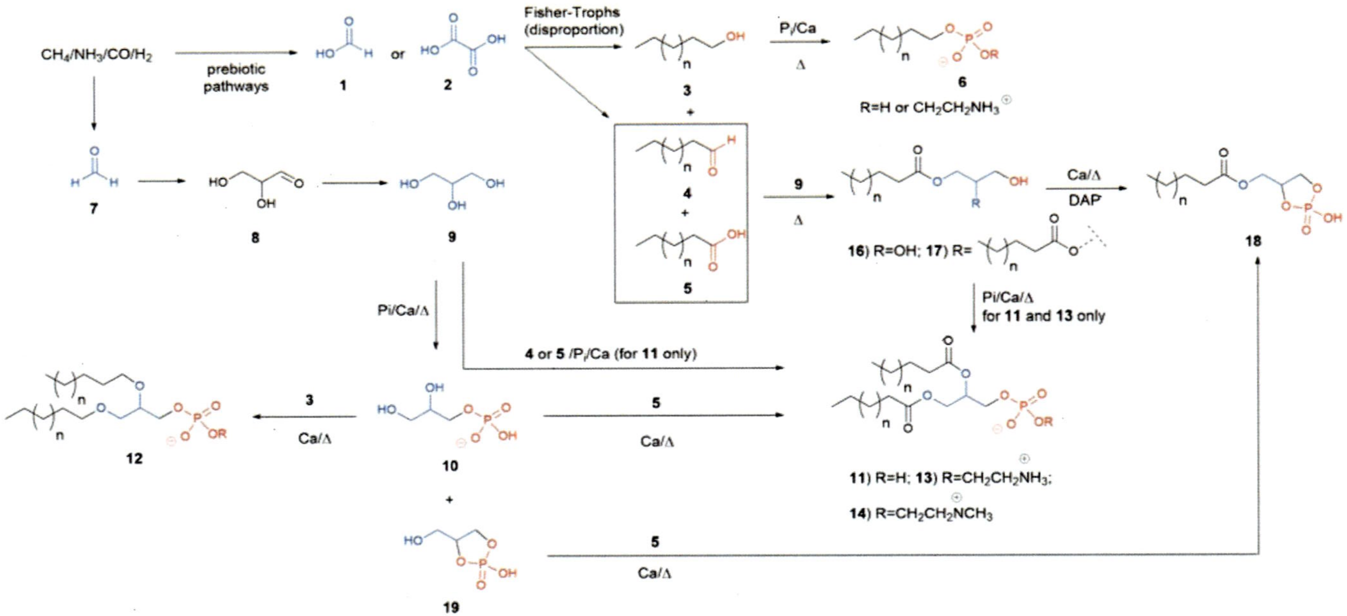

Fig. 6.4 A schematic synthetic pathway for prebiotic lipid-based amphiphiles. The synthesis of those compounds seems to have common building blocks precursors such as formamide, glyceraldehyde or glycerol (**7–9**). The phosphorylation and the acylation mainly occur in the presence of a phosphate source and a condensing agent. Δ stands for temperatures between 65 °C and 130 °C; DAP stands for diamidophosphate (Gibard et al. 2018)

The prebiotic origin of glyceraldehyde and glycerol was exhaustively reviewed in recent years (Patel et al. 2015) and will not be part of our discussion. Incomplete lipids could have been formed from this glycerol and single-chain amphiphiles (Fiore and Strazewski 2016). Simoneit and collaborators showed that C7 to C16 (except C8) alkanoic acids (**5**) could be condensed with glycerol (**9**) to form mono, di and triacylglycerol (not shown) in simulated hydrothermal conditions (Simoneit et al. 2007). The esterification between glycerol and fatty acids can also occur in environments such as hydrothermal pools where the heat and the evaporation favor the condensation reactions (Hargreaves et al. 1977; Eichberg et al. 1977).

Glycerol phosphate (**10**) seems to be the precursor of racemic glycerol-phosphate esters (**11**) and ethers (**12**) (Hargreaves et al. 1977; Eichberg et al. 1977; Epps et al. 1978, 1979). The formation of more complex phosphoethanol amines (**13**) and phosphocholines (**14**) occurred from phosphatidic acids (**11**), instead (Rao et al. 1982, 1987). Incomplete forms of phospholipids, such as mono- and di-acyl glycerols (**16, 17**) (Rushdi and Simoneit 2006) that probably played a crucial role in forming stable amphiphilic membranes (Fiore and Strazewski 2016) were successfully phosphorylated under plausible prebiotic conditions as well (Fayolle et al. 2017). More recently, glycerol based cyclic phosphates (**18**) were obtained in prebiotic conditions from glycerol or monoacyl glycerols or *via* alkylation of cyclic glycerol phosphate (Gibard et al. 2018; Toparlak et al. 2019; Bonfio et al. 2019).

6.5 Which Composition for Prebiotic Lipidic Compartments?

As highlighted previously, phospholipids are the most common modern membranogenic amphiphiles but they are not as convenient to form the first protocells. Indeed, they tend to form very stable membranes with slow dynamics. These features reduce the membrane exchange (e.g., K+ permeability coefficient: 10^{-10}–10^{-12} cm/s) (Paula et al. 1996) and the membrane growth upon supply with amphiphiles. Nevertheless, these two aspects are crucial for any living system allowing it to self-sustain and to self-reproduce. In modern cells this dynamic is enhanced by membrane proteins but one should not expect such complex membranes for the first protocells. Vesicles can also be formed from fatty acids with a hydrocarbon chain which is long enough (at least 8 carbons) in a narrow range of pH (Deamer 2017). These compartments show facilitated membrane exchange (e.g., K+ permeability coefficient: around 10–6 cm/s) (Paula et al. 1996) and growth upon supply with amphiphiles. This is presumably due to a higher mobility of the amphiphiles which are flipping more often between the leaflets (Chen and Walde 2010).

These interesting features added to their prebiotic availability make fatty acids appealing candidates to form the first protocells. Indeed, vesicles made of fatty acids allow the transport of ions (Chen and Szostak 2004) and small molecules such as

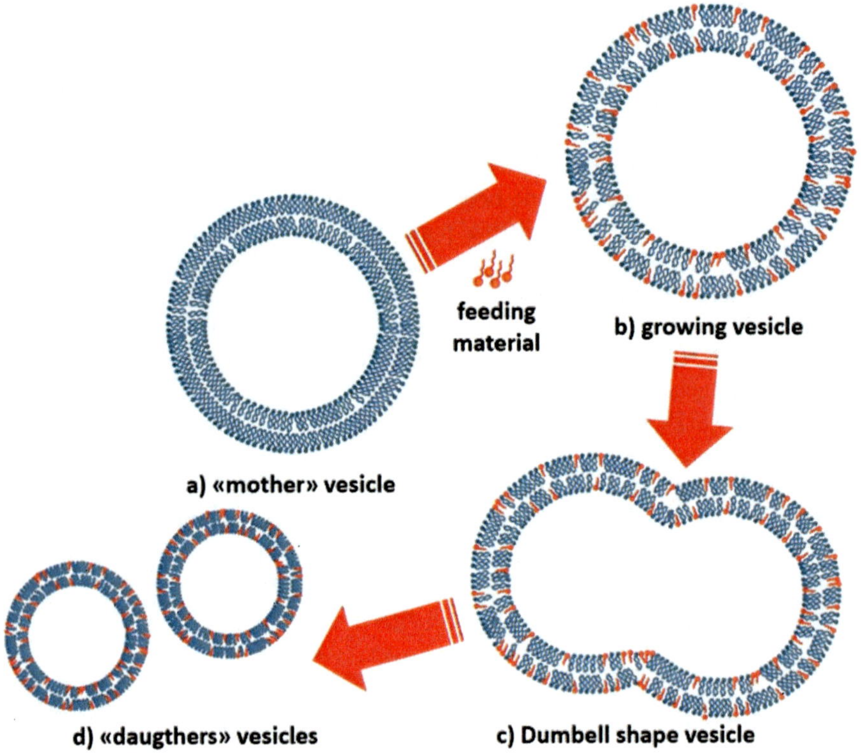

Fig. 6.5 A representative scheme of growth and division as described by Bozîč and Svetina (Božič and Svetina 2004). Vesicle self-reproduction is a process in which a growing vesicle (**a** → **b**) first transforms its spherical shape into a dumbbell shape (**b** → **c**), and then splits into two spherical daughter vesicles (**c** → **d**); "Mother" vesicles refers to the vesicles that were submitted to G&D and "daughter" vesicles to those that generate from this process

riboses (Sacerdote and Szostak 2005). Besides, these compartments can also retain longer biopolymers such as oligonucleotides (Chen et al. 2005). Moreover, vesicles made of pure fatty acids can grow and divide (Walde et al. 1994; Hanczyc et al. 2003) upon feeding with amphiphiles and partially conserve their content (Fig. 6.5) (Zhu and Szostak 2009; Berclaz et al. 2001).

Nevertheless, these compartments made of fatty acids also have their drawbacks. Whereas the self-assembly of phospholipids vesicles is mainly driven by the hydrophobic effect and the Van der Waals interaction, the contribution of the charged head group is mandatory for fatty acid liposomes (Hanczyc and Monnard 2017). That explains why these compartments are so fragile. Fatty acids vesicles can only be formed in a narrow range of pH near the apparent pKa of the monocarboxylic acids for which headgroups can establish electrostatic interactions. For that reason, these membranes are also easily disrupted by electrostatic interactions with aqueous salts and multivalents cations such as Mg^{2+} or Ca^{2+} which are inevitable in marine environments (Monnard et al. 2002). Besides, the cvc of fatty acids is tremendously higher (about 10^{-4} M to 10^{-1} M) than the one of phospholipids (less than 10^{-9} M) and highly concentrated media would be necessary to form fatty acid compartments (Monnard and Deamer 2003). Finally, the high temperatures which are common in hydrothermal environments are not conducive to the stability of fatty acid vesicles (Mansy and Szostak 2008). Nevertheless, membranes made of several fatty acids instead of a single one showed decreased cvc (Cape et al. 2011) and a pH range stability which is extended (Jordan et al. 2019).

Other types of simple lipids can form vesicles. For instance, mono-alkyl phosphates are protonable amphiphiles that can form vesicles able to encapsulate fluorescent probes such as pyranine. Thus, these vesicles could also encapsulate biomolecules such as short peptides and oligonucleotides. Even though mono-alkyl phosphates can form vesicles at relatively low concentration (the cvc of decyl-phosphate is about 0.67 mM at pH 2) they are only stable at very low pH around their first pKa (Albertsen et al. 2014).

It has been found almost two decades ago that using mixtures of lipids to form protocell membranes led to a higher stability (Apel et al. 2002). In this study, the conditions of formation of vesicles made of nonanoic acid, among others, were studied without or with a small amount of nonanol. In the first case the vesicles formed were not stable for pH higher than 7.1 whereas the addition of 10% mol of nonanol enhanced the stability until a pH of 11. Noteworthy, the addition of fatty alcohols also decreased the cvc from 85 to 20 mM. Contrarily to the carboxylic acid of the fatty acids, the alcohol function of the alkanoles would be protonated in a large range of pH enabling hydrogen bonding even for high pH and then the stabilization of mixed vesicles. Mono-alkyl glycerols are other prebiotic compounds able to stabilize fatty acid membranes since vesicles made of mono-alkyl glycerol and fatty acids get lower cvc and are stable at higher pH than pure fatty acid vesicles. The former are also more prone to retain macromolecules (such as carboxyfluorescein) for temperatures going up to 60 °C than the latter (Maurer et al. 2009). Besides, alkanols and mono-alkyl glycerols are interesting co-surfactants in mono-alkyl phosphates vesicles as well. Indeed, pure decyl phosphate vesicles are stable at pH 2 exclusively whereas supplementing these membranes with 4% mol of 1-decanol or 33% mol of 1-decanoyl-glycerol enabled their stabilization at pH 7 and 12 (Albertsen et al. 2014).

Many other compounds considered as prebiotic themselves could favour the formation of lipid bilayers and stabilize them. The addition of amphiphilic compounds such as isoprenoids, (Jordan et al. 2019) polycyclic aromatic hydrocarbons (Groen et al. 2012) or alkyl amines (Namani et al. 2007) are able to enhance the stability of protocells made of fatty acids. Nucleobases and riboses (Black et al. 2013; Black and Blosser 2016) as well as amino acids (Cornell et al. 2019a) can also bind to and stabilize fatty acid membranes. Noteworthy, the environment itself might play a role in the formation of lipid membranes since minerals (Hanczyc et al. 2007) and monovalent salts (Maurer and Nguyen 2016) can promote the self-assembly of amphiphiles in vesicles.

Some combinations of simple prebiotic lipids are quite relevant as the first protocells. For instance, vesicles made of short chain fatty acid:alkanol:mono-alkyl glycerol (4:1:1 molar ratios) are at the same time very permeable to small molecules as riboses (Mansy et al. 2008) and stable at high temperatures (Mansy and Szostak 2008). Lately, some complex protocells made of C10–15 fatty acid:alkanol:C10 isoprenoid (1:1:1 molar ratios) were also proved to be stable in the harsh conditions of hydrothermal vents (alkaline pH, high concentrations of salts and divalent cations, temperature of 70 °C) (Jordan et al. 2019).

As highlighted previously, phospholipids are poor candidates to form protocells because they are less available and because they form rigid membranes. That is the reason why they are more often used as models to study other phenomena linked to the origins of life such as the growth and division of vesicles (Lonchin et al. 1999). However, compartments made of phospholipids and fatty acids reveal interesting features. For instance, liposomes made of oleic acid:1-palmitoyl-2-oleyl-sn-glyero-3-phosphocholine (POPC) (1:1 molar ratios) are at the same time stable at high concentrations of divalent cations such as Mg^{2+} and permeable to small molecules such as Mg^{2+} or nucleotides (Jin et al. 2018). Besides, mixed membranes containing cyclophospholipids are also relevant protocells since decanoate cyclophospholipids: decanol vesicles (2:1 molar ratios) were stable for pH 5 to 10 and to high concentrations of Na^{+}, Mg^{2+} and Ca^{2+} (Toparlak et al. 2019).

6.6 Non-lipidic Prebiotic Compartments

In addition of the molecules found in modern organisms, prebiotic chemistry generated a tremendous variety of molecules. It is possible that the chemical evolution was strongly impacted by some molecules which are not found anymore in contemporary biochemistry. It could notably be the case of molecules able to encapsulate a chemical system such as α-hydroxy acids (Chandru et al. 2020). Such compartments may have played an important role and form protocells before the emergence of life as we know it (Monnard and Walde 2015).

Among others, α-hydroxy acids (αHAs, Fig. 6.6) are interesting candidates as prebiotic compartments. Indeed, αHAs are coproduced along with α-amino acids in prebiotic settings (Jia et al. 2019). The polymerization of αHAs occuring in drying pools gave polyesters that can assemble in membraneless microdroplets. These microdroplets can segregate and entrap fluorescent molecules, providing early compartments. Even if such compartments were not selected by chemical evolution, they could have been important players by protecting, exchanging, and encapsulating primitive components on the early Earth.

Otherwise, the presence in modern cells of non-membrane based organelles has also been taken as an example for non-lipidic prebiotic compartments. In particular, ribonucleotidic particles that present a wall of RNA and proteins are one of the most ancient systems. The protein parts are characterized by low sequence complexity (LSC) and intrinsecally disordered regions (IDR), significantly easier

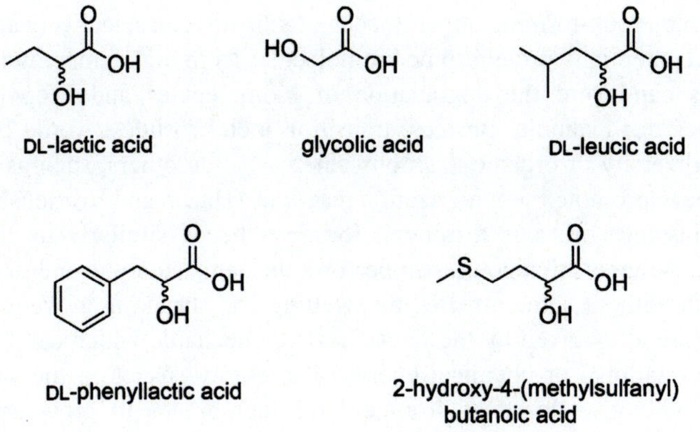

Fig. 6.6 The structure of the polymerizable α-hydroxy acids (αHAs) recently studied by Jia et al. (2019)

to form, and they typically bind RNA to form organelles (Hansma 2017). These coacervates are self-assembled membrane-free compartments. They could have also been present in a prebiotic world since coacervation can happen between a wide diversity oppositely charged polymers such as polypeptides and oligonucleotides. The protocells obtained can also sequester interesting compounds for a system such as enzymes (Koga et al. 2011). This form of compartmentalization is interesting especially since it maintains the functionality of the encapsulated compounds. Recently, coacervate protocells formed with the association of carboxymethyl dextran sodium salt (or poly-L-lysine) and simple ribozymes were used as models. These protocells supported the catalysis of hammerhead ribozymes and selectively kept or released RNAs depending on their sizes (Drobot et al. 2018).

6.7 The Peculiar Case of Minerals: The Case Study of Mica

The theory emphasizing the emergence of life through compartmentalization of substrates in a mixture has been widely supported up to this moment. For most researchers this means protocells composed of amphiphiles which bear an aqueous medium inside, resembling modern cells. Doubts raised on this kind of assemblies concerning the simultaneous syntheses of completely different structures (amphiphiles and encapsulated material) in a prebiotically compatible environment. On the molecular economy point of view, it is a useless waste. This is why other theories underlined the important role of minerals that can host and retain prebiotic compounds in a crowded system. Even though the role of minerals is likely to be limited to primitive prebiotic systems, one can not ignore the interesting features they offer. They could have ensured a link between a diluted prebiotic environment and a compartmentalized prebiotic system (Hansma 2017).

Common rock-forming minerals such as silicates can selectively adsorb, concentrate and potentially protect some biomolecules from high temperatures. Therefore, minerals can avoid the degradation of biomolecules and keep their catalytic properties, for example. Besides, transition metal sulfides, would have catalyzed a wide diversity of organic reactions whereas some others, such as clay minerals would have promoted polymerization reactions (Hazen and Sverjensky 2010).

For instance, mica is a mineral forming sheets, similarly to clay, adsorbing prebiotic polymers that could still perform the same catalytic and/or support activities with without being affected by swelling and shrinking in wet/dry conditions. The main energy given by the mica sheets is mechanic which can be transformed in conformational or chemical energy. The empty spaces in the structure is not only providing a closed environment for membranes to grow and divide, but also contribute to form covalent bonds making the polymers (by enhancing their proximity) and to modify their arrangement and positions. The mica sheets are furthermore compatible environments for most origin-of-life scenarios, such as RNA world or “metabolism” first (Hansma 2017).

6.8 How to Encapsulate?

Compartmentalized systems such as liposomes or vesicles are defined as protocells, given the fact that they resemble in dimension, composition and form to the simplest kind of cell imaginable. Their importance is indeed enhanced by the possibility to fill them through a method called encapsulation. The process of encapsulation could be thought as the introduction of external material inside the lumen of vesicles. In systems chemistry field, this material generally consists in small biomolecules (nucleobases, nucleotides, small RNA strands, peptides . . .) which are necessary to perform an evolutionary process (Szostak et al. 2001). The main effort is for them to cross the hydrophobic bilayer of vesicles, while these compounds are normally characterized by a high degree of polarity.

Two solutions can be imagined, which describe the two main methods to perform encapsulation. In the first one, the molecule that must be inserted is firstly dissolved in the same medium from which the bilayer is formed. To better explain, the membrane formation and the encapsulation processes are simultaneous: the arrangement of phospholipids forming the barrier results in the entrapment of water-soluble molecules. In this case, the same parameters which affect the membrane formation influence the encapsulation. The preparation of giant vesicles, intended as a form of compartments selected by their micrometrical dimension, was firstly introduced by Reeves and Dowben (1969). The method, generally known as natural swelling or gentle hydration method, is based on the controlled and stepwise hydration of dry lipids placed on a glass surface. In fact, supramolecular aggregations are kinetically (and not thermodynamically) controlled processes. This is why the production of vesicles which corresponds to a local minimum of the free energy is more accessible than the true energetic minimum. Thus, the production can be highly influenced by the conditions of vesicles formation. In the last years different methods of vesicles preparation have been introduced, and each one of them has been differently applied to produce lipid aggregates with specific characteristics (Table 6.1) (Walde et al. 2010).

Even though there is not a control on the vesicle dimensions, the gentle hydration method still has a pivotal role because it guarantees the formation of vesicles in a controlled pH range without any specific instrumentation. Furthermore, given the fact that it is the simplest method of vesicles production, it can be thought as the more prebiotically compatible. In the last years an example of vesicle formation has been explored by Damer and coworkers. In their work it is imagined that a thin layer of prebiotically plausible lipids (for example decanoic acid) could have formed on the bottom of hydrothermal pools. The alternation between day and night are thought as cycles of "wet/dry" or "cold/hot", in which the preformed dry film of lipid-like structures is hydrated and dehydrated (Damer and Deamer 2015). The hydration corresponds to vesicles formation, which randomly introduce in the lumen hydrophilic molecules present in the medium, while in the drying process the film is restored. When the hydration step happens again it will form vesicles encapsulating other randomly selected molecules. By repetition of cycles an evolving system should be produced.

A second encapsulation method is based on the preformation of vesicles and subsequent introduction of material inside them. In this autoamplification scenario the coexistence of membranes and early life relevant molecules, such as nucleotides or amino acids, is explained by their direct bond. To explain how the interaction between the lipid barrier and the encapsulated molecules has been obtained, the presence of nucleobases first, nucleotides and amino acids later (Cornell et al. 2019b) have been proved to stabilize membrane formation. In fact, empty pure fatty acids vesicles made of one type of fatty acids are usually not stable in high concentrations of salt or other cations (such as Mg^{2+} or Fe^{3+}). However, such conditions would have been common in prebiotically plausible environments like spring water pools. On the other hand, the presence of monomers (amino acids, nucleobases or nucleotides) has a stabilizing effect against salt-induced flocculation. Differences in the effect given by each nucleobase proved that the interaction between membranes and substrates is specific, and not related to general properties like viscosity or ionic strength. Furthermore, the presence of such biomolecules anchored to the membrane performs an autocatalytic process, in which their replication and polymerization is enhanced by their bond to the membrane (Black and Blosser 2016) (Fig. 6.7).

Encapsulation is interesting when it involves evolutionary necessary molecules. To produce protocells which are capable of Darwinian evolution two components are needed: self-replicating membranes and self-replicating genetic material such as nucleic acids. As previously mentioned, the main problem encountered by biomolecules during encapsulation is membrane crossing. Given their polarity, they have poor affinity with the long chained components of membrane. Though, the encapsulation in liposomes of polar structures, mainly RNA strands, has been previously obtained by hydrophobic modification of the polynucleotide (Pfeiffer

Table 6.1 Methods of vesicles preparations (Walde et al. 2010)

Method name	Explanation	Type of lipids	Vesicles obtained	Equipment	Encapsulation efficency	Pros	Cons
Natural swelling	A preformed dry film of lipids supported on a glass surface is gently dried in order to form vesicles.	Charged and uncharged	Variable	No special equipment	Low	Highly applicable	Not very performant, scarcely reproducible
Electroformation	A preformed dry film of lipids supported on a glass surface is gently dried in order to form vesicles. Furthermore, the addition of an external electronic fields allows the control on the quality of vesicles obtained.	Only uncharged	Uniform, unilamellar and spherical. Not high yields	Electrodes built in an observation chamber and an AC or DC device	Low	Controllable and reproducible	Need of specific equipment, not uniformely applicable
w/o emulsion transfer method or inverted-emulsion	An w/o emulsion is brought into contact with a less dense aqueous phase leading to vesicle formation as the lipid-coated water droplets migrate through the lipid-saturated o/w interface	Only bilayer-forming lipid (for example, POPC or egg PC/cholesterol (8:2))	Controllable (up to few micromiters). High yields	No special equipment	High	Easy and performant	Not GVs formed. Possible presence of oil in the membrane

Lipid-coated ice droplet hydration method	From an initial surfactant-stabilized w/o emulsion, the surfactant is replaced with lipids while keeping the dispersed water droplets in a frozen state, and the oil is replaced with an aqueous solution	Specific-egg PC/cholesterol/SA (5:5:1)	Uniform, reproducible, often oligolamellar. High yields.	Microfluidics	High	Controllable and reproducible	Need of specific equipment. Possible presence of oil in the membrane
Lipid stabilized w/o/w double emulsion method	In a lipid-stabilized w/o/w double emulsion an inner water droplet is localized within an oil droplet which again is in an outer bulk aqueous phase. Each internal water droplet is coated with lipid molecules. Removal of the volatile oil leads to the formation of giant vesicles.	Specific (DMPC or POPC or diblock copolymers)	Often not unilamellar. High yields.	Microfluidics	High	Precise	Delicate. Possible presence of oil in the membrane
Fusion of small vesicles	Preformed LUV in particular conditions undergo fusion in order to form bigger vesicles.	Specific (oleic acid/oleate (1:1) or egg PC)	No control on size and lamellarity	No special equipment.	Low	No organic solvents needed	Not performant, possible vesicle leakage.

(continued)

Table 6.1 (continued)

Method name	Explanation	Type of lipids	Vesicles obtained	Equipment	Encapsulation efficency	Pros	Cons
From planar bilayer	A planar lipid bilayer is first formed between two aqueous solutions in a double-well chamber. Vesicle formation is induced by detaching fragments of the planar bilayer by jet-blowing an aqueous solution onto the bilayer.	Specific (DPhPC)	Control on size and lamellarity by flow conditions	Microfluidics	High	Precise	Delicate, possible presence of oil in the membrane
Lipids in water-miscible solvents	Mechanism of vesicle formation still not known, reported for dropwise addition of water in an organic solution of graft copolymer COS-g-PCL.	Specific (COS-g-PCL)	No control on size and lamellarity	No special equipment	Low	Simple	Not well estabilished
From micellized bilayer-forming lipids	An aqueous solution of egg PC in presence of chaotropic ions forms mainly micelles. The removal of such ions produces vesicles.	Specific (egg PC)	No control on size	No special equipment	Not reported	Simple	Not well estabilished
From bilayers in in a w/o two-phase system	The starting state is an oil/water two-phase system, where the oil phase is chosed to be heavier, therefore forms the lower layer. By gently evaporation of the organic solvent, vesicles are formed.	Egg PC or DPPC	No control on size, mainly multilamellar vesicles	No special equipment	Low	Simple, fast	Not well estabilished

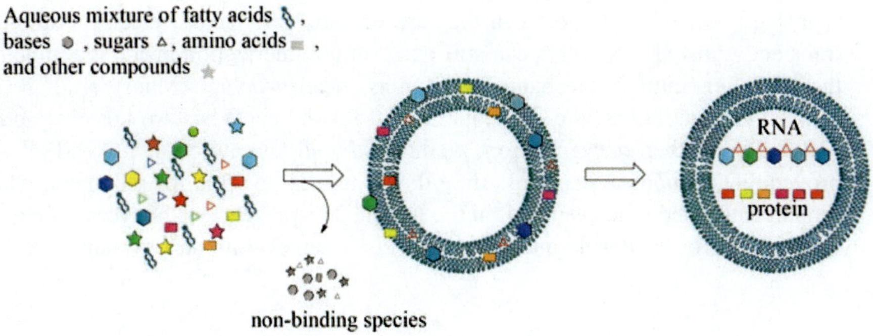

Fig. 6.7 Autoamplification process imagined by Black and coworkers (2016)

and Höök 2004) or bivalent cation complex formation (Vlassov et al. 2001). While effective, these methods could not be applied in the prebiotic field. For this reason, in recent years other ways to localize and/or introduce RNAs in membranes have been developed. In particular, the availability of amino acids in protocellular environments made peptides perfect candidates for nucleic acids anchoring. Firstly explored in 2014 thanks to Strazewski's group, (Isaad et al. 2014) the direct amide bond between a hydrophobic 20-mer peptide composed of leucine and alanine and a polar head formed of a 22-meric hairpin stem-loop molecule mimicking the aminoacyl stem of transfer RNA produced an amphiphilic peptidyl-RNA. By fluorescence microscopy it has been shown that its ambiguous polarity perfectly mixed with the components of GVs, while it could not permit the complete encapsulation. Other cases of peptide-anchoring to polynucleotides have given similar results (Kamat et al. 2015; Izgu et al. 2016). The attachment of RNA strands to the lipidic barrier is still considered important for the development of prebiotically relevant reactions, such as its replication.

Even if useful, the previously mentioned methods are generally not very effective, or can produce degradation of the polynucleotides (Monnard et al. 2002; Adamala and Szostak 2013; Blain and Szostak 2014). Another method to perform encapsulation in vesicles explored in this review is the freeze-thaw cycle (Qiao et al. 2017). Here, vesicles are subject to temperature shifts starting from room temperature, then freezed at $-100\ ^\circ$ C and finally heated up to $100\ ^\circ$ C. In this process the DNA strands (64 kb of length) are introduced through the phospholipid barrier thanks to small gaps formed by the ice crystals formation. Heating is not necessary to enhance encapsulation but rather to reform the barrier and to restore its spherical form. By reiterating the process, it is possible to encapsulate DNA in considerably higher concentrations than the outside solutions without modifying the integrity and the conformation of the DNA strand.

While the permeability is a necessary property to ensure the encapsulation, the obvious limit of the process is that the permanence of the molecules inside the vesicles can be only temporary. In fact, the difficulty for polar structure to interact with amphiphiles is surely a problem to overcome, but a membrane easy to cross represents as well an issue, as in the case of early fatty acids vesicles. Although, transiently hosted monomers can still perform polymerization inside the lumen of the GVs, generating a mechanism known as "heterotrophic" (Mansy et al. 2008). Here, it is thought that the nutrients needed for the synthesis are not obtained in situ, but are rather emerging from a complex external environment. While it was proved that monomers can cross the bilayer, it is impossible for polymers, which remain entrapped (encapsulated) in the lumen. The presence of bilayers once more is necessary for the development and selection of an evolutionary system.

6.9 Conclusions

Lipidic molecules, amphiphilic polymers and cavities inside minerals could have provided effective encapsulation or crowding of biotic molecules and promoted processes that were crucial steps for life emergence. Then, Darwinian evolution of biomolecules promoted the selection of lipids for the formation of capsula that have been dynamic in composition, sufficiently stable to retain macromolecules and semi-permeable allowing reactive small molecules to get across the membrane. Today this capsula is become even more complex and is called the cell membrane. Mixed vesicles made of a combination of prebiotic simple lipids such as fatty acids and alkanols, fatty acids and mono-alkyl glycerols (or more complex mixtures) show higher stabilities whereas keeping a good permeability to small compounds. It could be that the very first protocell membranes were made of such kind of lipids. Even though complete lipids don't seem at first sight plausible compounds for protocells, they can be synthesized in prebiotic conditions and mixed membranes containing complete lipids (phospholipids, cyclphospholipids . . .) and simple lipids (fatty acids, alkanols . . .) are at the same time very stable and prone to the transfer of small molecules. Such protocells made of a wide diversity of lipids (simple, incomplete and complete lipids) could have encapsulated biomolecules such as self-replicative RNA (Joyce and Szostak 2018), self-replicative peptides (Lee et al. 1996) and other related molecules. Even though conditions for the introductions of such polar structures inside protocells are still debated, it is generally accepted that compartmentalization may have helped the reproduction of biomolecules, by allowing condensation reaction to happen in water. It can be argued that complete lipids were progressively integrated in protocells membranes replacing simple lipids. It was notably illustrated with oleic acid: 1,2-dioleyl-*sn*-glyero-3-phosphate (DOPA) membranes growing faster than pure fatty acids membranes revealing an evolutive benefit for the incorporation of glycerol phospholipids (Budin and Szostak 2011). Over time, protocells made of simple lipids would have given cells made of complete lipids as we know them today passing by minimal autopoietic structures that gave birth to the cell's cenancestor (Betts et al. 2018).

References

Adamala, K., Szostak, J.W.: Nonenzymatic template-directed RNA synthesis inside model protocells. Science. **342**, 1098–1100 (2013). https://doi.org/10.1126/science.1241888

Alberts, B., Johnson, A., Lewis, J., Raff, M., Roberts, K., Walter, P.: Molecular Biology of the Cell. Garland Science, New York (2014)

Albertsen, A.N., Duffy, C.D.D., Sutherland, J.D.D., Monnard, P.-A.A.: Self-assembly of phosphate Amphiphiles in mixtures of prebiotically plausible surfactants. Astrobiology. **14**, 462–472 (2014). https://doi.org/10.1089/ast.2013.1111

Apel, C.L., Deamer, D.W., Mautner, M.N.: Self-assembled vesicles of monocarboxylic acids and alcohols: conditions for stability and for the encapsulation of biopolymers. Biochim. Biophys. Acta Biomembr. **1559**, 1–9 (2002). https://doi.org/10.1016/S0005-2736(01)00400-X

Berclaz, N., Müller, M., Walde, P., Luisi, P.L.: Growth and transformation of vesicles studied by ferritin labeling and cryotransmission electron microscopy. J. Phys. Chem. B. **105**, 1056–1064 (2001). https://doi.org/10.1021/jp001298i

Betts, H.C., Puttick, M.N., Clark, J.W., et al.: Integrated genomic and fossil evidence illuminates life's early evolution and eukaryote origin. Nat. Ecol. Evol. **2**, 1556–1562 (2018). https://doi.org/10.1038/s41559-018-0644-x

Black, R., Blosser, M.: A self-assembled aggregate composed of a fatty acid membrane and the building blocks of biological polymers provides a first step in the emergence of protocells. Life. **6**, 33 (2016). https://doi.org/10.3390/life6030033

Black, R.A., Blosser, M.C., Stottrup, B.L., et al.: Nucleobases bind to and stabilize aggregates of a prebiotic amphiphile, providing a viable mechanism for the emergence of protocells. Proc. Natl. Acad. Sci. **110**, 13272–13276 (2013). https://doi.org/10.1073/pnas.1300963110

Blain, J.C., Szostak, J.W.: Progress toward synthetic cells. Annu. Rev. Biochem. **83**, 615–640 (2014). https://doi.org/10.1146/annurev-biochem-080411-124036

Bonfio, C., Caumes, C., Duffy, C.D., et al.: Length-selective synthesis of acylglycerol-phosphates through energy-dissipative cycling. J. Am. Chem. Soc. **141**, 3934–3939 (2019). https://doi.org/10.1021/jacs.8b12331

Božič, B., Svetina, S.: A relationship between membrane properties forms the basis of a selectivity mechanism for vesicle self-reproduction. Eur. Biophys. J. **33**, 565–571 (2004). https://doi.org/10.1007/s00249-004-0404-5

Budin, I., Szostak, J.W.: Physical effects underlying the transition from primitive to modern cell membranes. Proc. Natl. Acad. Sci. **108**, 5249–5254 (2011). https://doi.org/10.1073/pnas.1100498108

Cape, J.L., Monnard, P.-A., Boncella, J.M.: Prebiotically relevant mixed fatty acid vesicles support anionic solute encapsulation and photochemically catalyzed trans-membrane charge transport. Chem. Sci. **2**, 661 (2011). https://doi.org/10.1039/c0sc00575d

Chandru, K., Mamajanov, I., Cleaves, H.J., Jia, T.Z.: Polyesters as a model system for building primitive biologies from non-biological prebiotic chemistry. Life. **10** (2020). https://doi.org/10.3390/life10010006

Chen, I.A., Szostak, J.W.: Membrane growth can generate a transmembrane pH gradient in fatty acid vesicles. Proc. Natl. Acad. Sci. USA. **101**, 7965–7970 (2004). https://doi.org/10.1073/pnas.0308045101

Chen, I.A., Walde, P.: From self-assembled vesicles to protocells. Cold Spring Harb. Perspect. Biol. **2**, a002170 (2010). https://doi.org/10.1101/cshperspect.a002170

Chen, I.A., Salehi-Ashtiani, K., Szostak, J.W.: RNA catalysis in model protocell vesicles. J. Am. Chem. Soc. **127**, 13213–13219 (2005). https://doi.org/10.1021/ja051784p

Chyba, C., Sagan, C.: Endogenous production, exogenous delivery and impact-shock synthesis of organic molecules: an inventory for the origins of life. Nature. **355**, 125–132 (1992). https://doi.org/10.1038/355125a0

Cornell, C.E., Black, R.A., Xue, M., et al.: Prebiotic amino acids bind to and stabilize prebiotic fatty acid membranes. Proc. Natl. Acad. Sci. USA. **116**, 17239–17244 (2019a). https://doi.org/10.1073/pnas.1900275116

Cornell, C.E., Black, R.A., Xue, M., et al.: Prebiotic amino acids bind to and stabilize prebiotic fatty acid membranes. Proc. Natl. Acad. Sci. **116**, 17239–17244 (2019b). https://doi.org/10.1073/pnas.1900275116

Damer, B., Deamer, D.: Coupled phases and combinatorial selection in fluctuating hydrothermal pools: a scenario to guide experimental approaches to the origin of cellular life. Life. **5**, 872–887 (2015). https://doi.org/10.3390/life5010872

Danger, G., D'Hendecourt, L.L.S., Pascal, R.: On the conditions for mimicking natural selection in chemical systems. Nat. Rev. Chem. **4**, 102–109 (2020). https://doi.org/10.1038/s41570-019-0155-6

Deamer, D.W.: Boundary structures are formed by organic components of the Murchison carbonaceous chondrite. Nature. **317**, 792–794 (1985). https://doi.org/10.1038/317792a0

Deamer, D.: The role of lipid membranes in life's origin. Life. **7**, 5 (2017). https://doi.org/10.3390/life7010005

Drobot, B., Iglesias-Artola, J.M., Le Vay, K., et al.: Compartmentalised RNA catalysis in membrane-free coacervate protocells. Nat. Commun. **9**, 1–9 (2018). https://doi.org/10.1038/s41467-018-06072-w

Eichberg, J., Sherwood, E., Epps, D.E., Oró, J.: Cyanamide mediated syntheses under plausible primitive earth conditions. J. Mol. Evol. **10**, 221–230 (1977). https://doi.org/10.1007/BF01764597

Epps, D.E., Sherwood, E., Eichberg, J., Or, J.: Cyanamide mediated syntheses under plausible primitive earth conditions: V. The synthesis of phosphatidic acids. J. Mol. Evol. **11**, 279–292 (1978)

Epps, D.E., Nooner, D.W., Eichberg, J., et al.: Cyanamide mediated synthesis under plausible primitive earth conditions: VI. The synthesis of glycerol and glycerophosphates. J. Mol. Evol. **14**, 235–241 (1979). https://doi.org/10.1007/BF01732490

Fayolle, D., Altamura, E., D'Onofrio, A., et al.: Crude phosphorylation mixtures containing racemic lipid amphiphiles self-assemble to give stable primitive compartments. Sci. Rep. **7**, 18106 (2017). https://doi.org/10.1038/s41598-017-18053-y

Fiore, M.: The synthesis of mono-alkyl phosphates and their derivatives: an overview of their nature, preparation and use, including synthesis under plausible prebiotic conditions. Org. Biomol. Chem. **16**, 3068–3086 (2018). https://doi.org/10.1039/C8OB00469B

Fiore, M., Strazewski, P.: Prebiotic lipidic amphiphiles and condensing agents on the early earth. Life. **6**, 17 (2016). https://doi.org/10.3390/life6020017

Fiore, M., Madanamoothoo, W., Berlioz-Barbier, A., et al.: Giant vesicles from rehydrated crude mixtures containing unexpected mixtures of amphiphiles formed under plausibly prebiotic conditions. Org. Biomol. Chem. **15**, 4231–4240 (2017). https://doi.org/10.1039/C7OB00708F

Gibard, C., Bhowmik, S., Karki, M., et al.: Phosphorylation, oligomerization and self-assembly in water under potential prebiotic conditions. Nat. Chem. **10**, 212–217 (2018). https://doi.org/10.1038/nchem.2878

Groen, J., Deamer, D.W., Kros, A., Ehrenfreund, P.: Polycyclic aromatic hydrocarbons as plausible prebiotic membrane components. Origin Life Evol. Biosph. **42**, 295–306 (2012). https://doi.org/10.1007/s11084-012-9292-3

Grommet, A.B., Feller, M., Klajn, R.: Chemical reactivity under nanoconfinement. Nat. Nanotechnol. **15**, 256–271 (2020). https://doi.org/10.1038/s41565-020-0652-2

Hanczyc, M.M., Monnard, P.-A.A.: Primordial membranes: more than simple container boundaries. Curr. Opin. Chem. Biol. **40**, 78–86 (2017). https://doi.org/10.1016/j.cbpa.2017.07.009

Hanczyc, M.M., Fujikawa, S.M., Szostak, J.W.: Experimental models of primitive cellular compartments: encapsulation, growth, and division. Science. **302**, 618–622 (2003). https://doi.org/10.1126/science.1089904

Hanczyc, M.M., Mansy, S.S., Szostak, J.W.: Mineral surface directed membrane assembly. Orig. Life Evol. Biosph. **37**, 67–82 (2007). https://doi.org/10.1007/s11084-006-9018-5

Hansma, H.: Better than membranes at the origin of life? Life. **7**, 28 (2017). https://doi.org/10.3390/life7020028

Hargreaves, W.R., Deamer, D.W.: Liposomes from ionic, single-chain Amphiphiles. Biochemistry. **17**, 3759–3768 (1978). https://doi.org/10.1021/bi00611a014

Hargreaves, W.R., Mulvil, S.J., Deamer, D.W.: Synthesis of phospholipids and membranes in prebiotic conditions. Nature. **266**, 78–80 (1977). https://doi.org/10.1038/266078a0

Hazen, R.M., Sverjensky, D.A.: Mineral surfaces, geochemical complexities, and the origins of life. Cold Spring Harb. Perspect. Biol. **2**, a002162–a002162 (2010). https://doi.org/10.1101/cshperspect.a002162

Huang, Y., Wang, Y., Alexandre, M.R., et al.: Molecular and compound-specific isotopic characterization of monocarboxylic acids in carbonaceous meteorites. Geochim. Cosmochim. Acta. **69**, 1073–1084 (2005). https://doi.org/10.1016/j.gca.2004.07.030

Isaad, A.L.C., Carrara, P., Stano, P., et al.: A hydrophobic disordered peptide spontaneously anchors a covalently bound RNA hairpin to giant lipidic vesicles. Org. Biomol. Chem. **12**, 6363–6373 (2014). https://doi.org/10.1039/c4ob00721b

IUPAC-IUB Commission on Biochemical Nomenclature: The nomenclature of lipids (recommendations 1976). J. Lipid. Res. **19**, 114–128 (1978)

Izgu, E.C., Björkbom, A., Kamat, N.P., et al.: N-Carboxyanhydride-mediated fatty acylation of amino acids and peptides for functionalization of protocell membranes. J. Am. Chem. Soc. **138**, 16669–16676 (2016). https://doi.org/10.1021/jacs.6b08801

Javaux, E.J.: Challenges in evidencing the earliest traces of life. Nature. **572**, 451–460 (2019). https://doi.org/10.1038/s41586-019-1436-4

Jia, T.Z., Chandru, K., Hongo, Y., et al.: Membraneless polyester microdroplets as primordial compartments at the origins of life. Proc. Natl. Acad. Sci. **116**, 15830–15835 (2019). https://doi.org/10.1073/pnas.1902336116

Jin, L., Kamat, N.P., Jena, S., Szostak, J.W.: Fatty acid/phospholipid blended membranes: a potential intermediate state in protocellular evolution. Small. **14**, 1–9 (2018). https://doi.org/10.1002/smll.201704077

Jordan, S.F., Rammu, H., Zheludev, I.N., et al.: Promotion of protocell self-assembly from mixed amphiphiles at the origin of life. Nat. Ecol. Evol. **3**, 1705–1714 (2019). https://doi.org/10.1038/s41559-019-1015-y

Joyce, G.F., Szostak, J.W.: Protocells and RNA self-replication. Cold Spring Harb. Perspect. Biol. **10** (2018). https://doi.org/10.1101/cshperspect.a034801

Kamat, N.P., Tobé, S., Hill, I.T., Szostak, J.W.: Electrostatic localization of RNA to protocell membranes by cationic hydrophobic peptides. Angew. Chem. Int. Ed. **54**, 11735–11739 (2015). https://doi.org/10.1002/anie.201505742

Kee, T.P., Monnard, P.-A.A.: Chemical systems, chemical contiguity and the emergence of life. Beilstein J. Org. Chem. **13**, 1551–1563 (2017). https://doi.org/10.3762/bjoc.13.155

Kitadai, N., Maruyama, S.: Origins of building blocks of life: a review. Geosci. Front. **9**, 1117–1153 (2018). https://doi.org/10.1016/j.gsf.2017.07.007

Klein, A.E., Pilpel, N.: Oxidation of n-alkanes photosensitized by 1-naphthol. J. Chem. Soc. Faraday Trans. 1 Phys. Chem. Condens. Phases. **69**, 1729 (1973). https://doi.org/10.1039/f19736901729

Koga, S., Williams, D.S., Perriman, A.W., Mann, S.: Peptide-nucleotide microdroplets as a step towards a membrane-free protocell model. Nat. Chem. **3**, 720–724 (2011). https://doi.org/10.1038/nchem.1110

Lee, D.H., Granja, J.R., Martinez, J.A., et al.: A self-replicating peptide. Nature. **382**, 525–528 (1996). https://doi.org/10.1038/382525a0

Lombard, J., López-García, P., Moreira, D.: The early evolution of lipid membranes and the three domains of life. Nat. Rev. Microbiol. **10**, 507–515 (2012). https://doi.org/10.1038/nrmicro2815

Lonchin, S., Luisi, P.L., Walde, P., Robinson, B.H.: A matrix effect in mixed phospholipid/fatty acid vesicle formation. J. Phys. Chem. B. **103**, 10910–10916 (1999). https://doi.org/10.1021/jp9909614

Lopez, A., Fiore, M.: Investigating prebiotic protocells for a comprehensive understanding of the origins of life: a prebiotic systems chemistry perspective. Life. **9**, 1–21 (2019). https://doi.org/10.3390/life9020049

Ma, W., Feng, Y.: Protocells: at the interface of life and non-life. Life. **5**, 447–458 (2015). https://doi.org/10.3390/life5010447

Mansy, S.S.: Membrane transport in primitive cells. Cold Spring Harb. Perspect. Biol. **2**, 1–14 (2010). https://doi.org/10.1101/cshperspect.a002188

Mansy, S.S., Szostak, J.W.: Thermostability of model protocell membranes. Proc. Natl. Acad. Sci. **105**, 13351–13355 (2008). https://doi.org/10.1073/pnas.0805086105

Mansy, S.S., Schrum, J.P., Krishnamurthy, M., et al.: Template-directed synthesis of a genetic polymer in a model protocell. Nature. **454**, 122–125 (2008). https://doi.org/10.1038/nature07018

Mariscal, C., Barahona, A., Aubert-Kato, N., et al.: Hidden concepts in the history and philosophy of origins-of-life studies: a workshop report. Origin Life Evol. Biosph. **49**, 111–145 (2019). https://doi.org/10.1007/s11084-019-09580-x

Maurer, S.E., Nguyen, G.: Prebiotic vesicle formation and the necessity of salts. Orig Life Evol. Biosph. **46**, 215–222 (2016). https://doi.org/10.1007/s11084-015-9476-8

Maurer, S.E., Deamer, D.W., Boncella, J.M., Monnard, P.A.: Chemical evolution of amphiphiles: glycerol monoacyl derivatives stabilize plausible prebiotic membranes. Astrobiology. **9**, 979–987 (2009). https://doi.org/10.1089/ast.2009.0384

Mccollom, T.M., Simoneit, B.R.T.: Abiotic formation of hydrocarbons and oxygenated compounds during thermal decomposition of iron oxalate. Orig. Life Evol. Biosph. **29**, 167–186 (1999). https://doi.org/10.1023/A:1006556315895

Mccollom, T.M., Ritter, G., Simoneit, B.R.T.: Lipid synthesis under hydrothermal conditions by Fischer-Tropsch-type reactions. Orig. Life Evol. Biosph. **29**, 153–166 (1999). https://doi.org/10.1023/A:1006592502746

Monnard, P.A., Deamer, D.W.: Preparation of vesicles from nonphospholipid amphiphiles. Methods Enzymol. **372**, 133–151 (2003). https://doi.org/10.1016/S0076-6879(03)72008-4

Monnard, P.-A.A., Walde, P.: Current ideas about prebiological compartmentalization. Life. **5**, 1239–1263 (2015). https://doi.org/10.3390/life5021239

Monnard, P.-A., Apel, C.L., Kanavarioti, A., Deamer, D.W.: Influence of ionic inorganic solutes on self-assembly and polymerization processes related to early forms of life: implications for a prebiotic aqueous medium. Astrobiology. **2**, 139–152 (2002). https://doi.org/10.1089/15311070260192237

Morigaki, K., Dallavalle, S., Walde, P., et al.: Autopoietic self-reproduction of chiral fatty acid vesicles. J. Am. Chem. Soc. **119**, 292–301 (1997). https://doi.org/10.1021/ja961728b

Murillo-Sánchez, S., Beaufils, D., González Mañas, J.M., et al.: Fatty acids' double role in the prebiotic formation of a hydrophobic dipeptide. Chem. Sci. **7**, 3406–3413 (2016). https://doi.org/10.1039/c5sc04796j

Namani, T., Ishikawa, T., Morigaki, K., Walde, P.: Vesicles from docosahexaenoic acid. Colloids Surf. B Biointerfaces. **54**, 118–123 (2007). https://doi.org/10.1016/j.colsurfb.2006.05.022

Ourisson, G., Nakatani, Y.: The terpenoid theory of the origin of cellular life: the evolution of terpenoids to cholesterol. Chem. Biol. **1**, 11–23 (1994). https://doi.org/10.1016/1074-5521(94)90036-1

Paltauf, F., Hermetter, A.: Strategies for the synthesis of glycerophospholipids. Prog. Lipid Res. **33**, 239–328 (1994). https://doi.org/10.1016/0163-7827(94)90028-0

Patel, B.H., Percivalle, C., Ritson, D.J., et al.: Common origins of RNA, protein and lipid precursors in a cyanosulfidic protometabolism. Nat. Chem. **7**, 301–307 (2015). https://doi.org/10.1038/nchem.2202

Paula, S., Volkov, A.G., Van Hoek, A.N., et al.: Permeation of protons, potassium ions, and small polar molecules through phospholipid bilayers as a function of membrane thickness. Biophys. J. **70**, 339–348 (1996). https://doi.org/10.1016/S0006-3495(96)79575-9

Peretó, J., López-García, P., Moreira, D.: Ancestral lipid biosynthesis and early membrane evolution. Trends Biochem. Sci. **29**, 469–477 (2004). https://doi.org/10.1016/j.tibs.2004.07.002

Pfeiffer, I., Höök, F.: Bivalent cholesterol-based coupling of oligonucletides to lipid membrane assemblies. J. Am. Chem. Soc. **126**, 10224–10225 (2004). https://doi.org/10.1021/ja048514b

Pizzarello, S., Shock, E.: The organic composition of carbonaceous meteorites: the evolutionary story ahead of biochemistry. Cold Spring Harb. Perspect. Biol. **2**, a002105–a002105 (2010). https://doi.org/10.1101/cshperspect.a002105

Powner, M.W., Sutherland, J.D.: Prebiotic chemistry: a new modus operandi. Philos. Trans. R. Soc. B. Biol. Sci. **366**, 2870–2877 (2011). https://doi.org/10.1098/rstb.2011.0134

Qiao, H., Hu, N., Bai, J., et al.: Encapsulation of nucleic acids into giant unilamellar vesicles by freeze-thaw: a way protocells may form. Origin Life Evol. Biosph. **47**, 499–510 (2017). https://doi.org/10.1007/s11084-016-9527-9

Rao, M., Eichberg, J., Oró, J.: Synthesis of phosphatidylcholine under possible primitive Earth conditions. J. Mol. Evol. **18**, 196–202 (1982). https://doi.org/10.1007/BF01733046

Rao, M., Eichberg, J., Oró, J.: Synthesis of phosphatidylethanolamine under possible primitive earth conditions. J. Mol. Evol. **25**, 1–6 (1987). https://doi.org/10.1007/BF02100033

Reeves, J.P., Dowben, R.M.: Formation and properties of thin-walled phospholipid vesicles. J. Cell. Physiol. **73**, 49–60 (1969). https://doi.org/10.1002/jcp.1040730108

Ruiz-Mirazo, K., Briones, C., De La Escosura, A.: Prebiotic systems chemistry: new perspectives for the origins of life. Chem. Rev. **114**, 285–366 (2014). https://doi.org/10.1021/cr2004844

Ruiz-Mirazo, K., Briones, C., de la Escosura, A.: Chemical roots of biological evolution: the origins of life as a process of development of autonomous functional systems. Open Biol. **7**, 170050 (2017). https://doi.org/10.1098/rsob.170050

Rushdi, A.I., Simoneit, B.R.T.T.: Abiotic condensation synthesis of glyceride lipids and wax esters under simulated hydrothermal conditions. Origin Life Evol. Biosph. **36**, 93–108 (2006). https://doi.org/10.1007/s11084-005-9001-6

Sacerdote, M.G., Szostak, J.W.: Semipermeable lipid bilayers exhibit diastereoselectivity favoring ribose. Proc. Natl. Acad. Sci. U. S. A. **102**, 6004–6008 (2005). https://doi.org/10.1073/pnas.0408440102

Saladino, R., Crestini, C., Pino, S., et al.: Formamide and the origin of life. Phys. Life Rev. **9**, 84–104 (2012). https://doi.org/10.1016/j.plrev.2011.12.002

Segre, D., Ben-Eli, D., Lancet, D.: Compositional genomes: prebiotic information transfer in mutually catalytic noncovalent assemblies. Proc. Natl. Acad. Sci. **97**, 4112–4117 (2000). https://doi.org/10.1073/pnas.97.8.4112

Shirt-Ediss, B., Murillo-Sánchez, S., Ruiz-Mirazo, K.: Framing major prebiotic transitions as stages of protocell development: three challenges for origins-of-life research. Beilstein J. Org. Chem. **13**, 1388–1395 (2017). https://doi.org/10.3762/bjoc.13.135

Simionescu, C.I., Dénes, F., Totolin, M.: The synthesis of some lipid-like structures in simulated primeval earth conditions. Biosystems. **13**, 149–156 (1981). https://doi.org/10.1016/0303-2647(81)90056-3

Simoneit, B.R.T.: Prebiotic organic synthesis under hydrothermal conditions: an overview. Adv. Space Res. **33**, 88–94 (2004). https://doi.org/10.1016/j.asr.2003.05.006

Simoneit, B.R.T., Rushdi, A.I., Deamer, D.W.: Abiotic formation of acylglycerols under simulated hydrothermal conditions and self-assembly properties of such lipid products. Adv. Space Res. **40**, 1649–1656 (2007). https://doi.org/10.1016/j.asr.2007.07.034

Szostak, J.W.: Systems chemistry on early Earth. Nature. **459**, 171–172 (2009). https://doi.org/10.1038/459171a

Szostak, J.W., Bartel, D.P., Luisi, P.L.: Synthesizing life. Nature. **409**, 387–390 (2001). https://doi.org/10.1038/35053176

Toparlak, D., Karki, M., Egas Ortuno, V., et al.: Cyclophospholipids increase protocellular stability to metal ions. Small. **1903381**, 1–8 (2019). https://doi.org/10.1002/smll.201903381

Vitas, M., Dobovišek, A.: Towards a general definition of life. Origin. Life Evol. Biosph. **49**, 77–88 (2019). https://doi.org/10.1007/s11084-019-09578-5

Vlassov, A., Khvorova, A., Yarus, M.: Binding and disruption of phospholipid bilayers by supramolecular RNA complexes. Proc. Natl. Acad. Sci. USA. **98**, 7706–7711 (2001). https://doi.org/10.1073/pnas.141041098

Walde, P.: Surfactant assemblies and their various possible roles for the origin(s) of life. Origin Life Evol. Biosph. ***36***(2), 109–150 (2006)

Walde, P., Wick, R., Fresta, M., et al.: Autopoietic self-reproduction of fatty acid vesicles. J. Am. Chem. Soc. **116**, 11649–11654 (1994). https://doi.org/10.1021/ja00105a004

Walde, P., Cosentino, K., Engel, H., Stano, P.: Giant vesicles: preparations and applications. ChemBioChem. **11**, 848–865 (2010). https://doi.org/10.1002/cbic.201000010

Zhu, T.F., Szostak, J.W.: Coupled growth and division of model protocell membranes. J. Am. Chem. Soc. **131**, 5705–5713 (2009). https://doi.org/10.1021/ja900919c

Chapter 7
First Steps Towards Molecular Evolution

Oliver Trapp

Abstract A fascinating question of abiogenesis is how simple organic molecules have developed into complex biological systems that are capable of being altered by evolutionary mechanisms. The fundamental question is what complexity is required to trigger evolution at the molecular level, i.e. molecules undergo their synthesis, selection and mutation leading to selectivity and replication. This question is closely linked to processes that lead to spontaneous symmetry breaking (chirogenesis) and ultimately to homochirality. This chapter focuses on mechanisms that lead to evolutionary systems at the level of organic molecules and open the possibility of chirogenesis. On the one hand, this is shown by the emergence of prebiotic organocatalysts and, on the other hand, a mechanism is discussed that leads naturally to the preferential formation of (deoxy)ribonucleosides and nucleotides. This creates the basis for self-sustaining and information-storing structures that can undergo dynamic alterations.

7.1 Introduction

The synthesis of biologically relevant molecules under probiotic conditions has developed rapidly in recent years and numerous advances have been made (Ruiz-Mirazo et al. 2014). Interestingly, numerous new pathways have been discovered that lead to amino acids, sugars, nucleobases, nucleosides, and nucleotides (Fialho et al. 2020; Yadav et al. 2020; Roy et al. 2016). The focus is on the prebiotic synthesis of cell components that are essential for metabolism (Muchowska et al. 2020; Preiner et al. 2020), information transfer and cell division.

The holy grail in this field is the combination of all partial components to a self-sustaining chemical system, which can undergo the Darwinian evolution (Woese 1968; Lincoln and Joyce 2009). This definition implies energy consumption and

O. Trapp (✉)
Department of Chemistry, Ludwig-Maximilians-Universität München, Munich, Germany
e-mail: Oliver.trapp@cup.uni-muenchen.de

A. Neubeck, S. McMahon (eds.), *Prebiotic Chemistry and the Origin of Life*,
Advances in Astrobiology and Biogeophysics,
https://doi.org/10.1007/978-3-030-81039-9_7

conversion from the environment for self-preservation and reproduction. Already in the middle of the twentieth century the idea arose that the formation of life forms from simple molecules would be possible if suitable reaction conditions were present (Oparin 1953; Damer and Deamer 2019). Indeed, Miller and Urey succeeded in synthesizing important biomolecules such as urea and proteinogenic amino acids by electrical discharges of water vapor and gas mixtures (Miller and Urey 1953). This "primordial soup" strategy, in which reaction mixtures are subjected to certain reaction conditions and then examined for the formation of biomolecules, looks plausible. However, the product mixtures obtained in such a way are extremely difficult to analyze due to the abundance of organic substances. The "bottom-up" approach is based on simple organic molecules but is limited to a small selection of reactive substrates. Conversely, the "top-down" approach can reduce the complexity of existing organisms to arrive at cells that are as primitive and prebiotically plausible as possible. Hutchison has succeeded in reducing a bacterial genome to a number of only 531 kilobases while maintaining life functions (Hutchison et al. 2016). But even such a simplified synthetic bacterium is chemically extraordinarily complex and far from the desired shape of the unicellular organism.

Besides the debate about the place of origin of the first functional cell (Mulkidjanian et al. 2012), there is controversy about the order of origin of replication and metabolic processes. Although the two processes are strongly intertwined in their present form, their simultaneous development from simple precursors and their timely merging are difficult to imagine. At the end of the 1960s, the approach was therefore to decouple the two processes and to investigate the independent survival strategies. Initially, it was speculated that nucleic acids were created first because they have the advantage of enzyme-free replication by means of elongation after Watson-Crick base pairing and subsequent denaturation. The concept was expanded by formulating precise ideas about RNA-based coenzymes and ribosomes. This theory was supported by the discovery of the catalytic, enzymatic function of RNA units in living cells (Joyce 1989; Benner et al. 2019). For the discovery of these ribozymes (Kruger et al. 1982; Guerrier-Takada et al. 1983), Cech and Altman were awarded the Nobel Prize in 1989. Due to the wide distribution of RNA, its multifunctionality in the field of information storage and its catalytic properties, the initial hypotheses were considered to be correct. Subsequently, the "RNA world" hypothesis was defined in 1986, implying that RNA alone is involved at the interface between abiotic and biotic processes (Gilbert 1986; Joyce 2002). It is also assumed that the development of DNA and proteins took place via natural selection during evolution. With the identification of a ribosome consisting of RNA, this hypothesis of a protein-free origin was confirmed. In addition, the thesis gained increasing attention, so that numerous studies on prebiotic, enzyme-free RNA synthesis followed. The latter can be divided into three reaction sequences, including the presentation of nucleoside building blocks (A, G, C, U), phosphorylation to nucleotides and subsequent polymerization to a sequence.

Since the existence of nucleobases on meteorites has been proven, their natural formation via prebiotic mechanisms is obvious. A possible synthetic approach to purine bases has been demonstrated starting from hydrogen cyanide. In 1914 the

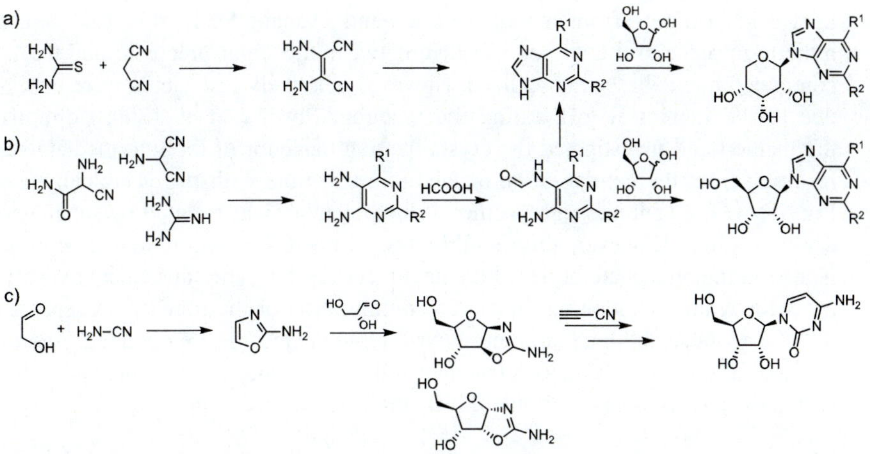

Fig. 7.1 Pathways to RNA nucleosides. (**a**) Fischer-Orgel and (**b**) Carell pathways utilizing purines or formamidopyrimidines (FaPys) in a reaction with D-ribose, respectively. (**c**) The Sutherland pathway via 2-aminooxazole as connecting building block between sugar and pyrimidine base. A: $R^1 = NH_2$, $R^2 = H$; G: $R^1 = OH$, $R^2 = NH_2$; C: $R^3 = NH_2$, $R^4 = H$; T: $R^3 = OH$, $R^4 = CH_3$

first successful purine nucleoside synthesis was achieved by glycosylation of the nucleobase demonstrated by Emil Fischer in Munich (Fischer and Helferich 1914; Kim and Benner 2017). This early discovery was taken up by Orgel and investigated in more detail (cf. Fig. 7.1a) (Fuller et al. 1972). The condensation of ribose with adenine in a melt resulted in a mixture of the mainly formed *N*-6-ribosyladenine and traces of the 6,9 disubstituted adenine. Starting from guanidine, a yield of 9% of the naturally occurring β-guanosine was obtained. It can be stated that the direct linkage of bases with sugars has the problem of the lack of reactivity of the *N*-9 atom as well as the low regioselectivity due to several reactive nitrogen centers. An alternative approach of Carell (cf. Fig. 7.1b) starts from formamide pyrimidines (FaPy), which in turn can be traced back to ammonium cyanide as a precursor molecule. FaPy have a high regioselectivity at the *N*-9 position in the reaction with ribose, resulting in a remarkable yield of 20% of the canonical β-adenoside (Becker et al. 2016, 2018). In contrast, the binding linkage is non-selective with respect to ribose, resulting in mixtures of pyranoses and furanoses and their respective anomers. It would be interesting to continue the investigation of possible selection mechanisms, showing the evolutionary preference of the β-furanose nucleoside over the other isomers (α-furanosides and α/β-pyranosides) resulting from this process.

Synthesis strategies for the canonical pyrimidine nucleosides uridine and cytidine were developed (Kruse et al. 2020). These are based on simple starting materials such as malonic acid with urea, acrylonitrile with urea, or cyanate with cyanoacetylene. Previous attempts to directly couple the pyrimidine bases to sugar have failed. Therefore, the investigations are focused on ring-closure reactions involving substituted sugars. Orgel demonstrated the first successful preparation of pyrimidine nucleoside isomers from aminoxazolines and cyanamide. In this key step, the hitherto unsuccessful C-N bond linkage between the sugar precursor and the later completed nucleobase was realized. However, relatively low yields were obtained due to the increasing pH during the reaction. Powner et al. (2009) optimized the method and investigated the crystallization behavior of the intermediate stage resulting from the condensation of the aminoxazoline with D-glyceraldehyde (cf. Fig. 7.1c). The canonic sugar isomer of the precursor was isolated by spontaneous crystallization. However, this equilibrates at the C2 position to the arabinose isomer in the phosphate buffer of the next reaction step. The subsequent hydrolysis proceeds while maintaining the stereo configuration of the four stereo centers, so that the α ribose and the β arabinose pyrimidine nucleoside are obtained. Avoiding ribose as a substrate is a major advantage of these reaction routes, as the low stability of pentose and its low probability of formation classify it as a limiting factor on early earth.

Inspired by this approach, Carell developed an alternative pyrimidine nucleoside synthesis in 2019 (Becker et al. 2019). Aminoxazoline is also of central importance here, as it is part of the target molecule structure. In contrast to previous work, the aminoxazoline does not link to the sugar but forms the backbone of the cytidine. Activation is achieved by a metal-catalyzed reaction with urea, creating a reactive intermediate for the subsequent addition of sugar. For further transformation, the presence of a redox-active metal complex is required, which accomplishes the ring opening and the reduction of the oxazole. Finally, condensation leads to the formation of the pyrimidine nucleoside. Furthermore, in this study the parallel and site-bound formation of all RNA nucleosides is plausible by environmental influences and alternating wet and dry periods. It was demonstrated, how different pH values and mineral deposits influence the solubility and reactivity of the intermediates and thus enable a simultaneous development from different starting molecules.

Another method for prebiotic synthesis of all four DNA nucleosides was introduced by Trapp (Teichert et al. 2019). Similar to Sutherland, direct condensation with ribose is bypassed and the sugar is built up from D-glyceraldehyde. In contrast to previous work, the strategy is based on an enamine formation between the *N*-9 atom and the acetaldehyde and a subsequent aldol addition with the triose. The ring closure to deoxyribose is kinetically controlled by a 5-exo-trig cyclization, so that the stereochemistry of the entire target molecule is defined by only one stereo center of the substrate. This is the first example with an exclusive formation of the β-furanose nucleoside. In addition, the one-pot reaction opens the first possible access to pyrimidine nucleosides directly from the bases. In summary, this represents a new perspective on a possible "DNA" world, which is quite likely due to the higher stability of DNA compared to its RNA analogue.

7.2 Evolution on a Molecular Level

Prebiotic organocatalysis might have played a key role in the simultaneous generation of molecular complexity at the very beginning of fundamental processes leading to prebiotically relevant molecules. Prebiotic organocatalysis offers many advantages, namely not only the induction of modification to selectivity but also the formation of larger molecules and a broad variability of chemical functionalities. The most important advantage is that no metals or metal ions are required for catalysis. In contrast to classical metal catalysis, organocatalysis is also possible in the presence of water, which was present everywhere on the early earth. An important advantage is that organocatalysis is able to modify simple organic molecules in a simple way, for example in combination with photoorganocatalysis CH activations. In the context of organocatalysis in organic synthesis, mainly amino acids (List 2002; Mukherjee et al. 2007) and short peptides have been investigated in extensive studies (Wiesner et al. 2008; Wennemers 2011). However, these transformations are limited to certain organic transformations and are often not able to create the molecular diversity we find today. It has been shown that proline and proline derivatives have extremely high catalytic activities and often show high enantioselectivity especially in aldol-like condensation but also in Mannich and Michael additions (Yang et al. 2008; Patora-Komisarska et al. 2011). The problem is that proline is rarely found in meteorites but also in experiments that simulate prebiotic conditions.

Organocatalysis is interesting because of very important reactions that can lead to a high variability on the one hand but also to a high selectivity in the formation of amino acids. This is the α-alkylation of aldehydes. This carbon-carbon bond forming transformation would extend small carbonyl compounds obtained in extraterrestrial material or in the presence of electrical discharges. Such a transformation expands the pool of prebiotically accessible molecular structures but also increases the insolubility of organic compounds in water, which inevitably leads to compartmentalization through micelles or lipid layers. A tremendous advantage is that these organocatalysts could form anywhere on the early Earth, making very dedicated geochemical conditions irrelevant.

It has been shown that especially in modern organic chemistry the intermolecular α-alkylation is dominated by photo redox organocatalysis.

If you consider one step further, you will come to the conclusion that when organocatalysts are formed from aldehydes or ketones, for example, so the organocatalysts themselves are capable of modifying their own chemical building blocks by α-alkylation. It is also likely that from a generated library of organocatalysts, one of the catalysts is selected on the basis of its activity and selectivity, thus influencing the formation of the new generation of catalysts. Logically, this selectivity is influenced by the chemical environment, i.e. the concentration of the available substrates, the concentration itself and the available water will influence the selection and reactivity or selectivity of the organocatalysts. It is also obvious whether the reaction takes place under photochemical conditions

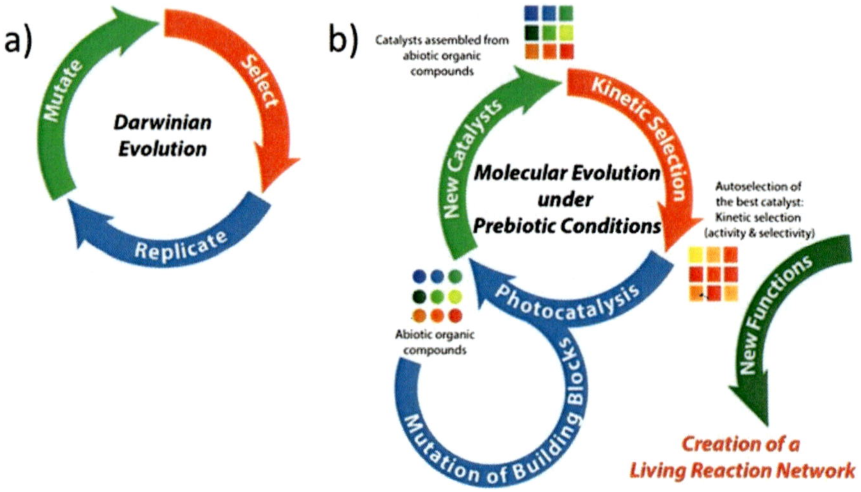

Fig. 7.2 Comparison of Darwinian evolution and evolution on a (small) molecular level. Darwinian evolution can be traced back to chemical evolution using the biochemical machinery to generate new functions

or in the dark because this influences the mechanism of the reaction, specifically whether an α-alkylation or an aldol reaction is preferred. This corresponds to a Darwinian evolution on a molecular level (cf. Fig. 7.2).

In the course of investigations into prebiotic chemistry, the Trapp Research Laboratory discovered a simple and direct access to a whole class of catalysts that allow intermolecular α-alkylations of aldehydes under plausible prebiotic conditions at an early Earth, i.e. reducing atmospheric conditions with water as oxygen source, volcanic activity, wet dry, and day and night cycles (Closs et al. 2020). Such a scenario is illustrated schematically in Fig. 7.3.

If one considers the currently best organocatalysts, cyclic structures with secondary amines and the modular implementation of sterically demanding residues can be identified as promising potential prebiotic catalysts. Paventi and Edward (1987) already identified a reaction which starts by the formation of α-amino nitriles in a Strecker-type reaction of ketones, hydrocyanic acid and ammonia, which react in the presence of hydrogen sulfide and another molecule of ketone to imidazolidine-4-thiones (cf. Fig. 7.4).

This reaction forming imidazolidine-4-thiones is not limited to ketones, but is broadly applicable to any carbonyl compound, such as aldehydes. The reaction can be performed in basic aqueous solution and thus all reaction components can be regarded as prebiotically plausible. This reaction works very well and imidazolidine-4-thiones can be obtained in yields of up to 76%, thus including aldehydes as a class of prebiotically relevant molecules.

In this way, a large number of potential organocatalysts can be formed (cf. Fig. 7.5). Only aldehydes or ketones are needed as organic feedstock, which shows its

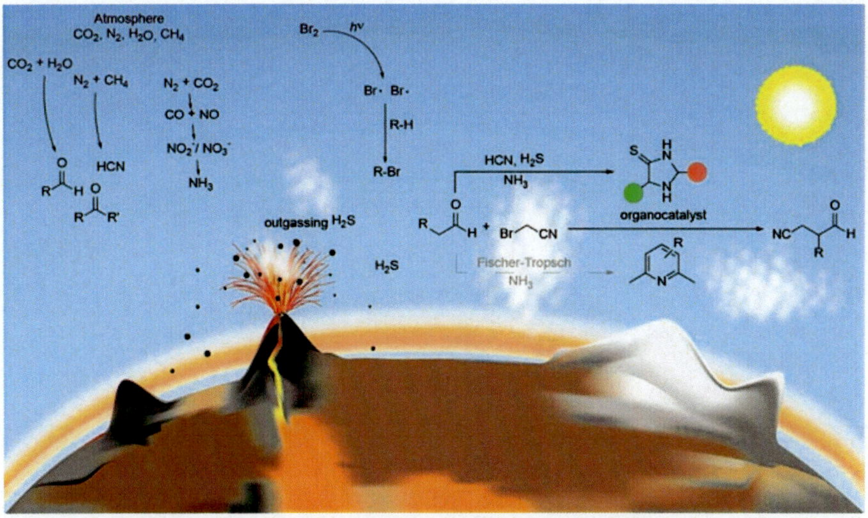

Fig. 7.3 Schematic representation of the early Earth with possibly occurring reactions for the formation of aldehydes and ketones, which form imidazolidine-4-thiones in a reaction with hydrocyanic acid, hydrogen sulfide and ammonia. These photoredox organocatalysts are able to modify aldehydes and ketones in α-position

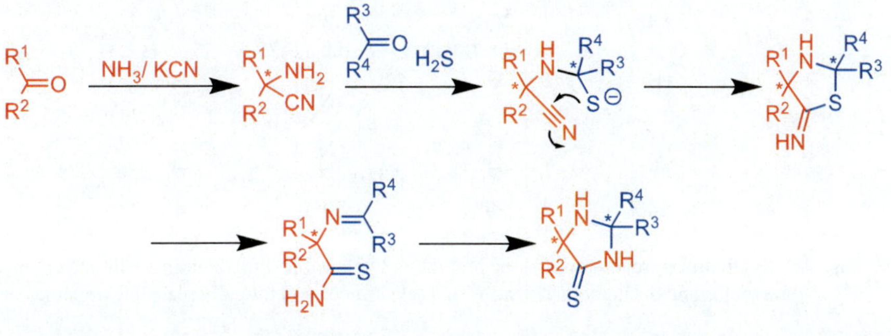

Fig. 7.4 Synthesis of imidazolidine-4-thiones, starting from carbonyl compounds which react with ammonia and hydrocyanic acid in a Strecker-type reaction to form the corresponding aminonitrile. These react in a second step with a further molecule of a carbonyl compound and hydrogen sulfide to form an adduct which first closes the ring and then converts under rearrangement into the imidazolidine-4-thiones

potential and relevance for the emergence of the first organocatalysts on the Early Earth.

This class of organocatalysts shows an initially unexpected behavior; since not only an incorporation of the second carbonyl compound takes place, but also an exchange of the aminonitrile already incorporated in the structure can take place. This is remarkable in that a mutation - quite analogous to a mutation in Darwinian evolution - can occur. This is shown in Fig. 7.6a. A dynamic exchange of the

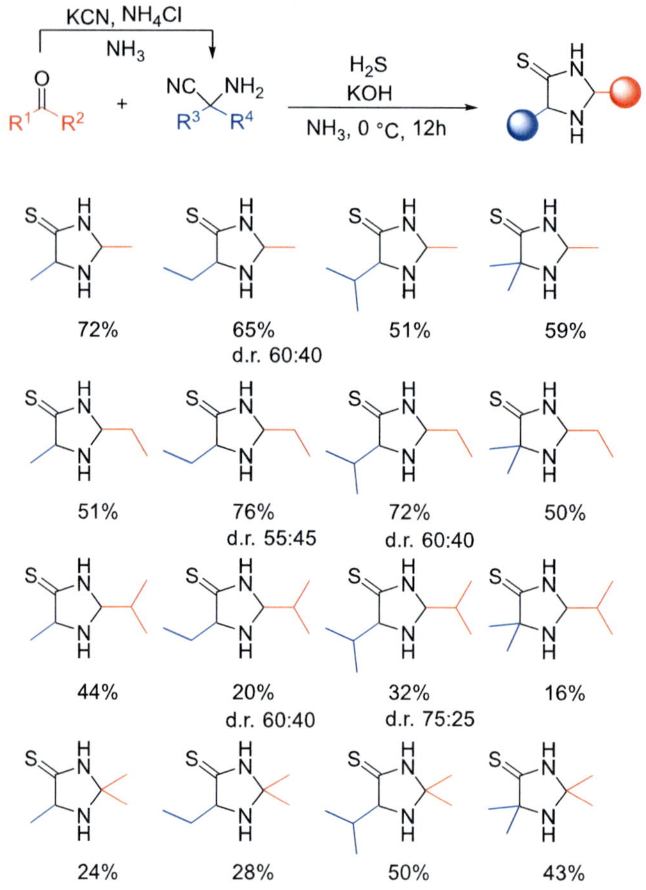

Fig. 7.5 Synthetic approach towards the prebiotically plausible imidazolidine-4-thione organocatalysts with yields and diastereomeric ratios of a selection of prebiotically plausible organocatalysts

aminonitrile occurs. If only aldehydes are used in this reaction, all possible reaction products are obtained and there is no significant preference for a specific structure. However, if a secondary aminonitrile is used, it is preferentially incorporated into the backbone of the imidazolidine-4-thione (Fig. 7.6b). Furthermore, a preferred formation of the anti-substituted diastereomer is observed, especially when sterically demanding aldehydes are incorporated. These features are of particular interest as they underline the dynamics of these systems and the potential enrichment of single diastereomers, thus culminating in a scenario of a first evolutionary selection on a molecular level on the Early Earth.

Considering the potential of the obtained imidazolidine-4-thiones a prebiotically plausible catalytic system for the α-alkylation of aldehydes can be envisioned. α-Cyanomethylation is an interesting reaction, because the reaction product gives

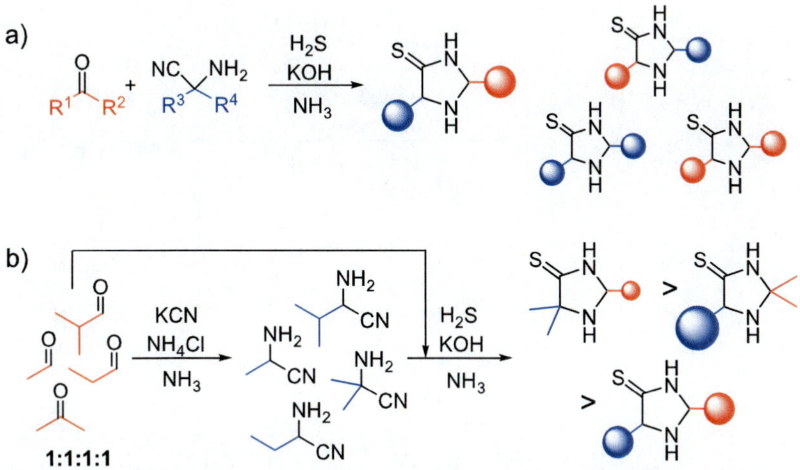

Fig. 7.6 (**a**) Products formed by reaction of the primary reactants (aminonitriles and aldehydes) in a dynamic equilibrium between them (R^3, $R^4 \neq H$). (**b**) Selectivity in the formation of imidazolidine-4-thione organocatalysts starting from a small reactant library

access to prebiotically relevant, more complex molecules by hydrolysis of the nitrile group and further transformation of the aldehyde functionality. The existence of 2-bromoacetonitrile is plausible under the conditions on the Early Earth, formed in a photoinduced radical recombination of interstellar abundant bromine and acetonitrile (Wennemers 2011). In addition, the sun is considered as an important energy supplier in the resource-poor environment at that time, whereby photoreactions have been highly probable. In our photocatalytic setup, the 365 nm LED lamp accounts for the high content of UV-A radiation (320–400 nm). Indeed, imidazolidine-4-thione organocatalysts are active in the photoredox organocatalysis of the α-alkylation of aldehydes, e.g. *n*-octanal resulting in 81% yield. Surprisingly, when this reaction was performed without any photosensitizer, the α-cyanomethylated *n*-propanal is obtained without any erosion in the reaction yield, thus overcoming the need for an additional reagent as well as the limits of metal complexes. The α-cyanomethylation of *n*-propanal can further be performed in acetonitrile or even in an excess of *n*-propanal without any additional solvent (Fig. 7.7).

Typically, a base is needed for these transformations (Fuks et al. 2020). Bases structurally related to 2,6-lutidine are 2,4,6-trimethylpyridine (collidine) and quinolines, such as quinoline, isoquinoline and 2-methylquinoline, which were detected in the Murchison meteorite, and are suggested to be formed by Fischer-Tropsch (Anders et al. 1973; Plows et al. 2003) type catalysis of aldehydes and ammonia or photochemically from benzene and naphthalene derivatives in H_2O/NH_3 ices, respectively. All these pathways are compatible with the here presented synthesis of the imidazolidine-4-thiones. Indeed, collidine and 2-methylquinoline were suitable prebiotic substitutes for 2,6-lutidine. This illustrates that the α-alkylation of aldehydes is feasible under prebiotic conditions using solar radiation as primary

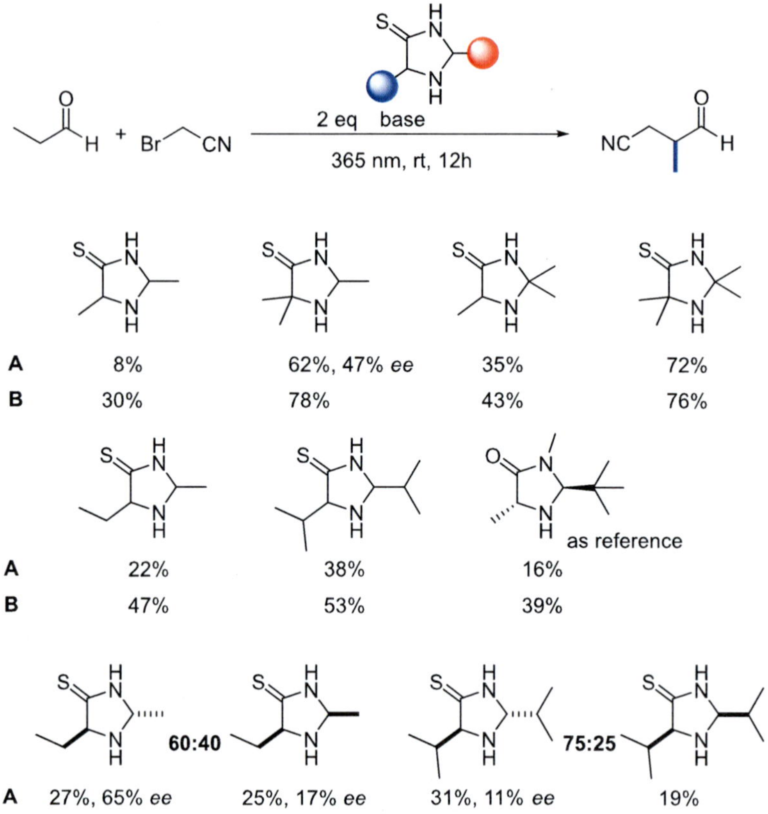

Fig. 7.7 α-Cyanomethylation of *n*-propanal for selected catalysts. A: reaction performed in DMSO with 2,6-lutidine under aerobic conditions; B: reaction performed in acetonitrile with collidine under aerobic conditions. The enantiomeric excess (*ee*) was determined by enantioselective GC measurements (heptakis(2,3-di-O-methyl-6-O-TBDMS)-β-cyclodextrin in PS 086 column, 7 m × 0.25 mm I.D.; film thickness 0.25 μm) of the corresponding alcohol. Yields were determined by ^{1}H-NMR analysis

energy source leading to modified carbonyl compounds, which can again be used as building blocks in the formation of the next generation organocatalysts.

These organocatalysts potentially also provide a platform for symmetry breaking leading to homochirality. Hydrolysis of the building blocks (α-aminonitrile) leads to amino acids according to the Strecker synthesis. It is remarkable, that several imidazolidine-4-thiones crystallize as an enantiomerically pure crystals out of their racemic solution, e.g. (2*R*,5*S*)-5-ethyl-2-methylimidazolidine-4-thione. In contrast to the formation of a racemic compound, this conglomerate can lead to spontaneous resolution, a phenomenon which is comprehensively proposed as a plausible hypothesis for the origin of molecular chirality. This feature is rather rare as most of the organocatalysts studied so far, including nearly all the proteinogenic amino acids, crystallize as racemic compounds.

Indeed, these prebiotically plausible catalysts can provide enantiomeric excesses (*ee*) of up to 64% in α-alkylation reactions.

7.3 Selective Formation of Deoxyribonucleosides Directed by Activation with Nucleobases

Due to the catalytic activity of RNA (Ferris 1993), it is assumed that RNA may have played a key role in the emergence of life. The hypothesis of the RNA world is that the transition to more complex but stable DNA for continuous information storage and replication (Lazcano et al. 1988), the development of the enzyme ribonucleotide reductase (Poole et al. 2002) was necessary to obtain deoxyribonucleotides from ribonucleotides. For the synthesis of DNA building blocks for a DNA-based life, there are so far only two described pathways that are prebiotically plausible. Sutherland and coworker have proposed a longer synthetic route that starts from the synthesis route to RNA nucleosides and uses an intermediate stereoisomer. Trapp and colleagues recently discovered a very short and universal pathway to DNA nucleosides, which also plays a role in the a priori synthesis of DNA nucleosides in nature. Furthermore, this path can be used to explain why nature has chosen ribose or deoxyribose as the sugar unit for nucleosides. This shows that a thorough examination of possible prebiotic reaction paths provides important insights into the intrinsic properties of these molecules and their reactivities under the given environmental conditions.

One of the first chemists to work on nucleoside synthesis was Emil Fischer (cf. Fig. 7.1a) (Roy et al. 2016). He retro-synthetically disassembled the nucleosides into their respective nucleobase and the sugar unit, just as we still explain the systematic structure of heterocycles and sugars to our students in organic chemistry. It has to be pointed out that the formation of RNA nucleosides and nucleotides from D-ribose seemed to be easier. This pathway was also pursued by Orgel (Kruse et al. 2020; Sanchez and Orgel 1970; Oró and Kimball 1961). The problem here is that under these drastic reaction conditions, the thermodynamically favored pyranose is formed instead of the kinetically favored furanose. To avoid this problem, a modified pathway was proposed by Carell, based on formamidopyrimidines (FaPys) as purine precursors (Preiner et al. 2020), which are easier to condense with aldoses (cf. Fig. 7.1b). However, the challenge to prepare D-ribose or D-deoxyribose remains, because there is no selective pathway to D-ribose (Pizzarello and Weber 2010; Meinert et al. 2016) for the formation of ribonucleosides. D-Ribose is subject to rapid decomposition in aqueous solution (Larralde et al. 1995) and the selectivity of the formose reaction (Butlerow 1861; Haas et al. 2020) is negligibly low with respect to D-ribose. Sutherland worked around this challenge for the preparation of pyrimidine ribonucleosides by introducing 2-aminooxazole as the connecting building block between sugar and pyrimidine base (cf. Fig. 7.1c) (Lincoln and Joyce 2009; Powner et al. 2010). The ribose unit is synthesized by reaction with D-glyceraldehyde.

The prebiotic access to the deoxyribonucleosides is even more difficult, since the previously described strategies require D-deoxyribose (Nuevo et al. 2018) or a dehydroxylation step, respectively. Furthermore, for these strategies there is no uniform pathway for both purine and pyrimidine nucleosides. A chemical prebiotic pathway via photoreduction has been considered by Sutherland and co-workers (Xu et al. 2019, 2020). involving a multiple step synthesis with low yields.

Deoxyribonucleosides can also form directly from the corresponding nucleobases and prebiotic aldehydes (Wiesner et al. 2008). Such a scenario is even more interesting since no complicated steps or even changing chemical reaction conditions are required and the reaction path is more plausible if the target molecules are formed due to their intrinsic reactivity. If a retro-synthetic analysis of the probiotic reaction network is performed and possible reaction pathways in aqueous solution are investigated, it can be concluded that the deoxyribonucleosides can be formed in a reaction analogous to organocatalysts via enamine intermediates from the nucleobases and acetaldehyde as well as D-glyceraldehyde or the precursors formaldehyde and glycol aldehyde.

The nucleophilic character of the nucleobases at nitrogen atom *N*-1 in pyrimidine and *N*-9 in purine bases allows a reaction with acetaldehyde in analogy to aminocatalysis (Knoevenagel 1898; List 2010) giving the corresponding enamines, the *N*-9-vinylpurine bases and *N*-1-vinylpyrimidine bases in water at 50 °C (cf. Fig. 7.8). The obtained vinyl nucleobases react stereoselectively with D-glyceraldehyde or even with formaldehyde (37 wt.% in water) to the deoxyribonucleosides dA, dG, dC, and dT. The nucleophilic attack of the vinyl nucleobase on the carbonyl group of the D-glyceraldehyde is stereoelectronically controlled in an unambiguous manner (Anh and Eisenstein 1976) (Fig. 7.8). The formed open chained deoxyribose system undergoes kinetically favored *5-exo-trig* cyclization (Baldwin 1976) to the furanose form under formation of the β-deoxyribonucleoside in defined stereochemistry, which makes this reaction highly plausible in a prebiotic reaction setting.

These deoxyribonucleoside forming reactions proceed even in mixtures selectively and under strictly prebiotic and mild conditions in water between 40 and 70 °C with the optimum at 50 °C. This reaction is very robust and proceeds in a broad neutral to alkaline pH range. Furthermore, prebiotically difficult explainable modulations of the pH are not necessary.

The question arose, which deoxynucleosides are obtained when other sugar precursors and formaldehyde are used in the reaction with the vinyl nucleobases. The reaction of the nucleobases with acetaldehyde, glycolaldehyde and formaldehyde yields deoxythreonucleoside (dTNA). This deoxynucleoside is complementary to Eschenmoser's threonucleosides (TNA) (Yu et al. 2012) as RNA progenitor (Schöning et al. 2000). Because the corresponding dTNA nucleoside is unable to polymerize to a polynucleic acid due to the lack of a second hydroxy group, it can be assumed that this is an evolutionary dead end (Fig. 7.9). 2*C*-hydroxymethylerythronucleoside (HMENA) is formed, which has also to be

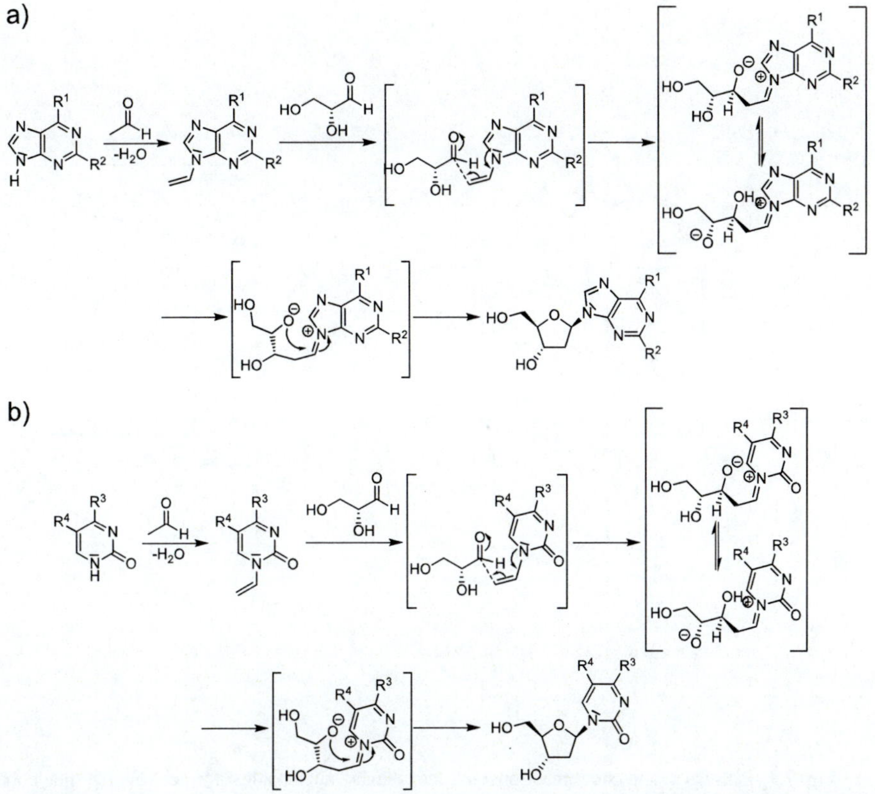

Fig. 7.8 Reaction mechanism to DNA nucleosides via nucleobase activation of acetaldehyde in combination with sugar forming aldol reactions under prebiotic reaction conditions. Stereoselectively controlled mechanism leading to deoxyribonucleosides (dA, dG, dC, dT) via vinyl nucleobases and reaction with D-glyceraldehyde. dA: $R^1 = NH_2$, $R^2 = H$; dG: $R^1 = OH$, $R^2 = NH_2$; dC: $R^3 = NH_2$, $R^4 = H$; dT: $R^3 = OH$, $R^4 = CH_3$

regarded as an evolutionary dead end, because of the formation of eight possible stereoisomers.

A side reaction of vinyl nucleobases and formaldehyde gives nucleosides derived from 2-hydroxymethyl-1,3-propanediol and pentaerythrol (PentaNA). The Cannizzaro reaction (Cannizzaro 1853), is the key step in both reactions. Potentially both compounds could form a polynucleic acid, but such a structure would be difficult to replicate due to branching and ambiguous sequencing. In addition, by polymerization the central carbon atom of the sugar type moiety becomes stereogenic leading to enantiomers and diastereomers in the polymer.

The C3-ketose dihydroxyacetone which is isomeric to glyceraldehyde reacts with the vinyl nucleobases to deoxyapionucleosides. Remarkably, the carbohydrate apiose occurs in pectins in the cell wall of plants and plays an important role in the glycosylation of secondary metabolites (Vongerichten 1902; Pičmanová and

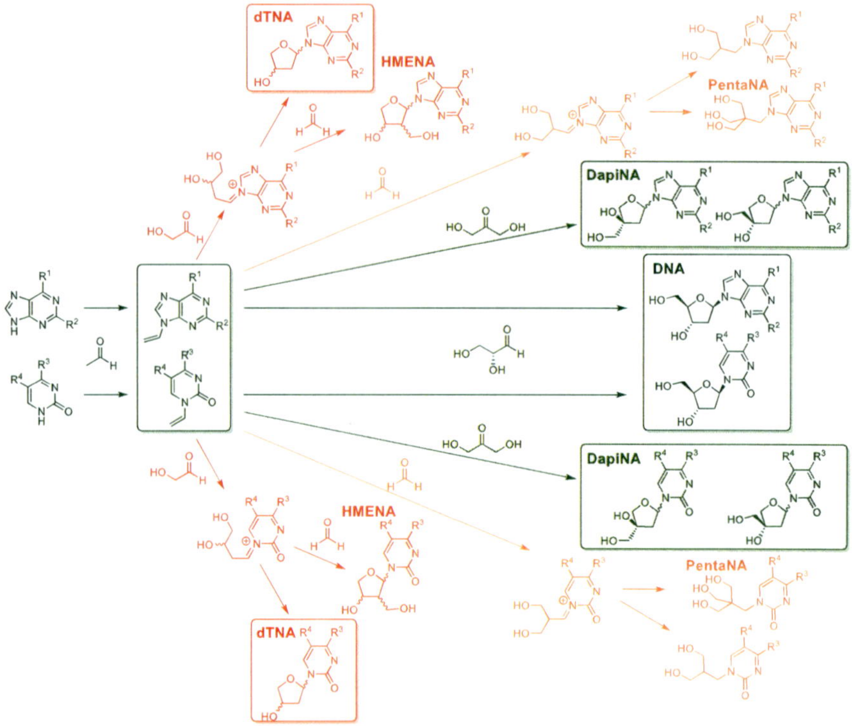

Fig. 7.9 Pathways and side reactions to deoxysugar nucleosides formed by the nucleobase activation of acetaldehyde. The initial step is the vinyl nucleobase formation starting from the nucleobase and acetaldehyde. Reaction with glycolaldehyde and formaldehyde leads to deoxythreonucleosides (dTNA) and 2C-hydroxymethylerythronucleoside (HMENA), which are unable to polymerize in a defined structure. In a side reaction under Cannizzaro conditions with formaldehyde two deoxynucleosides are formed derived from 2-hydroxymethyl-1,3-propanediol and pentaerythrol (PentaNA). Reaction of the vinyl nucleosides with dihydroxyacetone results in the deoxyapionucleosides (DApiNA), and with D-glyceraldehyde the deoxyribonucleosides (DNA) are obtained, respectively. A: $R^1 = NH_2$, $R^2 = H$; G: $R^1 = OH$, $R^2 = NH_2$; C: $R^3 = NH_2$, $R^4 = H$; T: $R^3 = OH$, $R^4 = CH_3$

Møller 2016). It is noteworthy that these DApiNA nucleosides are also capable to polymerize to a polynucleic acid and can therefore be regarded as a DNA progenitor. But it has to be pointed out that the formation of glyceraldehyde in the formose reaction is favored and therefore also DNA should be favored over DApiNA. In addition, while DNA nucleosides are built in a defined reaction from D-glyceraldehyde, two additional configurational isomers of DApiNA nucleosides are possible, summing up to a total of four epimers.

7.4 Summary and Outlook

In summary, a class of prebiotic plausible photoredox organocatalysts was presented in this chapter. In particular the formation under prebiotic plausible conditions was shown. The special feature of such organocatalysts is that they are able to modify their own molecular building blocks in such a way that new catalysts can be created in another generation. This means that such a catalyst system is able to mutate and to select the best catalyst by activity and selectivity. Such a system meets all the criteria of evolution, namely reproduction (synthesis), selection, and mutation. The preferred organocatalysts isomers show improved catalytic performance in terms of activity and enantioselectivity. Of particular interest in this context is the enantiomerically pure crystallization of single enantiomers thereby opening a plausible pathway for symmetry breaking, leading to homochirality. In addition to the access to substituted aldehydes, which is necessary for compartmentalization to form cellular structures, also the formation of the canonical amino acids is plausible. Due to the similarity to established organocatalysts, many prebiotic functionalizations via organocatalysis are conceivable.

Reactions related to organocatalysis play an important role in prebiotic chemistry. This can be impressively demonstrated with the highly selective formation of β-deoxynucleosides by activation of acetaldehyde and an *in-situ* sugar-forming cascade. The stereochemical course, which has led to the preferred formation of deoxyribose over other deoxysugars is striking. The molecular evolutionary selection principle is clearly the minimization of possible stereoisomers. This corresponds to a minimization of entropy and, by definition, to a shift from equilibrium, which can be understood as a criterion for Life on a molecular basis.

In a nucleobase activated aldol reaction, a transition from all canonical nucleobases to deoxyribonucleosides occurs. The reaction proceeds under mild reaction conditions, in water, basic mineral additives enhance the reaction. When glycolaldehyde, formaldehyde or dihydroxyacetone, which can equilibrate with glyceraldehyde, deoxyapionucleosides are formed. However, these potential DNA nucleoside progenitors are ruled out because of their inability to polymerize or their stereochemical variability.

The here presented logical prebiotic pathway proves the ingenuity of Nature to minimize the number of possible stereoisomers and explains why D-deoxyribose is favored over all other sugars and thus opens a non-enzymatic pathway towards DNA nucleosides. It can be concluded that RNA and DNA evolved simultaneously.

References

Anders, E., Hayatsu, R., Studier, M.H.: Organic compounds in meteorites. Science. **182**, 781–790 (1973)

Anh, N.T., Eisenstein, O.: Induction asymetrique 1–2: comparaison ab initio des modeles de Cram, de Cornforth, de Karabatsos et de Felkin. Tetrahedron Asymmetry. **17**, 155–158 (1976)

Baldwin, J.E.: Rules for ring closure. J. Chem. Soc. Chem. Commun. **1976**, 734–736 (1976)

Becker, S., Thoma, I., Deutsch, A., Gehrke, T., Mayer, P., Zipse, H., Carell, T.: A high-yielding, strictly regioselective prebiotic purine nucleoside formation pathway. Science. **352**, 833–836 (2016)

Becker, S., Schneider, C., Okamura, H., Crisp, A., Amatov, T., Dejmek, M., Carell, T.: Wet-dry cycles enable the parallel origin of canonical and non-canonical nucleosides by continuous synthesis. Nat. Commun. **9**, 163 (2018)

Becker, S., Feldmann, J., Wiedemann, S., Okamura, H., Schneider, C., Iwan, K., Crisp, A., Rossa, M., Amatov, T., Carell, T.: Unified prebiotically plausible synthesis of pyrimidine and purine rna ribonucleotides. Science. **366**, 76–82 (2019)

Benner, S.A., Bell, E.A., Biondi, E., Brasser, R., Carell, T., Kim, H.-J., Mojzsis, S.J., Omran, A., Pasek, M.A., Trail, D.: When did life likely emerge on earth in an RNA-first process? ChemSystemsChem. (2019). https://doi.org/10.1002/syst.202000009

Butlerow, A.: Bildung einer zuckerartigen substanz durch synthese. Liebigs Ann. **120**, 295–298 (1861)

Cannizzaro, S.: Über den der Benzoesäure entsprechenden Alkohol. Justus Liebigs Ann. Chem. **88**, 129–130 (1853)

Closs, A.C., Fuks, E., Bechtel, M., Trapp, O.: Prebiotically plausible organocatalysts enabling a selective photoredox α-alkylation of aldehydes on the early earth. Chem. Eur. J. **26**, 10702–10706 (2020)

Damer, B., Deamer, D.: The hot spring hypothesis for an origin of life. Astrobiology. **20**(4), 429–452 (2019)

Ferris, J.P.: Catalysis and prebiotic RNA synthesis. Orig. Life Evol. Biosph. **23**, 307–315 (1993)

Fialho, D.M., Roche, T.P., Hud, N.V.: Prebiotic syntheses of noncanonical nucleosides and nucleotides. Chem. Rev. **120**, 4806–4830 (2020)

Fischer, E., Helferich, B.: Synthetische glucoside der purine. Ber. Dtsch. Chem. Ges. **47**, 210–235 (1914)

Fuks, E., Huber, L., Schinkel, T., Trapp, O.: Investigation of straightforward, photoinduced alkylations of electron-rich heterocompounds with electron-deficient alkyl bromides in the sole presence of 2,6-Lutidine. Eur. J. Org. Chem. (2020). https://doi.org/10.1002/ejoc.202001003

Fuller, W.D., Sanchez, R.A., Orgel, L.E.: Studies in prebiotic synthesis: vi. Synthesis of purine nucleosides. J. Mol. Biol. **67**, 25–33 (1972)

Gilbert, W.: Origin of life: the RNA world. Nature. **319**, 618 (1986)

Guerrier-Takada, C., Gardiner, K., Marsh, T., Pace, N., Altman, S.: The rna moiety of ribonuclease P is the catalytic subunit of the enzyme. Cell. **35**, 849–857 (1983)

Haas, M., Lamour, S., Christ, S.B., Trapp, O.: Mineral-mediated carbohydrate synthesis by mechanical forces in a primordial geochemical setting. Commun. Chem. (2020). https://doi.org/10.1038/s42004-020-00387-w

Hutchison, C.A., Chuang, R.-Y., Noskov, V.N., Assad-Garcia, N., Deerinck, T.J., Ellisman, M.H., Gill, J., Kannan, K., Karas, B.J., Ma, L., Pelletier, J.F., Qi, Z.-Q., Richter, R.A., et al.: Design and synthesis of a minimal bacterial genome. Science. **351**, aad6253 (2016)

Joyce, G.F.: RNA evolution and the origins of life. Nature. **338**, 217 (1989)

Joyce, G.F.: The antiquity of RNA-based evolution. Nature. **418**, 214 (2002)

Kim, H.-J., Benner, S.A.: Prebiotic stereoselective synthesis of purine and noncanonical pyrimidine nucleotide from nucleobases and phosphorylated carbohydrates. PNAS. **114**, 11315 (2017)

Knoevenagel, E.: Ueber den chemismus der condensierenden Wirkung des Ammoniaks und organischer Amine bei Reactionen zwischen Aldehyden und Acetessigester. Ber. Dtsch. Chem. Ges. **31**, 738–748 (1898)

Kruger, K., Grabowski, P.J., Zaug, A.J., Sands, J., Gottschling, D.E., Cech, T.R.: Self-splicing RNA: autoexcision and autocyclization of the ribosomal rna intervening sequence of tetrahymena. Cell. **31**, 147–157 (1982)

Kruse, F.M., Teichert, J.S., Trapp, O.: Prebiotic nucleoside synthesis - the selectivity of simplicity. Chem. Eur. J. **26** (2020). https://doi.org/10.1002/chem.202001513

Larralde, R., Robertson, M.P., Miller, S.L.: Rates of decomposition of ribose and other sugars: implications for chemical evolution. PNAS. **92**, 8158–8160 (1995)

Lazcano, A., Guerrero, R., Margulis, L., Oró, J.: The evolutionary transition from RNA to DNA in early cells. J. Mol. Evol. **27**, 283–290 (1988)

Lincoln, T.A., Joyce, G.F.: Self-sustained replication of an RNA enzyme. Science. **323**, 1229 (2009)

List, B.: Proline-catalyzed asymmetric reactions. Tetrahedron. **58**, 5573–5590 (2002)

List, B.: Emil Kknoevenagel and the roots of aminocatalysis. Angew. Chem. Int. Ed. **49**, 1730–1734 (2010)

Meinert, C., Myrgorodska, I., Marcellus, P.D., Buhse, T., Nahon, L., Hoffmann, S.V., D'Hendecourt, L.L.S., Meierhenrich, U.J.: Ribose and related sugars from ultraviolet irradiation of interstellar ice analogs. Science. **352**, 208–212 (2016)

Miller, S.L., Urey, H.C.: Organic compound synthes on the primitive earth. Science. **117**, 245–251 (1953)

Muchowska, K.B., Varma, S.J., Moran, J.: Nonenzymatic metabolic reactions and life's origins. Chem. Rev. **120**, 7708–7744 (2020)

Mukherjee, S., Yang, J.W., Hoffmann, S., List, B.: Asymmetric enamine catalysis. Chem. Rev. **107**, 5471–5569 (2007)

Mulkidjanian, A.Y., Bychkov, A.Y., Dibrova, D.V., Galperin, M.Y., Koonin, E.V.: Origin of first cells at terrestrial, anoxic geothermal fields. PNAS. **109**, E821–E830 (2012)

Nuevo, M., Cooper, G., Sandford, S.A.: Deoxyribose and deoxysugar derivatives from photoprocessed astrophysical ice analogues and comparison to meteorites. Nat. Commun. **9**, 5276 (2018)

Oparin, A.I.: The Origin of Life, VIII. The Macmillan Company, New York (1953)

Oró, J., Kimball, A.P.: Synthesis of purines under possible primitive earth conditions. I. Adenine from hydrogen cyanide. Arch. Biochem. Biophys. **94**, 217–227 (1961)

Patora-Komisarska, K., Benohoud, M., Ishikawa, H., Seebach, D., Hayashi, Y.: Organocatalyzed michael addition of aldehydes to nitro alkenes – generally accepted mechanism revisited and revised. Helv. Chim. Acta. **94**, 719–745 (2011)

Paventi, M., Edward, J.T.: Preparation of α-aminothioamides from aldehydes. Can. J. Chem. **65**, 282–289 (1987)

Pičmanová, M., Møller, B.L.: Apiose: one of nature's witty games. Glycobiology. **26**, 430–442 (2016)

Pizzarello, S., Weber, A.L.: Stereoselective syntheses of pentose sugars under realistic prebiotic conditions. Orig. Life Evol. Biosph. **40**, 3–10 (2010)

Plows, F.L., Elsila, J.E., Zare, R.N., Buseck, P.R.: Evidence that polycyclic aromatic hydrocarbons in two carbonaceous chondrites predate parent-body formation. Geosci. Front. **67**, 1429–1436 (2003)

Poole, A.M., Logan, D.T., Sjöberg, B.-M.: The evolution of the ribonucleotide reductases: much ado about oxygen. J. Mol. Evol. **55**, 180–196 (2002)

Powner, M.W., Gerland, B., Sutherland, J.D.: Synthesis of activated pyrimidine ribonucleotides in prebiotically plausible conditions. Nature. **459**, 239–242 (2009)

Powner, M.W., Sutherland, J.D., Szostak, J.W.: Chemoselective multicomponent one-pot assembly of purine precursors in water. J. Am. Chem. Soc. **132**, 16677–16688 (2010)

Preiner, M., Igarashi, K., Muchowska, K.B., Yu, M., Varma, S.J., Kleinermanns, K., Nobu, M.K., Kamagata, Y., Tüysüz, H., Moran, J., Martin, W.F.: A hydrogen-dependent geochemical analogue of primordial carbon and energy metabolism. Nat. Ecol. Evol. **4**, 534–542 (2020)

Roy, B., Depaix, A., Périgaud, C., Peyrottes, S.: Recent trends in nucleotide synthesis. Chem. Rev. **116**, 7854–7897 (2016)

Ruiz-Mirazo, K., Briones, C., Escosura, A.D.L.: Prebiotic systems chemistry: new perspectives for the origins of life. Chem. Rev. **114**, 285–366 (2014)

Sanchez, R.A., Orgel, L.E.: Studies in prebiotic synthesis. J. Mol. Biol. **47**, 531–543 (1970)

Schöning, K.-U., Scholz, P., Guntha, S., Wu, X., Krishnamurthy, R., Eschenmoser, A.: Chemical etiology of nucleic acid structure: the α-threofuranosyl-($3' \rightarrow 2'$) oligonucleotide system. Science. **290**, 1347 (2000)

Teichert, J.S., Kruse, F.M., Trapp, O.: Direct prebiotic pathway to DNA nucleosides. Angew. Chem. Int. Ed. **58**, 9944–9947 (2019)

Vongerichten, E.: Ueber Apiose, eine β-Oxymethylerythrose. Justus Liebigs Ann. Chem. **321**, 71–83 (1902)

Wennemers, H.: Asymmetric catalysis with peptides. Chem. Commun. **47**, 12036–12041 (2011)

Wiesner, M., Revell, J.D., Tonazzi, S., Wennemers, H.: Peptide catalyzed asymmetric conjugate addition reactions of aldehydes to nitroethylene—a convenient entry into γ^2-amino acids. J. Am. Chem. Soc. **130**, 5610–5611 (2008)

Woese, C.R.: The fundamental nature of the genetic code: prebiotic interactions between polynucleotides and polyamino acids or their derivatives. Proc. Natl. Acad. Sci. USA. **59**, 110 (1968)

Xu, J., Green, N.J., Gibard, C., Krishnamurthy, R., Sutherland, J.D.: Prebiotic phosphorylation of 2-thiouridine provides either nucleotides or dna building blocks via photoreduction. Nat. Chem. **11**, 457–462 (2019)

Xu, J., Chmela, V., Green, N., Russell, D., Janicki, M., Góra, R., Szabla, R., Bond, A., Sutherland, J.: Selective prebiotic formation of rna pyrimidine and dna purine nucleosides. Nature. **582**, 60–66 (2020)

Yadav, M., Kumar, R., Krishnamurthy, R.: Chemistry of abiotic nucleotide synthesis. Chem. Rev. **120**, 4766–4805 (2020)

Yang, J.W., Chandler, C., Stadler, M., Kampen, D., List, B.: Proline-catalysed mannich reactions of acetaldehyde. Nature. **452**, 453–455 (2008)

Yu, H., Zhang, S., Chaput, J.C.: Darwinian evolution of an alternative genetic system provides support for tna as an rna progenitor. Nat. Chem. **4**, 183 (2012)

Chapter 8
Virus Origins and the Origin of Life

Donald Pan

Abstract Viruses are enigmatic, with scientists debating, often heatedly, about the nature of their existence, such as whether viruses should be categorized as living or non-living. They are not even represented in most conventional trees of life. However, viruses have played a driving role in the co-evolution of life on Earth since its origins. Viruses can serve as molecular fossils, providing a possible glimpse at a pool of genes that may have been present from a time before the Last Universal Common Ancestor (LUCA). The study of virology and the origin of life have often been intertwined. Since the earliest studies of viruses, viruses have been hypothesized to be ancient, possibly anteceding cell-based life. Other hypotheses propose that cell-based life preceded viruses, with viruses "escaping" from cellular life forms. This chapter will introduce viruses and their relationship with the origin of life, including the various proposed origin scenarios involving viruses and virus-like elements.

8.1 Introduction

Any discussion on the origin and nature of viruses inevitably leads down a path toward a discussion of the origin and nature of life itself. Viruses have challenged the definitions of life and how scientists think about it, occupying the hazy boundary between living and non-living. Ever since the earliest days of virology, they have been hypothesized to play a role in the origin of life, predating cells. Soon after bacteriophage were discovered in the 1910s, they were proposed to be related to the earliest life forms, being given the name "protobe" (first life) by their French-

D. Pan (✉)
School of Oceanography, Shanghai Jiao Tong University, Shanghai, China
e-mail: donald.pan@sjtu.edu.cn

A. Neubeck, S. McMahon (eds.), *Prebiotic Chemistry and the Origin of Life*, Advances in Astrobiology and Biogeophysics,
https://doi.org/10.1007/978-3-030-81039-9_8

Canadian discoverer Felix D'Herelle[1] (Vaughan 1927). One of the foundational figures in the modern study of the origin of life, J.B.S. Haldane, in his 1929 essay "The Origin of Life" suggested that life had existed in a "virus stage for many millions of years" before the first cell formed (Haldane 1929).

Viruses have played an outsized role in the history of humanity, for example smallpox, polio, and more recently SARS-CoV-2. For much of the past century, virology had focused on human and agricultural diseases of economic impact. Around the beginning of this century, discoveries of novel viruses like giant viruses (La Scola et al. 2003) and new families of archaeal viruses (Prangishvili et al. 2017), as well as the development of sequencing technologies and metagenomics have revealed the global impact and importance of viruses to life on Earth. In this new era, there has been a revival of interest in the evolutionary origins of viruses and the roles that viruses may have played at the origin of life on Earth. There has also been increasing interest in viruses for astrobiology and the search for extraterrestrial life (Janjic 2018; Berliner et al. 2018; Griffin 2013).

Viruses are intertwined with life on Earth. They are known to infect all domains of life. As far as we know, there are apparently no cellular organisms that do not harbor viruses or related genetic parasites. Their fingerprints are found within the genomes of all cellular organisms. Viruses and related selfish replicating genetic elements have been locked in an arms race with cellular life likely since its origins. Viruses are also abundant in extreme environments that may have been the setting for the origin of life, for example hydrothermal vents (Lossouarn et al. 2015; Ortmann and Suttle 2005) and terrestrial hot springs (Schoenfeld et al. 2008; Munson-McGee et al. 2018). Thus, viruses, genetic parasites, and other selfish replicating elements offer clues to early life. In addition, viruses have been implicated in various major evolutionary events in the history of the Earth (Koonin 2016).

There are a multitude of hypotheses regarding virus origins and their possible involvement in the origin of life, far too many to cover within a chapter. However, the principles and key concepts used in virological origin of life thinking will be reviewed here.

8.2 The Place of Viruses in Origin of Life Studies

Since their discovery, viruses sparked debate over whether they were living entities and what the nature of life was. Coincidentally, viruses were discovered around the time when modern efforts began to tackle questions surrounding the origin of life on Earth. In 1859, Charles Darwin's *On the Origin of Species* had been published, and Louis Pasteur had disproven spontaneous generation. Soon after,

[1] D'Herelle was appointed "Professor of Protobiology" at Yale in 1928. While it had originally referred to the study of bacteriophage, interestingly, nowadays protobiology more often refers to the study of the origin of life.

viruses were discovered when Ivanovsky and Beijerinck independently discovered that an agent smaller than cells ("filterable viruses") could cause disease. The subsequent discoveries of bacteriophage by Frederick Twort and Felix D'Herelle (Duckworth 1976) prompted early thinkers of origin of life studies like J.B.S Haldane to consider whether viruses may be a transitionary step between the prebiotic soup and the emergence of the first cells (the "virus-first" hypothesis) (Lazcano 2016; Tirard 2017). In 1935, the crystallization of the tobacco mosaic virus by Wendell Stanley seemed to confirm that simple crystallizable chemicals could have properties of life (Porcar and Peretó 2018). This sparked a heated debate over whether viruses are living matter or not, one which continues to this day (Haldane 1929).

Another foundational figure in origin of life studies, A.I. Oparin, proposed a metabolism-first model for the origin of life. Being obligate parasites without any metabolic capability, viruses were excluded from this model (Kostyrka 2016). With the recognition that viruses were obligate parasites unable to independently reproduce, the virus-first hypothesis fell out of favor until more recently (Podolsky 1996; López-García 2012). In its place were hypotheses such as the "regression" hypothesis which suggested that viruses were once parasitic cells that had regressed into a simpler form and the "escape" hypothesis in which viruses were originally selfish host genes that escaped the confines of the cell. Refer to the Sect. 8.4.

8.3 What Are Viruses? Evolution of the Virus Concept

> Viruses should be considered as viruses because viruses are viruses.
> -André Lwoff (Lwoff 1957)

What are viruses? This seemingly simple question has been fraught with ambiguities and debate from the earliest days of virology to today. Even the definition of virus has evolved over time. The word "virus" derives from Latin *vira*, meaning "poison", and had originally referred to anything causing disease.[2]

Modern virology started in the late 1800s when Ivanovsky and Beijerinck found that an agent (now known to be tobacco mosaic virus), which had passed through Chamberland filters[3] that no known bacteria could pass, was capable of causing disease in tobacco leaves. Viruses were redefined by their ability to pass through filters and were initially called "*contagium vivum fluidum*" (Beijerinck 1898) (contagious living fluid) and later "filterable viruses". Viruses infecting bacteria,

[2] This association with disease likely formed an initial bias in the field of virology towards the study of disease-causing viruses infecting humans and other animals and plants of economic importance.

[3] An early type of porcelain filter developed in 1884 by Charles Chamberland, who worked with Louis Pasteur.

which were coined "bacteriophage",[4] were discovered by Twort and D'Herelle (Duckworth 1976) in the early 1900s. In 1926, Thomas Rivers found that viruses are obligate parasites. Wendell Stanley's 1935 crystallization of the tobacco mosaic virus confirmed that viruses are particulate, not fluid, in nature, and in 1937 Bawdan and Pirie found that virus particles are composed of protein and nucleic acid (Pennazio and Roggero 2000). During this early period of virology, the virus particle (virion) and the virus were considered one and the same.

However, the virus concept shifted with the discovery of lysogeny. Initially discovered by Burnet & MacKie in 1929 and further developed by André Lwoff, some bacteriophage were found to be incorporated into the host as prophage (Lwoff 1966). The virus concept could no longer be limited to the virus particle, it must also include the virus integrated with the host cell. Some viruses can also reside in host cells without integrating into the host genome. It became clear that viruses can have a distinct intracellular and extracellular phase. This necessitated a shift in the virus concept to not only be centered on the virion, but to also include all the phases of the virus replication cycle, extracellular as well as intracellular. It was no longer appropriate to confuse the virion for the virus (Forterre 2016). This was recognized early on by Lwoff (1957), however this confusion remains a major trapping nowadays with much of the public (as well as many scientists).

Raoult and Forterre proposed a definition of viruses that would divide life into two categories: capsid-encoding (viruses) and ribosome-encoding organisms (cells) (Raoult and Forterre 2008). Under this definition, viruses necessarily encode capsids which serve to propagate the virus extracellularly. Cellular life forms all encode ribosomes, while there are no known viruses that encode ribosomes. This further led to the formation of the "virocell concept" (Forterre 2013a) which asserts that the infected virion-producing cell is in fact the living phase of the virus (Fig. 8.1). Cells that do not produce virions are termed 'ribocells'. The virus in the course of infection converts the cell into a 'virion factory'.[5] A virus co-existing with its host cell is termed a 'ribovirocell'. Framing the virus-infected cell as a cellular form of the virus can offer a perspective on phenonema such as the infection of other cells by an infected cell via direct cell-to-cell contact rather than through extracellular virion release (Cifuentes-Muñoz et al. 2018; Zhong et al. 2013; Mothes et al. 2010; Dupont and Sattentau 2020). The virocell concept also aligns with why a virus may encode genes for metabolism (auxiliary metabolic genes) that are seemingly useless for the metabolically inert virion (e.g. such as genes related to photosynthesis or fermentation) (Rosenwasser et al. 2016; Thompson et al. 2011; Schvarcz and Steward 2018).

[4] This initiated a distinction between the viruses of prokaryotes and eukaryotes. The prokaryote/eukaryote division divided virology between the study of "bacteriophage" and the study of eukaryotic "viruses", a division that is still palpable today (Forterre 2010). This distinction was also formed before the discovery of archaea, creating a confusing situation in which many had referred to archaeal viruses as archaeal phage (Abedon and Murray 2013).

[5] Interestingly, in his 1963 Nobel lecture, André Lwoff had also referred to the infected cell as being converted into a "virus factory" by the virus (Lwoff 1966).

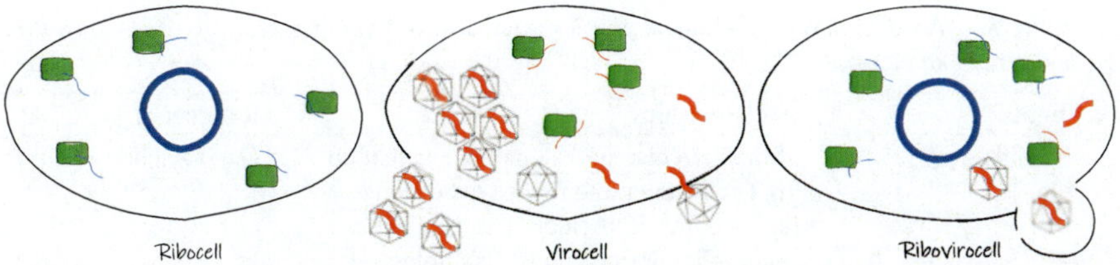

Fig. 8.1 The virocell concept. Ribocells are ribosome encoding cells (ribosomes represented in green). Virocells are virion-producing cells infected by a virus. Ribovirocells are virion-producing cells in which the host is still able to divide—the replicating virus co-exists with the host. Host genome is represented in blue and virus genome in red

Under this definition, a bona fide virus encodes the genes necessary for genome replication as well as structural genes for the capsid. However this definition will not include viruses that do not form virions such as narnaviruses (Hillman and Cai 2013) and hypoviruses (Suzuki et al. 2018) nor extracellular membrane vesicles that encapsulate viral genomes (Gill et al. 2018; Gaudin et al. 2014; Soler et al. 2015). Numerous other selfish genetic elements possibly related to bona fide viruses would also be excluded, such as plasmids, satellite viruses, transposons, introns, viroids, and other related genetic elements (Koonin and Dolja 2013). Table 8.1 provides an overview of these virus-like biological entities. Because viruses recombine with other genetic elements like transposons and plasmids (Toussaint and Merlin 2002) and can transition into other genetic elements through the gain or loss of genes, there is no hard border between categories of selfish elements.

For the purpose of examining the origin of viruses and their roles in the origin of life, I will utilize the replicator paradigm coined by Koonin which includes phylogenetically related replicons as part of a wider "viral world" (Koonin and Starokadomskyy 2016; Koonin and Dolja 2014) in which the virus concept is "genetic, informational parasitism" (Koonin and Dolja 2014; Rohwer and Barott 2013). The reason for utilizing this paradigm is that capsidless selfish elements likely preceded bona fide viruses in evolution (Koonin and Dolja 2014; Krupovic et al. 2019; Krupovic and Koonin 2017).

As more convention-breaking viruses were discovered and vast amounts of genomic data became available, a rethinking of traditional notions of viruses became necessary. Recently, the views of viruses have shifted dramatically. The discovery of giant viruses eliminated the notion that virions are necessarily small. Further investigations of viruses have also shown that they play a more complex role than mere disease-causing agents. They also engage in mutualistic interactions with their hosts (Roossinck and Bazán 2017).

It is still challenging to rigorously define what a virus is. There is also debate about whether viruses should be considered organisms and what the definition of an organism should be (Pepper and Herron 2008). Many have resorted to referring to viruses as "biological entities" to avoid this. However, it still suffers from the same lack of a rigorous definition: what is a biological entity or biological individual? If

Table 8.1 An overview of biological entities related to or hypothesized to be related to viruses and virus-like elements

Entity	Relationship	References
Viroids	Small circular ssRNA pathogens that do not code any proteins. A type of long non-coding RNA(lncRNA). Can act as self-splicing ribozymes. Examples: *Pospiviroidae* and *Avsunviroidae*	Gago-Zachert (2016)
Virusoid/satellite RNA	Viroid-like pathogen that can be packaged into virions with a helper virus. Example: *Deltavirus*	Flores et al. (2011)
Group I introns	May be ancestral to viroids	Hadidi (1986), Lewin (1986)
Satellite viruses	Does not encode a capsid but requires a helper virus for encapsidation. Example: Phage P4	Briani et al. (2001)
Plasmids	Some phage may spend part of their life cycle as plasmids (phasmids/phagemids). May transition into/from bona fide viruses by recruitment/loss of capsid genes. Example: Phage N15	Lagos and Goldstein (1987), Rybchin and Svarchevsky (1999), Krupovic et al. (2011)
Pathogenicity islands	Some pathogenicity islands can be packaged into virions. May have derived from phage. Example: *S. aureus* pathogenicity islands (SaPI), phage-inducible chromosomal islands (PICI)	Fillol-Salom et al. (2018)
Group II introns	May be ancestral to retroelements, RNA viruses, and the spliceosome	Koonin et al. (2015), (2020), Zimmerly and Semper (2015)
Transposons	Some phage behave as transposons (encoding a transposase) or are directly related to transposons. Example: Phage Mu	Harshey (2012), Toussaint and Rice (2017)
LTR retrotransposons	Share a common ancestry with retroviruses. Lost the ability to form virions, but some may potentially retain the ability. Some can form virus-like particles.	Koonin and Dolja (2014)
Endogenous viral elements (EVE)	A type of LTR retrotransposon. Derived from retroviruses. Integrated into the genomes of germ-line cells.	Patel et al. (2011)
SINEs/ LINEs	Non-LTR retrotransposons. May be related to retroviruses.	Koonin et al. (2015)
Polintons/Mavericks	Self-synthesizing transposons in eukaryotes. Related to virophages. Predicted to form virions, but not yet observed. May have evolved from bacteriophage (*Tectiviridae*).	Koonin and Krupovic (2017), Krupovic and Koonin (2015)
Transpovirons	Plasmid-like element. May have derived from polintons or polinton-like viruses	Koonin and Krupovic (2017)

(continued)

Table 8.1 (continued)

Entity	Relationship	References
Casposons	A type of self-synthesizing transposon. May have evolved from viruses. Likely ancestors of CRISPR-Cas system.	Krupovic and Koonin (2016)
Grounded/defective/cryptic prophage	Prophage that are unable to excise from the host genome or produce virions.	Ramisetty and Sudhakari (2019)
Gene transfer agents (GTAs)	Exapted from a prophage, but now serves to transfer host DNA rather than viral DNA.	Redfield and Soucy (2018)
Virus-like particles (VLPs)	Structures made from viral capsid proteins but do not contain viral genetic material (non-infectious). May be defective virions. Examples: Arc, gene transfer agents	Budnik and Thomson (2020)
Exapted proteins from viruses	Example: Tailocin, Type VI secretion system, encapsulin, telomerase, proteins from endogenous retroviruses (synctin).	Roossinck and Bazán (2017), Koonin and Krupovic (2018), Bobay et al. (2014)
Membrane vesicles	May transmit viral genomes or entire virion particles between cells.	Gill et al. (2018), Gaudin et al. (2014)

viruses can integrate into hosts, is an endogenized virus still a virus or biological entity? The question of the nature of viruses will likely continue to be debated as it may belong more to the realm of philosophy (Pradeu et al. 2016).

Terms

Bona fide virus—a virus that can produce virions
Virion—virus particle
Bacteriophage/phage—virus infecting bacteria
Provirus/prophage—a virus or phage integrated into the host genome
Lysogenic/latent—a stage of a viral replication cycle in which the viral genome replicates together with the host, but produces no virions
Lytic—a stage of the viral replication cycle that results in death of the cell
Endogenous/endogenized—embedded into a host genome, sometimes permanently as a "domesticated" viral sequences

8.3.1 Are Viruses Alive?

When one discusses the definition of life, the question of whether or not viruses are alive is invariably raised. This question has been asked since the beginnings of virology. NASA's definition of life is a "self-sustaining chemical system capable of

Darwinian evolution" (Benner 2010). Under this definition, a virion, reliant on host cellular systems for energy transduction, would not be considered alive. However, inside a living system, as a virocell, the virus has all the characteristics of a living entity (Forterre 2016). Others have taken the view that the entire virus life cycle should be considered a living process (Dupré and Guttinger 2016). Because there is not yet any widely accepted formal definition of life nor is there any widely accepted rigorous definition of virus, it is a question that belongs to the philosophical domain. No matter how one defines life or viruses, viruses are intertwined with cellular living systems.

8.3.2 *Evidence for Primordial Viruses*

8.3.2.1 Viruses Co-evolved with Cellular Life on Earth

The modern era of viral ecology likely started with the discovery of the enormous quantity of virus particles in the ocean, exceeding the number of cells (Bergh et al. 1989). Since then, virions have been revealed to be ubiquitous in every environment wherever cells exist. The development of viral metagenomics and the availability of massive sequence data have revealed the enormous genetic diversity of viruses, far greater than that of cellular life forms (Rohwer and Barott 2013). Most of the genetic diversity on Earth resides in the virosphere.

Viruses have shaped the genomes of all cellular life forms and have co-evolved with cellular life for billions of years (Forterre and Prangishvili 2009). There does not seem to be any organism that does not harbor viruses or related genetic parasites. Organisms for which no viruses are known seem to more likely reflect a lack of effort in searching rather than an actual absence of viruses. However even in their genomes, there can still be remnants of past viral infections such as integrated proviruses and other parasitic genetic elements. Many eukaryotic genomes are filled with selfish elements, for instance, the human genome in which parasitic genetic elements make up a significant portion (Koonin and Dolja 2013). From the perspective of the virus, cells can be seen as a habitat and rare resource necessary for replication. Viruses engage in complex interactions of competition and cooperation in order to compete for space in the cellular habitat (Díaz-Muñoz et al. 2017; Secor and Dandekar 2020; Domingo-Calap et al. 2020). And given the abundance and diversity of virions in the environment, it is no wonder that all cellular life forms on Earth have evolved systems to resist the constant invasion of genetic parasites for example CRISPR-Cas, restriction modification, and RNAi (Broecker and Moelling 2019). However, virus-host co-evolution is not only an arms race. Viruses also engage in mutualistic symbioses (Roossinck and Bazán 2017). Viruses can mediate horizontal gene transfer and can shape cellular genomes through integration. Some viruses can integrate into host genomes as proviruses which may become endogenized, losing their ability to produce virions. The viral components can then become exapted by the host for other uses (Koonin and Krupovic 2018). These interactions between virus and host have shaped all life on Earth.

8.3.2.2 Viruses Share Homologous Genes Not Found in Cellular Life Forms

Viral sequence data has allowed the identification of virus "hallmark genes" shared across major viral lineages (Koonin et al. 2020). The viruses of the 3 domains of life share hallmark genes that do not share any homology with those from cellular life forms (Koonin et al. 2006, 2020), suggesting that viruses are an ancient lineage, deriving from before the Last Universal Cellular Ancestor (LUCA, also referred to as Last Universal Common Ancestor). Whereas the core genes of all cellular life forms are monophyletic, deriving from the same lineage as LUCA, core viral replication genes such as RNA-dependent RNA polymerase, reverse transcriptase, protein primed DNA polymerase, and rolling-circle replication initiation endonuclease do not share homology with those from cellular life forms (Krupovic et al. 2019; Koonin 2006a). In addition, homologous viral capsid proteins containing the double jelly roll fold and HK97 fold are encoded by diverse viruses infecting all 3 domains of life, but are not found in cellular life forms (Abrescia et al. 2012; Krupovic and Bamford 2011) suggesting that viruses infecting the 3 domains had evolved before the 3 domains diverged from LUCA (Bamford 2003; Krupovič and Bamford 2008; Krupovic et al. 2020).

8.3.2.3 Viruses Utilize Diverse Modes of Replication

Several features of viruses make them attractive as models for pre-LUCA life forms. One primary feature is the utilization of a greater variety of genetic information storage and replication strategies than extant cellular life forms. Viruses can encode their genomes in linear or circular form, as double-stranded DNA (dsDNA), single-stranded DNA (ssDNA), double-stranded RNA (dsRNA), or single-stranded RNA (ssRNA), whereas extant cells only utilize dsDNA. This is illustrated by the Baltimore classification of viruses (Baltimore 1971), which groups viruses by their genome types and modes of transcription and replication (Fig. 8.2). All cells follow the "Central Dogma", coined by Francis Crick (1958), in which genomic information is stored as dsDNA, transcribed into RNA, and translated into protein. However, viruses can store their genomic information in myriad forms and can replicate in ways that defy the Central Dogma, suggesting that they did not derive from LUCA. Furthermore, viruses do not only rely on host cell machinery, but many also encode their own polymerases (Choi 2012). The many modes of replication and the diversity of replication machineries encoded by viruses offer tantalizing clues to possible alternative modes of replication on Earth before the evolution of DNA-based genomes. Take for example reverse transcribing viruses (Group VI and VII on the Baltimore classification, Fig. 8.2) that use the enzyme reverse transcriptase to produce DNA from RNA genomes or templates. Viruses encoding reverse transcriptase may have been instrumental during the transition from the RNA World to DNA World via RNA-dependent DNA synthesis (Forterre 2005).

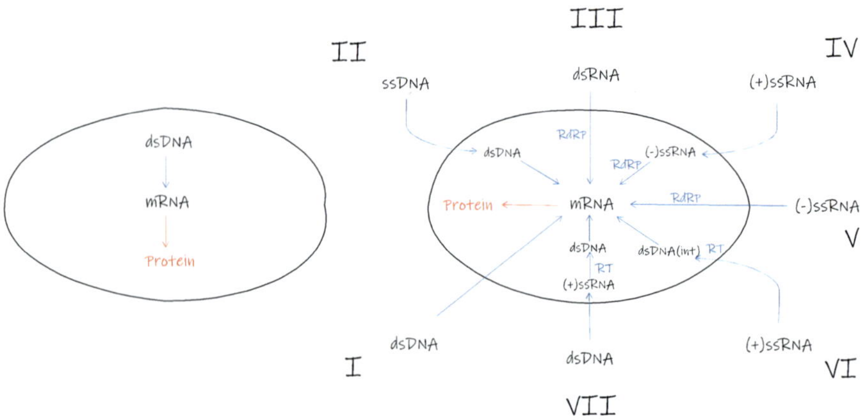

Fig. 8.2 "The Central Dogma" of cellular life forms (left) and the diversity of viral genomes based on the Baltimore Classification (right). I: double-stranded DNA viruses, II: single-stranded DNA viruses, III: double-stranded RNA viruses, IV: positive-sense single-stranded RNA viruses, V: negative-sense single-stranded RNA viruses, VI: reverse transcribing single-stranded RNA viruses, VII: reverse transcribing double-stranded DNA viruses, RdRP: RNA-dependent RNA polymerase, RT: Reverse transcriptase, int: integrated into host genome. The ovals delineate the interior and exterior of a cell

8.3.2.4 Giant Viruses

While large viruses infecting algae have been known since the 1980s (Van Etten and Meints 1999; Van Etten et al. 1991), the discovery of giant viruses with genomes of over 1 million basepairs in length (as large as some prokaryotic genomes), has challenged the conception of what viruses are. Visible by light microscopy, the mimivirus was originally thought to be a bacterium (initially named "*Bradfordcoccus*") when it was first observed in 1992. It remained classified as a bacterium until 2003 when it was found to actually be an enormous virus (La Scola et al. 2003). Since then, numerous giant viruses have been discovered including Pandoravirus (Philippe et al. 2013), Pithovirus (Legendre et al. 2014), and Tupanvirus (Abrahão et al. 2018). Unlike other viruses, giant viruses encode thousands of genes including many involved in translation (Abrahão et al. 2018), leading some to claim that giant viruses represent an intermediate stage between a parasitic cell and typical smaller viruses or that giant viruses belong to a fourth domain of life that may have preceded LUCA (Boyer et al. 2010; Raoult 2013; Colson et al. 2012; Legendre et al. 2012). Some have also hypothesized that the giant viruses are an ancient lineage that arose before the origin of eukaryotes (Guglielmini et al. 2019).

Giant viruses were also found to be susceptible to infection by other viruses (virophage). Based on this evidence, some have suggested that viruses should be considered living organisms (Pearson 2008). More recently, "huge phages" with genomes as large as 735 kbp were detected based on metagenomic sequencing (Al-

Shayeb et al. 2020), reigniting the question of whether they may have risen from a pre-cellular stage of evolution.

8.4 Virus Origin Scenarios

Three major scenarios predominate discourse on the origin of viruses and related selfish elements—the "Virus-first", "Cell reduction/Regression", and "Escape/Endogenous" hypotheses (Fig. 8.3). Although the 3 scenarios have been around for nearly a century, they still survive to present day albeit with updates and revisions. Because viruses are polyphyletic in origin, the scenarios need not be mutually exclusive.

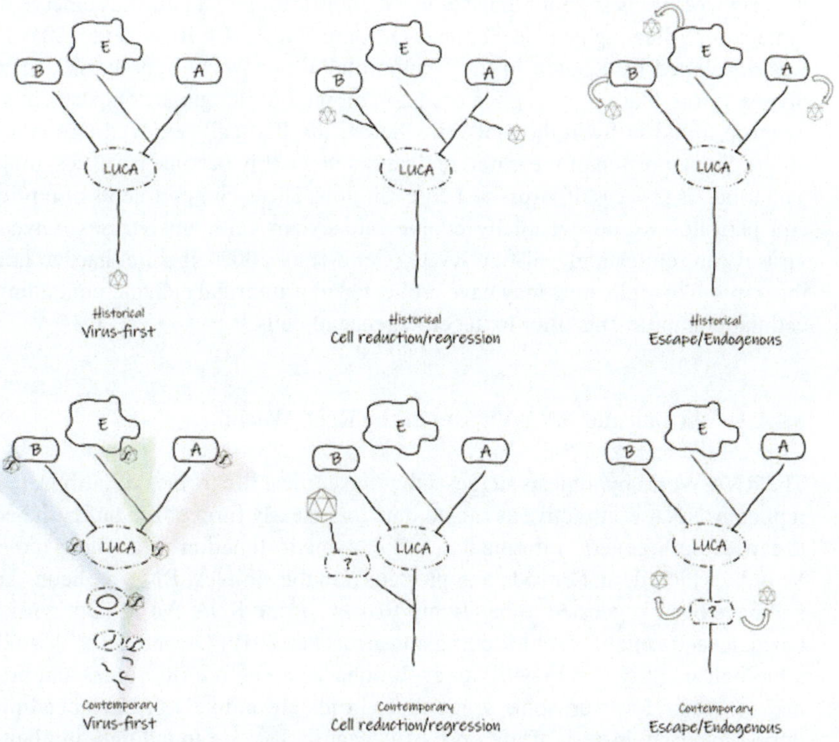

Fig. 8.3 Schemas of the various virus origin scenarios: the Virus-first, Cell reduction/regression, and Escape/Endogenous hypotheses. Although the original, historical hypotheses have largely been superseded by contemporary interpretations, they are displayed here for comparison. Details are described in the Sect. 8.4. *LUCA* last universal cellular ancestor; *B* Bacteria; *E* Eukarya; *A* Archaea

8.4.1 Virus-First Hypothesis

The idea that viruses are older than cellular life forms is as old as virology itself, with some of the foundational thinkers in virology and origin of life studies suggesting that viruses were the first forms of life on Earth (Haldane 1929; Falk and Lazcano 2012). This "Virst-first hypothesis" asserts that viruses emerged before cellular life forms evolved. This idea had been disputed because viruses are obligate intracellular parasites of cells—it would be impossible for bona fide viruses to precede cells in evolution. Another potential problem is that most virus capsid proteins seem to originate from cellular proteins (Krupovic et al. 2019; Krupovic and Koonin 2017). If one were to utilize a virus concept that strictly limits viruses to capsid-encoding genetic parasites, the virus-first hypothesis would have to be rejected. However, this is avoided with a more expansive virus concept by including related virus-like genetic replicators (Koonin et al. 2006). The appearance of parasitic replicators at the origin of life is supported by studies demonstrating that parasites emerge in any system of replicating genetic elements (Koonin et al. 2017; Iranzo et al. 2016). This was developed by Koonin in the "Primordial Virus World hypothesis", in which diverse, virus-like selfish replicators first emerged in inorganic compartments such as those found in hydrothermal vents before the first cells evolved (Koonin et al. 2006). Some portion of the initial replicators inevitably become parasites, initiating a continuous process of virus-host co-evolution. These compartments of replicators and parasites would eventually evolve into extant cells and viruses. Given that viruses can replicate in cell-free lysate (Cello et al. 2002), it is not hard to imagine that virus-like replicators may have replicated in primordial enzyme-containing pre-cellular compartments prior to the emergence of cells.

8.4.1.1 Viroids and RNA Viruses in the RNA World

The RNA World hypothesis suggests that the earliest life forms on Earth were RNA replicons. RNA is attractive as a candidate for an early form of life on Earth because it carries both genetic information and enzymatic function in a single molecule. Viroids, which do not encode any proteins, are the simplest RNA replicons known and have been suggested to be "living fossils" of the RNA World (Durzyńska and Goździcka-Józefiak 2015; Moelling and Broecker 2019; Diener 2016). Viroids are a biological entity that may fill an evolutionary gap before ribosomes and proteins had evolved. However, some argue that viroids are unlikely to have been present during the early history of life on Earth because they are found only in plants and not in other domains of life, suggesting that they may be a more recent evolutionary development (Diener 2016). Regardless, viroids and RNA viruses in general serve as models of replicators in the RNA World (Fisher 2010). In fact, the viral quasispecies model, which has been utilized extensively to study RNA viruses, was originally developed to model the behavior of autonomous prebiotic replicons on early Earth

(Eigen 1971). The high error-rates of RNA virus and viroid replication make them fit to study as quasispecies (Holmes 2011; Wilke 2005).

8.4.2 Cell Reduction or Regression Hypothesis

The regression hypothesis asserts that viruses evolved from parasitic cells that underwent reductive evolution into viruses, akin to the reductive evolution of endosymbionts like the mitochondria. Robert Green and Sir Patrick Laidlaw had first proposed this idea in the 1930s (Kostyrka 2016; Podolsky 1996). However, criticisms of this hypothesis focus on the lack of any plausible mechanism by which a cell would regress into a virion form. The discovery of giant viruses reignited contemporary discourse regarding this hypothesis. The virions of giant viruses, upon entry into the cell, form "virion factories" within the cytoplasm, reminiscent of endosymbiotic organelles. The genomes of giant viruses are as large as some prokaryotes and code for thousands of genes, including many of those involved in translation. The recently discovered Tupanvirus was found to encode a near-complete set of translation proteins with the exception of ribosomes (Abrahão et al. 2018). Some have claimed that the giant viruses represent the remnants of a fourth domain (later reassessed "Fourth TRUC—Things Resisting Uncompleted Classification") that had existed prior to LUCA (Colson et al. 2012; Nasir et al. 2012; Aherfi et al. 2016). However, this had been refuted by phylogenetic analyses suggesting that giant viruses had evolved from smaller viruses (Moreira and López-García 2015; Yutin et al. 2014). Instead, giant virus lineages had recruited genes from their hosts, expanding their genomes.

8.4.3 Escape or Endogenous Hypothesis

The escape hypothesis asserts that viruses emerged from cellular genes that became selfish and evolved into viruses capable of escaping the cell. This idea had first been proposed in the 1920s and had become widely accepted among virologists (Lwoff 1966; Forterre and Krupovic 2012). The strong specificity between viruses and their hosts seem to support this idea—there do not seem to be any viruses that can infect hosts across domains (Nasir et al. 2014). However, the escape hypothesis does not explain the apparent homologies between viruses infecting different domains of life (Koonin 2006a; Abrescia et al. 2012; Bamford 2003). To account for this, a modified version of the escape hypothesis had been proposed, with viruses escaping pre-LUCA protocells rather than modern post-LUCA cells (Forterre and Krupovic 2012). In this modified scenario, selfish genes within the genomes of RNA World protocells became viruses by recruiting host proteins for encapsidation. The escape hypothesis was further modified by merging it with the virus-first hypothesis in which early pre-cellular replicators emerged first, symbiotically associating with the first protocells after they emerged. The intracellular replicators may then evolve into bona-fide viruses by recruiting capsid proteins from those early protocells (Krupovic et al. 2019).

8.5 Other Early Evolutionary Scenarios

Origin of the DNA World The transition from the RNA World to DNA World has been proposed to have been mediated by viruses. The innovation of reverse transcription may have been key for this transition, possibly carried out by retroviruses. Forterre had hypothesized that DNA based cells may have evolved from viral infection of RNA based cells (Forterre 2002, 2005; Durzyńska and Goździcka-Józefiak 2015). DNA replication machineries may also have been viral innovations (Forterre 2013b). However, some reject the possibility of RNA-based cells (Koonin 2006b; Koonin and Martin 2005).

Vesicular Origin of Life Vesicles have been proposed as early precursors to the first protocell (Chen and Walde 2010; Luisi et al. 1999; Tessera 2009). Recently, extracellular membrane vesicles produced by all 3 domains of life have been found to package viral genomes as well as virions (Gill et al. 2018; Gaudin et al. 2014). It is conceivable that viral genomes may have first been packaged in vesicles in a precellular world before capsids were developed. Viral vesicles may also have been formed by ancient cells predating LUCA (Forterre and Krupovic 2012; Mughal et al. 2020).

Origin of Eukaryotes Viruses and other virus-like elements have also been hypothesized to have driven the evolution of eukaryotes. The eukaryotic nucleus was hypothesized to have derived from a large dsDNA virus (Bell 2009). The viral factories created by giant viruses have been compared to eukaryotic nuclei (Forterre 2017). Furthermore, jumbo phages (unusually large-sized bacteriophage with genomes above 200 kbp) have recently been found to form structures similar to nuclei upon infection of a host cell for protection from host defense mechanisms (Mendoza et al. 2020). In another hypothesis, invasions of introns into proto-eukaryotic cells may also have pushed the development of the nucleus as well as the spliceosome and telomerase (Koonin 2006c).

8.6 Conclusion

Ever since their discovery over a century ago, viruses have been hypothesized to play a role in the origin of life. The study of virology and the origin of life have had a remarkable history with both fields sharing and exchanging ideas such as the quasispecies model and virus-first hypothesis. The seemingly contradictory qualities of viruses as being very simple as well as exquisitely complex in their diversity have made viruses very attractive as models of early life on Earth. They have been utilized to fill numerous gaps and transitions in the origin and evolution of life such as the transition from the RNA World to DNA World and the origin of eukaryotes. Because viruses have always been present and co-evolving with their hosts, it is not possible to know the origin and evolution of life on earth without asking how viruses have shaped its course. Given the resilience of these ideas, it is likely that future thinkers and scientists will continue to ponder the role of viruses in the origin of life. While some questions of life, viruses, and origins may lean towards philosophical inquiry, they are still a useful exercise to help guide thinking and expand the conceptions of what life may be.

Acknowledgments The author would like to thank the reviewers for the time it took to check content and provide helpful suggestions for this chapter.

References

Abedon, S. T., Murray KL Archaeal viruses, not archaeal phages: an archaeological dig. In: Archaea. https://www.hindawi.com/journals/archaea/2013/251245/ (2013). Accessed 24 Oct 2020

Abrahão, J., Silva, L., Silva, L.S., et al.: Tailed giant Tupanvirus possesses the most complete translational apparatus of the known virosphere. Nat. Commun. **9**, 749 (2018). https://doi.org/10.1038/s41467-018-03168-1

Abrescia, N.G.A., Bamford, D.H., Grimes, J.M., Stuart, D.I.: Structure unifies the viral universe. Annu. Rev. Biochem. **81**, 795–822 (2012). https://doi.org/10.1146/annurev-biochem-060910-095130

Aherfi, S., Colson, P., La Scola, B., Raoult, D.: Giant viruses of amoebas: an update. Front. Microbiol. **7** (2016). https://doi.org/10.3389/fmicb.2016.00349

Al-Shayeb, B., Sachdeva, R., Chen, L.-X., et al.: Clades of huge phages from across Earth's ecosystems. Nature. **578**, 425–431 (2020). https://doi.org/10.1038/s41586-020-2007-4

Baltimore, D.: Expression of animal virus genomes. Bacteriol. Rev. **35**, 235–241 (1971)

Bamford, D.H.: Do viruses form lineages across different domains of life? Res. Microbiol. **154**, 231–236 (2003). https://doi.org/10.1016/S0923-2508(03)00065-2

Beijerinck, M.W.: Ueber ein Contagium vivum fluidum als Ursache der Fleckenkrankheit der Tabaksblatter. Verh Kon Akad Wetensch. **5**, 3–21 (1898)

Bell, P.J.L.: The viral eukaryogenesis hypothesis: a key role for viruses in the emergence of eukaryotes from a prokaryotic world environment. Ann. N. Y. Acad. Sci. **1178**, 91–105 (2009). https://doi.org/10.1111/j.1749-6632.2009.04994.x

Benner, S.A.: Defining Life. Astrobiology. **10**, 1021–1030 (2010). https://doi.org/10.1089/ast.2010.0524

Bergh, Ø., BØrsheim, K.Y., Bratbak, G., Heldal, M.: High abundance of viruses found in aquatic environments. Nature. **340**, 467–468 (1989). https://doi.org/10.1038/340467a0

Berliner, A.J., Mochizuki, T., Stedman, K.M.: Astrovirology: viruses at large in the universe. Astrobiology. **18**, 207–223 (2018). https://doi.org/10.1089/ast.2017.1649

Bobay, L.-M., Touchon, M., Rocha, E.P.C.: Pervasive domestication of defective prophages by bacteria. PNAS. **111**, 12127–12132 (2014). https://doi.org/10.1073/pnas.1405336111

Boyer, M., Madoui, M.-A., Gimenez, G., et al.: Phylogenetic and phyletic studies of informational genes in genomes highlight existence of a 4th domain of life including giant viruses. PLoS One. **5**, e15530 (2010). https://doi.org/10.1371/journal.pone.0015530

Briani, F., Dehò, G., Forti, F., Ghisotti, D.: The plasmid status of satellite bacteriophage P4. Plasmid. **45**, 1–17 (2001). https://doi.org/10.1006/plas.2000.1497

Broecker, F., Moelling, K.: Evolution of immune systems from viruses and transposable elements. Front. Microbiol. **10** (2019). https://doi.org/10.3389/fmicb.2019.00051

Budnik, V., Thomson, T.: Structure of an Arc-ane virus-like capsid. Nat. Neurosci. **23**, 153–154 (2020). https://doi.org/10.1038/s41593-019-0580-3

Cello, J., Paul, A.V., Wimmer, E.: Chemical synthesis of poliovirus cDNA: generation of infectious virus in the absence of natural template. Science. **297**, 1016–1018 (2002). https://doi.org/10.1126/science.1072266

Chen, I.A., Walde, P.: From self-assembled vesicles to protocells. Cold Spring Harb. Perspect. Biol., 2 (2010). https://doi.org/10.1101/cshperspect.a002170

Choi, K.H.: Viral polymerases. In: Rossmann, M.G., Rao, V.B. (eds.) Viral Molecular Machines, pp. 267–304. Springer, Boston, MA (2012)

Cifuentes-Muñoz, N., Dutch, R.E., Cattaneo, R.: Direct cell-to-cell transmission of respiratory viruses: the fast lanes. PLoS Pathog. **14** (2018). https://doi.org/10.1371/journal.ppat.1007015

Colson, P., de Lamballerie, X., Fournous, G., Raoult, D.: Reclassification of giant viruses composing a fourth domain of life in the new order megavirales. INT. **55**, 321–332 (2012). https://doi.org/10.1159/000336562

Crick, F.H.: On protein synthesis. Symp. Soc. Exp. Biol. **12**, 138–163 (1958)

Díaz-Muñoz, S.L., Sanjuán, R., West, S.: Sociovirology: conflict, cooperation, and communication among viruses. Cell Host Microbe. **22**, 437–441 (2017). https://doi.org/10.1016/j.chom.2017.09.012

Diener, T.O.: Viroids: "living fossils" of primordial RNAs? Biol. Direct. **11**, 15 (2016). https://doi.org/10.1186/s13062-016-0116-7

Domingo-Calap, P., Mora-Quilis, L., Sanjuán, R.: Social bacteriophages. Microorganisms. **8**, 533 (2020). https://doi.org/10.3390/microorganisms8040533

Duckworth, D.H.: Who discovered bacteriophage? Bacteriol. Rev. **40**, 793–802 (1976)

Dupont, M., Sattentau, Q.J.: Macrophage cell-cell interactions promoting HIV-1 infection. Viruses. **12** (2020). https://doi.org/10.3390/v12050492

Dupré, J., Guttinger, S.: Viruses as living processes. Stud. Hist. Philos. Biol. Biomed. Sci. **59**, 109–116 (2016). https://doi.org/10.1016/j.shpsc.2016.02.010

Durzyńska, J., Goździcka-Józefiak, A.: Viruses and cells intertwined since the dawn of evolution. Virol. J. **12**, 169 (2015). https://doi.org/10.1186/s12985-015-0400-7

Eigen, M.: Selforganization of matter and the evolution of biological macromolecules. Naturwissenschaften. **58**, 465–523 (1971). https://doi.org/10.1007/BF00623322

Falk, R., Lazcano, A.: The forgotten dispute: A.I. Oparin and H.J. Muller on the origin of life. Hist. Philos. Life Sci. **34**, 373–390 (2012)

Fillol-Salom, A., Martínez-Rubio, R., Abdulrahman, R.F., et al.: Phage-inducible chromosomal islands are ubiquitous within the bacterial universe. ISME J. **12**, 2114–2128 (2018). https://doi.org/10.1038/s41396-018-0156-3

Fisher, S.: Are RNA viruses vestiges of an RNA world? J. Gen. Philos. Sci. **41**, 121–141 (2010). https://doi.org/10.1007/s10838-010-9119-8

Flores, R., Grubb, D., Elleuch, A., et al.: Rolling-circle replication of viroids, viroid-like satellite RNAs and hepatitis delta virus: variations on a theme. RNA Biol. **8**, 200–206 (2011). https://doi.org/10.4161/rna.8.2.14238

Forterre, P.: The origin of DNA genomes and DNA replication proteins. Curr. Opin. Microbiol. **5**, 525–532 (2002). https://doi.org/10.1016/S1369-5274(02)00360-0

Forterre, P.: The two ages of the RNA world, and the transition to the DNA world: a story of viruses and cells. Biochimie. **87**, 793–803 (2005). https://doi.org/10.1016/j.biochi.2005.03.015

Forterre, P.: Defining life: the virus viewpoint. Orig. Life Evol. Biosph. **40**, 151–160 (2010). https://doi.org/10.1007/s11084-010-9194-1

Forterre, P.: The virocell concept and environmental microbiology. ISME J. **7**, 233–236 (2013a). https://doi.org/10.1038/ismej.2012.110

Forterre, P.: Why are there so many diverse replication machineries? J. Mol. Biol. **425**, 4714–4726 (2013b). https://doi.org/10.1016/j.jmb.2013.09.032

Forterre, P.: To be or not to be alive: how recent discoveries challenge the traditional definitions of viruses and life. Stud. Hist. Philos. Biol. Biomed. Sci. **59**, 100–108 (2016). https://doi.org/10.1016/j.shpsc.2016.02.013

Forterre, P.: Viruses in the 21st century: from the curiosity-driven discovery of Giant viruses to new concepts and definition of life. Clin. Infect. Dis. **65**, S74–S79 (2017). https://doi.org/10.1093/cid/cix349

Forterre, P., Krupovic, M.: The origin of virions and virocells: the escape hypothesis revisited. In: Witzany, G. (ed.) Viruses: Essential Agents of Life, pp. 43–60. Springer Netherlands, Dordrecht (2012)

Forterre, P., Prangishvili, D.: The great billion-year war between ribosome- and capsid-encoding organisms (cells and viruses) as the major source of evolutionary novelties. Ann. N. Y. Acad. Sci. **1178**, 65–77 (2009). https://doi.org/10.1111/j.1749-6632.2009.04993.x

Gago-Zachert, S.: Viroids, infectious long non-coding RNAs with autonomous replication. Virus Res. **212**, 12–24 (2016). https://doi.org/10.1016/j.virusres.2015.08.018

Gaudin, M., Krupovic, M., Marguet, E., et al.: Extracellular membrane vesicles harbouring viral genomes. Environ. Microbiol. **16**, 1167–1175 (2014). https://doi.org/10.1111/1462-2920.12235

Gill, S., Catchpole, R., Forterre, P.: Extracellular membrane vesicles in the three domains of life and beyond. FEMS Microbiol. Rev. **43**, 273–303 (2018). https://doi.org/10.1093/femsre/fuy042

Griffin, D.W.: The quest for extraterrestrial life: what about the viruses? Astrobiology. **13**, 774–783 (2013). https://doi.org/10.1089/ast.2012.0959

Guglielmini, J., Woo, A.C., Krupovic, M., et al.: Diversification of giant and large eukaryotic dsDNA viruses predated the origin of modern eukaryotes. Proc. Natl. Acad. Sci. USA. **116**, 19585–19592 (2019). https://doi.org/10.1073/pnas.1912006116

Hadidi, A.: Relationship of viroids and certain other plant pathogenic nucleic acids to group I and II introns. Plant Mol. Biol. **7**, 129–142 (1986). https://doi.org/10.1007/BF00040139

Haldane, J.B.: The origin of life. Ration. Annu. **148** (1929)

Harshey, R.M.: The Mu story: how a maverick phage moved the field forward. Mob. DNA. **3**, 21 (2012). https://doi.org/10.1186/1759-8753-3-21

Hillman, B.I., Cai, G.: Chapter 6 - The Family Narnaviridae: simplest of RNA viruses. In: Ghabrial, S.A. (ed.) Advances in Virus Research, pp. 149–176. Academic Press, New York (2013)

Holmes, E.C.: What does virus evolution tell us about virus origins? J. Virol. **85**, 5247–5251 (2011). https://doi.org/10.1128/JVI.02203-10

Iranzo, J., Puigbò, P., Lobkovsky, A.E., et al.: Inevitability of genetic parasites. Genome Biol. Evol. **8**, 2856–2869 (2016). https://doi.org/10.1093/gbe/evw193

Janjic, A.: The need for including virus detection methods in future Mars missions. Astrobiology. **18**, 1611–1614 (2018). https://doi.org/10.1089/ast.2018.1851

Koonin, E.V.: Temporal order of evolution of DNA replication systems inferred by comparison of cellular and viral DNA polymerases. Biol. Direct. **1**, 39 (2006a). https://doi.org/10.1186/1745-6150-1-39

Koonin, E.V.: On the origin of cells and viruses: a comparative-genomic perspective. Israel J. Ecol. Evol. **52**, 299–318 (2006b). https://doi.org/10.1560/IJEE_52_3-4_299

Koonin, E.V.: The origin of introns and their role in eukaryogenesis: a compromise solution to the introns-early versus introns-late debate? Biol. Direct. **1**, 22 (2006c). https://doi.org/10.1186/1745-6150-1-22

Koonin, E.V.: Viruses and mobile elements as drivers of evolutionary transitions. Philos. Trans. R. Soc. B Biol. Sci. **371**, 20150442 (2016). https://doi.org/10.1098/rstb.2015.0442

Koonin, E.V., Dolja, V.V.: A virocentric perspective on the evolution of life. Curr. Opin. Virol. **3**, 546–557 (2013). https://doi.org/10.1016/j.coviro.2013.06.008

Koonin, E.V., Dolja, V.V.: Virus world as an evolutionary network of viruses and Capsidless selfish elements. Microbiol. Mol. Biol. Rev. **78**, 278–303 (2014). https://doi.org/10.1128/MMBR.00049-13

Koonin, E.V., Krupovic, M.: Polintons, virophages and transpovirons: a tangled web linking viruses, transposons and immunity. Curr. Opin. Virol. **25**, 7–15 (2017). https://doi.org/10.1016/j.coviro.2017.06.008

Koonin, E.V., Krupovic, M.: The depths of virus exaptation. Curr. Opin. Virol. **31**, 1–8 (2018). https://doi.org/10.1016/j.coviro.2018.07.011

Koonin, E.V., Martin, W.: On the origin of genomes and cells within inorganic compartments. Trends Genet. **21**, 647–654 (2005). https://doi.org/10.1016/j.tig.2005.09.006

Koonin, E.V., Starokadomskyy, P.: Are viruses alive? The replicator paradigm sheds decisive light on an old but misguided question. Stud. Hist. Philos. Biol. Biomed. Sci. **59**, 125–134 (2016). https://doi.org/10.1016/j.shpsc.2016.02.016

Koonin, E.V., Senkevich, T.G., Dolja, V.V.: The ancient virus world and evolution of cells. Biol. Direct. **1**, 29 (2006). https://doi.org/10.1186/1745-6150-1-29

Koonin, E.V., Dolja, V.V., Krupovic, M.: Origins and evolution of viruses of eukaryotes: the ultimate modularity. Virology. **479–480**, 2–25 (2015). https://doi.org/10.1016/j.virol.2015.02.039

Koonin, E.V., Wolf, Y.I., Katsnelson, M.I.: Inevitability of the emergence and persistence of genetic parasites caused by evolutionary instability of parasite-free states. Biol. Direct, 12 (2017). https://doi.org/10.1186/s13062-017-0202-5

Koonin, E.V., Dolja, V.V., Krupovic, M., et al.: Global organization and proposed megataxonomy of the virus world. Microbiol. Mol. Biol. Rev., 84 (2020). https://doi.org/10.1128/MMBR.00061-19

Kostyrka, G.: What roles for viruses in origin of life scenarios? Stud. Hist. Philos. Biol. Biomed. Sci. **59**, 135–144 (2016). https://doi.org/10.1016/j.shpsc.2016.02.014

Krupovič, M., Bamford, D.H.: Virus evolution: how far does the double β-barrel viral lineage extend? Nat. Rev. Microbiol. **6**, 941–948 (2008). https://doi.org/10.1038/nrmicro2033

Krupovic, M., Bamford, D.H.: Double-stranded DNA viruses: 20 families and only five different architectural principles for virion assembly. Curr. Opin. Virol. **1**, 118–124 (2011). https://doi.org/10.1016/j.coviro.2011.06.001

Krupovic, M., Koonin, E.V.: Polintons: a hotbed of eukaryotic virus, transposon and plasmid evolution. Nat. Rev. Microbiol. **13**, 105–115 (2015). https://doi.org/10.1038/nrmicro3389

Krupovic, M., Koonin, E.V.: Self-synthesizing transposons: unexpected key players in the evolution of viruses and defense systems. Curr. Opin. Microbiol. **31**, 25–33 (2016). https://doi.org/10.1016/j.mib.2016.01.006

Krupovic, M., Koonin, E.V.: Multiple origins of viral capsid proteins from cellular ancestors. PNAS. **114**, E2401–E2410 (2017). https://doi.org/10.1073/pnas.1621061114

Krupovic, M., Prangishvili, D., Hendrix, R.W., Bamford, D.H.: Genomics of bacterial and archaeal viruses: dynamics within the prokaryotic virosphere. Microbiol. Mol. Biol. Rev. **75**, 610–635 (2011). https://doi.org/10.1128/MMBR.00011-11

Krupovic, M., Dolja, V.V., Koonin, E.V.: Origin of viruses: primordial replicators recruiting capsids from hosts. Nat. Rev. Microbiol. **17**, 449–458 (2019). https://doi.org/10.1038/s41579-019-0205-6

Krupovic, M., Dolja, V.V., Koonin, E.V.: The LUCA and its complex virome. Nat. Rev. Microbiol. **18**, 661–670 (2020). https://doi.org/10.1038/s41579-020-0408-x

La Scola, B., Audic, S., Robert, C., et al.: A giant virus in amoebae. Science. **299**, 2033 (2003). https://doi.org/10.1126/science.1081867

Lagos, R., Goldstein, R.: Evolutionary relationship between plasmids and phages: phasmid P4 as a model. Arch. Biol. Med. Exp. **20**, 325–332 (1987)

Lazcano, A.: Alexandr I. Oparin and the origin of life: a historical reassessment of the heterotrophic theory. J. Mol. Evol. **83**, 214–222 (2016). https://doi.org/10.1007/s00239-016-9773-5

Legendre, M., Arslan, D., Abergel, C., Claverie, J.-M.: Genomics of megavirus and the elusive fourth domain of life. Commun. Integr. Biol. **5**, 102–106 (2012)

Legendre, M., Bartoli, J., Shmakova, L., et al.: Thirty-thousand-year-old distant relative of giant icosahedral DNA viruses with a pandoravirus morphology. PNAS. **111**, 4274–4279 (2014). https://doi.org/10.1073/pnas.1320670111

Lewin, R.: Viroids may be escaped introns. Science. **233**, 1384 (1986). https://doi.org/10.1126/science.3749883

López-García, P.: The place of viruses in biology in light of the metabolism-versus-replication-first debate. Hist. Philos. Life Sci. **34**, 391–406 (2012)

Lossouarn, J., Dupont, S., Gorlas, A., et al.: An abyssal mobilome: viruses, plasmids and vesicles from deep-sea hydrothermal vents. Res. Microbiol. **166**, 742–752 (2015). https://doi.org/10.1016/j.resmic.2015.04.001

Luisi, P.L., Walde, P., Oberholzer, T.: Lipid vesicles as possible intermediates in the origin of life. Curr. Opin. Colloid Interface Sci. **4**, 33–39 (1999). https://doi.org/10.1016/S1359-0294(99)00012-6

Lwoff, A.: The concept of virus. J. Gen. Microbiol. **17**, 239–253 (1957). https://doi.org/10.1099/00221287-17-2-239

Lwoff, A.: Interaction among virus, cell, and organism. Science. **152**, 1216–1220 (1966). https://doi.org/10.1126/science.152.3726.1216

Mendoza, S.D., Nieweglowska, E.S., Govindarajan, S., et al.: A bacteriophage nucleus-like compartment shields DNA from CRISPR nucleases. Nature. **577**, 244–248 (2020). https://doi.org/10.1038/s41586-019-1786-y

Moelling, K., Broecker, F.: Viroids-first—a model for life on Earth, Mars and exoplanets. Geosciences. **9**, 241 (2019). https://doi.org/10.3390/geosciences9050241

Moreira, D., López-García, P.: Evolution of viruses and cells: do we need a fourth domain of life to explain the origin of eukaryotes? Philos. Trans. R. Soc. B Biol. Sci. **370**, 20140327 (2015). https://doi.org/10.1098/rstb.2014.0327

Mothes, W., Sherer, N.M., Jin, J., Zhong, P.: Virus cell-to-cell transmission. J. Virol. **84**, 8360–8368 (2010). https://doi.org/10.1128/JVI.00443-10

Mughal, F., Nasir, A., Caetano-Anollés, G.: The origin and evolution of viruses inferred from fold family structure. Arch. Virol. (2020). https://doi.org/10.1007/s00705-020-04724-1

Munson-McGee, J.H., Peng, S., Dewerff, S., et al.: A virus or more in (nearly) every cell: ubiquitous networks of virus–host interactions in extreme environments. ISME J. **12**, 1706–1714 (2018). https://doi.org/10.1038/s41396-018-0071-7

Nasir, A., Kim, K.M., Caetano-Anolles, G.: Giant viruses coexisted with the cellular ancestors and represent a distinct supergroup along with superkingdoms archaea, Bacteria and Eukarya. BMC Evol. Biol. **12**, 156 (2012). https://doi.org/10.1186/1471-2148-12-156

Nasir, A., Forterre, P., Kim, K.M., Caetano-Anollés, G.: The distribution and impact of viral lineages in domains of life. Front. Microbiol. **5** (2014). https://doi.org/10.3389/fmicb.2014.00194

Ortmann, A.C., Suttle, C.A.: High abundances of viruses in a deep-sea hydrothermal vent system indicates viral mediated microbial mortality. Deep Sea Res. I Oceanogr. Res. Pap. **52**, 1515–1527 (2005). https://doi.org/10.1016/j.dsr.2005.04.002

Patel, M.R., Emerman, M., Malik, H.S.: Paleovirology – ghosts and gifts of viruses past. Curr. Opin. Virol. **1**, 304–309 (2011). https://doi.org/10.1016/j.coviro.2011.06.007

Pearson, H.: "Virophage" suggests viruses are alive. Nature. **454**, 677–677 (2008). https://doi.org/10.1038/454677a

Pennazio, S., Roggero, P.: The discovery of the chemical nature of tobacco mosaic virus. Riv. Biol. **93**, 253–282 (2000)

Pepper, J.W., Herron, M.D.: Does biology need an organism concept? Biol. Rev. Camb. Philos. Soc. **83**, 621–627 (2008). https://doi.org/10.1111/j.1469-185X.2008.00057.x

Philippe, N., Legendre, M., Doutre, G., et al.: Pandoraviruses: amoeba viruses with genomes up to 2.5 Mb reaching that of parasitic eukaryotes. Science. **341**, 281–286 (2013). https://doi.org/10.1126/science.1239181

Podolsky, S.: The role of the virus in origin-of-life theorizing. J. Hist. Biol. **29**, 79–126 (1996). https://doi.org/10.1007/BF00129697

Porcar, M., Peretó, J.: Creating life and the media: translations and echoes. Life Sci. Soc. Policy. **14**, 19 (2018). https://doi.org/10.1186/s40504-018-0087-9

Pradeu, T., Kostyrka, G., Dupré, J.: Understanding viruses: philosophical investigations. Stud. Hist. Philos. Biol. Biomed. Sci. **59**, 57–63 (2016). https://doi.org/10.1016/j.shpsc.2016.02.008

Prangishvili, D., Bamford, D.H., Forterre, P., et al.: The enigmatic archaeal virosphere. Nat. Rev. Microbiol. **15**, 724–739 (2017). https://doi.org/10.1038/nrmicro.2017.125

Ramisetty, B.C.M., Sudhakari, P.A.: Bacterial 'grounded' prophages: hotspots for genetic renovation and innovation. Front. Genet. **10** (2019). https://doi.org/10.3389/fgene.2019.00065

Raoult, D.: TRUC or the need for a new microbial classification. INT. **56**, 349–353 (2013). https://doi.org/10.1159/000354269

Raoult, D., Forterre, P.: Redefining viruses: lessons from Mimivirus. Nat. Rev. Microbiol. **6**, 315–319 (2008). https://doi.org/10.1038/nrmicro1858

Redfield, R.J., Soucy, S.M.: Evolution of bacterial gene transfer agents. Front. Microbiol. **9** (2018). https://doi.org/10.3389/fmicb.2018.02527

Rohwer, F., Barott, K.: Viral information. Biol. Philos. **28**, 283–297 (2013). https://doi.org/10.1007/s10539-012-9344-0

Roossinck, M.J., Bazán, E.R.: Symbiosis: viruses as intimate partners. Annu. Rev. Virol. **4**, 123–139 (2017). https://doi.org/10.1146/annurev-virology-110615-042323

Rosenwasser, S., Ziv, C., van Creveld, S.G., Vardi, A.: Virocell metabolism: metabolic innovations during host–virus interactions in the ocean. Trends Microbiol. **24**, 821–832 (2016). https://doi.org/10.1016/j.tim.2016.06.006

Rybchin, V.N., Svarchevsky, A.N.: The plasmid prophage N15: a linear DNA with covalently closed ends. Mol. Microbiol. **33**, 895–903 (1999). https://doi.org/10.1046/j.1365-2958.1999.01533.x

Schoenfeld, T., Patterson, M., Richardson, P.M., et al.: Assembly of viral metagenomes from Yellowstone Hot Springs. Appl. Environ. Microbiol. **74**, 4164–4174 (2008). https://doi.org/10.1128/AEM.02598-07

Schvarcz, C.R., Steward, G.F.: A giant virus infecting green algae encodes key fermentation genes. Virology. **518**, 423–433 (2018). https://doi.org/10.1016/j.virol.2018.03.010

Secor, P.R., Dandekar, A.A.: More than simple parasites: the sociobiology of bacteriophages and their bacterial hosts. mBio. (2020). https://doi.org/10.1128/mBio.00041-20

Soler, N., Krupovic, M., Marguet, E., Forterre, P.: Membrane vesicles in natural environments: a major challenge in viral ecology. ISME J. **9**, 793–796 (2015). https://doi.org/10.1038/ismej.2014.184

Suzuki, N., Ghabrial, S.A., Kim, K.-H., et al.: ICTV virus taxonomy profile: Hypoviridae. J. Gen. Virol. **99**, 615–616 (2018). https://doi.org/10.1099/jgv.0.001055

Tessera, M.: Life began when evolution began: a lipidic vesicle-based scenario. Orig. Life Evol. Biosph. **39**, 559 (2009). https://doi.org/10.1007/s11084-009-9175-4

Thompson, L.R., Zeng, Q., Kelly, L., et al.: Phage auxiliary metabolic genes and the redirection of cyanobacterial host carbon metabolism. PNAS. **108**, E757–E764 (2011). https://doi.org/10.1073/pnas.1102164108

Tirard, S.: J. B. S. Haldane and the origin of life. J. Genet. **96**, 735–739 (2017). https://doi.org/10.1007/s12041-017-0831-6

Toussaint, A., Merlin, C.: Mobile elements as a combination of functional modules. Plasmid. **47**, 26–35 (2002). https://doi.org/10.1006/plas.2001.1552

Toussaint, A., Rice, P.A.: Transposable phages, DNA reorganization and transfer. Curr. Opin. Microbiol. **38**, 88–94 (2017). https://doi.org/10.1016/j.mib.2017.04.009

Van Etten, J.L., Meints, R.H.: Giant viruses infecting algae. Annu. Rev. Microbiol. **53**, 447–494 (1999). https://doi.org/10.1146/annurev.micro.53.1.447

Van Etten, J.L., Lane, L.C., Meints, R.H.: Viruses and viruslike particles of eukaryotic algae. Microbiol. Mol. Biol. Rev. **55**, 586–620 (1991)

Vaughan, V.C.: A chemical concept of the origin and development of life. A preliminary presentation. Chem. Rev. **4**, 167–188 (1927). https://doi.org/10.1021/cr60014a001

Wilke, C.O.: Quasispecies theory in the context of population genetics. BMC Evol. Biol. **5**, 44 (2005). https://doi.org/10.1186/1471-2148-5-44

Yutin, N., Wolf, Y.I., Koonin, E.V.: Origin of giant viruses from smaller DNA viruses not from a fourth domain of cellular life. Virology. **466–467**, 38–52 (2014). https://doi.org/10.1016/j.virol.2014.06.032

Zhong, P., Agosto, L.M., Munro, J.B., Mothes, W.: Cell-to-cell transmission of viruses. Curr. Opin. Virol. **3**, 44–50 (2013). https://doi.org/10.1016/j.coviro.2012.11.004

Zimmerly, S., Semper, C.: Evolution of group II introns. Mob. DNA. **6**, 7 (2015). https://doi.org/10.1186/s13100-015-0037-5

Chapter 9
Reconstructing the Last Universal Common Ancestor

Anthony M. Poole

Abstract There is general agreement that bacteria, archaea, and eukarya share common ancestry. However, tracing back extant lineages to reconstruct the ancestral gene set of the three domains has proven to be non-trivial, as there is little unambiguous signal this far back in time. In this chapter, I explain the basic principles behind reconstruction of the Last Universal Common Ancestor (LUCA) and summarise a few of the challenges associated with reconstruction. Finally, I consider whether a mid-resolution LUCA might be the most achievable goal, particularly from the perspective of the classes of chemistry available to early life.

9.1 Introduction

The concept of a Last Universal Common Ancestor (LUCA) has been central to efforts to reconstruct the early evolution of life (Penny and Poole 1999; Koonin 2003; Glansdorff et al. 2008; Weiss et al. 2016; Kyrpides et al. 1999). LUCA is commonly assumed to be a specific cellular ancestor at the root of the tree of life, but it is more correct to consider it the ancestral state at the root of the tree.

From an evolutionary point of view, the basic principles behind reconstruction of ancestral states, and therefore LUCA, are straightforward in principle, if not in practice. These largely derive from phylogenetic methods for reconstruction of evolutionary trees and the application of cladistics to assess antiquity. Together, these can be used to frame the problem of reconstructing LUCA as follows: given a set of extant species, related in a known way on a rooted phylogenetic tree, it is possible to apply a simple model of evolution to reconstruct character states in the

A. M. Poole (✉)
Bioinformatics Institute, University of Auckland, Auckland, New Zealand

School of Biological Sciences, University of Auckland, Auckland, New Zealand

Te Ao Mārama Centre for Fundamental Inquiry, University of Auckland, Auckland, New Zealand
e-mail: a.poole@auckland.ac.nz

A. Neubeck, S. McMahon (eds.), *Prebiotic Chemistry and the Origin of Life*,
Advances in Astrobiology and Biogeophysics,
https://doi.org/10.1007/978-3-030-81039-9_9

ancestor. It is essentially this general approach that researchers working on LUCA reconstruction employ. This has been a successful approach for reconstruction of the Last Eukaryotic Common Ancestor (LECA) (Koumandou et al. 2013; Neumann et al. 2010), but there are a number of complexities that make reconstruction of LUCA much more challenging. To provide some background, I will outline the basic concepts behind reconstructing ancestral states, before discussing some of the challenges with reconstructing LUCA.

The most straightforward example of using a tree to determine an ancestral state is where the character states are contained within the data used to build the phylogenetic tree. This approach, called ancestral sequence reconstruction, has been successfully employed to predict the ancestral temperatures that LUCA likely inhabited, with independent data all pointing to a non-hyperthermophilic LUCA (Galtier et al. 1999; Boussau et al. 2008). The principle works as follows: by reconstructing sequences at the internal nodes of a tree, it is possible to predict the ancestral state. An illustrative example is provided in Fig. 9.1, and shows the basic principles by which Galtier et al. (1999) and Bossau et al. (2008) have predicted the optimal growth temperature for LUCA, using rRNA and ribosomal protein sequence data. In both cases, the relationships between the three domains, Archaea, Bacteria and Eukarya, can be ascertained using these sequence data (Woese and Fox 1977; Ciccarelli et al. 2006; Forterre 2015). The assumptions here are that the inferred relationships between species on the tree are correct and that one can accurately reconstruct the ancestral state from the available data.

A final detail that is significant here is that the location of the root of the tree cannot be ascertained using these particular genes. Instead, researchers have used ancestral gene duplication events that occurred prior to the divergence of the three

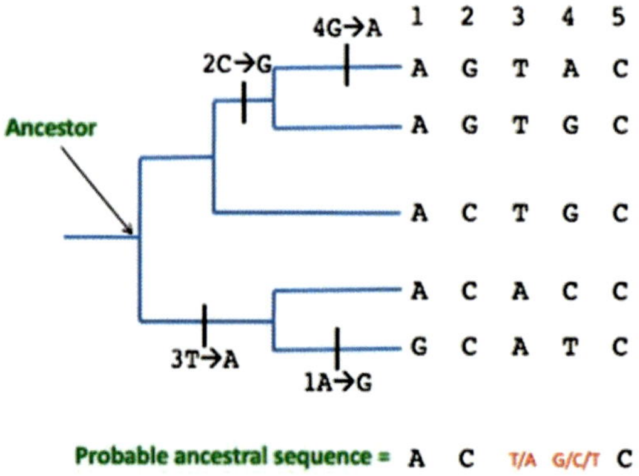

Fig. 9.1 The basic premise of ancestral sequence reconstruction. In the toy example shown, a probable ancestral sequence is generated by parsimony (reconstruction via the fewest possible evolutionary events) given a known tree of five sequences. The tree is rooted and the ancestral state is derived from reconstructing the most probable sequence at the root. For each column in the ancestor the most parsimonious ancestral state is established. Note however that for column 3, two ancestral states are equally probable, and for column 4 three sequence states are equally probable

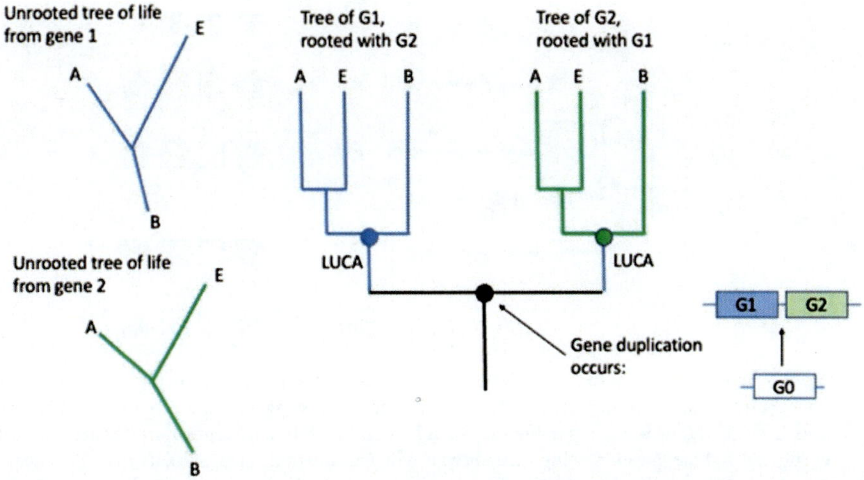

Fig. 9.2 Rooting the tree of life using an ancestral gene duplication. This figure illustrates the way in which gene duplications that predate the divergence of the three domains (Bacteria, Archaea, Eukarya) have been used to root the tree of life. Two trees at left show how building a tree from only one of the duplicate genes yields trees that do not carry sufficient information to determine the relationships between the three domains. Each single gene tree is a 'star' tree. To establish how the domains are related, it is possible to build a tree from sequences from both gene duplicates (Gene 1 and Gene 2). The gene duplicates are similar in sequence and evolved from a common ancestor (Gene 0), so a tree can be built from both genes simultaneously. The root of the overall tree is on the branch connecting the two genes (black dot) and the point at which the blue tree (from G1) connects with the green tree (from G2), and vice versa, is the root of the tree of life. In this example, the root places the archaea and eukarya together to the exclusion of bacteria

domains of life. The logic here is that a duplication which occurred prior to the divergence of the species one is interested in can be used to root the tree containing those species, since the tree of each duplicate can root the other tree (Fig. 9.2). Two independent studies (Gogarten et al. 1989; Iwabe et al. 1989) both placed the root between bacteria and the other two domains, though a range of other rootings have been proposed, including rooting the tree within the bacteria (Skophammer et al. 2007).

Assuming one correctly rooted the tree, it is possible in principle to place specific features in LUCA by using the tree to infer how those traits have evolved. There are two assumptions being made here. First, we apply some model of evolution for the traits, and our assumption is that we have identified a model that accurately captures how the traits have evolved. In Fig. 9.3, the principle of parsimony is applied, which favours the simplest explanation of the available data. (Note that, for sequence data in particular, we often use more sophisticated evolutionary models.) The second assumption is that those features share the same pattern of evolution as the reference tree.

For features that exist in all three domains, it is in principle straightforward to place features in LUCA. Where a feature is not present in all lineages, it becomes

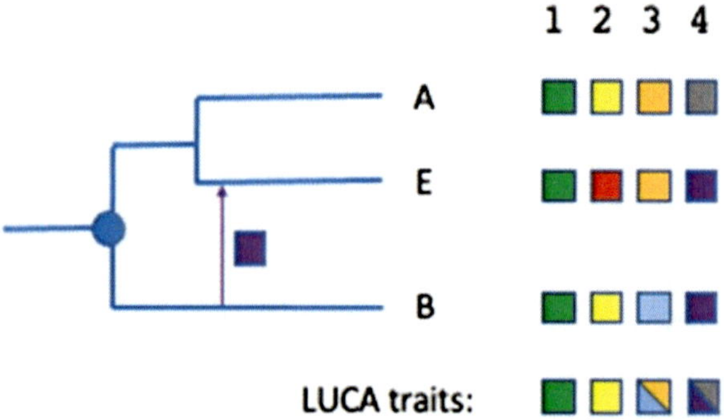

Fig. 9.3 Placement of traits in LUCA. This figure uses the same principle as the ancestral sequence reconstruction depicted in Fig. 9.1, but here the sequences are replaced by traits. The traits in each column are comparable (depicted by number) but non-identical (differences depicted by colour), and the question being asked here is which of the traits is most likely to be the ancestral trait. Again, for simplicity, a simple case using parsimony is illustrated. Trait 1 is universal (green box carried by all three domains) so, if there are no confounding factors, the green trait can be placed in LUCA. For trait 2, A and B have a yellow box, while E has red. Assuming the position of the root is correct and there are no confounding factors, then yellow is inferred to be ancestral because it is on both sides of the root and the simplest interpretation is that yellow changed to red once in the evolution of E. For trait 3, the ancestor could be either blue or orange; even though two of three lineages carry orange, only a single change is required for a blue ancestor to switch to orange, and vice versa. Finally, for trait 4, a horizontal gene transfer event has occurred (from B to E, arrow). This obscures the ancestral state of E, and, again, it is not possible to establish if purple or grey is ancestral

trickier, as one may need to factor in both the origin and the possibility of loss of a feature. Add to that the fact that there is not a consensus on the placement of the root (which can have multiple possible locations (Fig. 9.4)) (Skophammer et al. 2007; Forterre and Philippe 1999; Dagan et al. 2010), and this kind of reconstruction starts to become challenging.

The other major complication is that species and genes do not always follow the same evolutionary path. Thus, the assumption that a trait has evolved along the tree does not always hold. There are two reasons this may not be the case. The first is a loss of signal—in this case, the gene might have evolved along the same patterns as the tree, but there is insufficient signal in the gene sequences for that evolutionary history to be reliably recovered. This is hardly surprising given the enormity of the evolutionary timescales. Sequences change over time and there is no law of evolution mandating that molecules maintain a complete and accurate record of their evolutionary history!

The second challenge is that the evolutionary history of genes does not match the relationships outlined in the tree of life. The major mechanism here is horizontal gene transfer, a process by which genes move from a donor to a recipient organism without a cellular reproductive event (Fig. 9.5a). Horizontal gene transfer events enable spread of genes between species. This may in turn lead to some genes having a distribution consistent with placement in LUCA, even though they are in fact

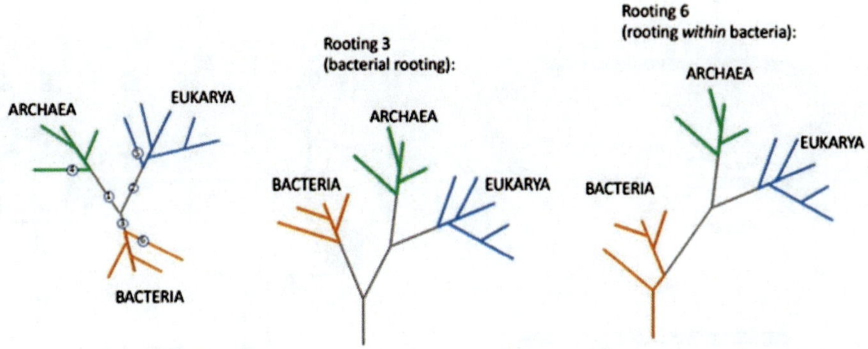

Fig. 9.4 Possible locations for the root of the tree of life. The left panel shows an unrooted 3-domain tree. Locations 1–3 show possible between-domain rootings, and numbers 4–6 depict the situation where the root falls within one of the three domains. The central panel (rooting position 3 on left panel) depicts the classical 'bacterial' rooting of the tree, which places Bacteria on one side of the root and Archaea and Eukarya on the other side. In this rooting, Archaea and Eukarya are monophyletic (i.e. they share a common ancestor). The right panel illustrates within-domain rooting, in this case rooting 6, which places the root within the diversity of modern bacteria. This has the effect of making Bacteria paraphyletic and implies that the monophyletic group encompassing Archaea and Eukarya evolved from within the Bacteria. For simplicity, this illustration does not include the 2-domain case, where Eukarya group within the diversity of modern Archaea. Under that model, the root could either be between Bacteria and Archaea or within Bacteria or within Archaea (but not within Eukarya)

younger genes that have spread via more recent horizontal gene transfer events (Fig. 9.5b) (Koonin 2003).

To try and avoid such false positive inferences, a strict criterion that has been applied is therefore that the evolutionary history of a particular gene underlying that trait, when reconstructed, shows the same evolutionary relationships as the gene or genes used to build the tree of life (Harris et al. 2003). The upshot is that, using this conservative criterion, it has only been possible to place a very small number of genes in LUCA (Harris et al. 2003); these are primarily associated with transcription and translation, and it is worth noting that the placement of pathways central to DNA synthesis and replication is uncertain (Leipe et al. 1999; Poole and Logan 2005; Forterre 1999; Lundin et al. 2010), despite the ubiquitous use of DNA as genetic material in all cellular life.

9.2 LUCA and Lost Signal: An Insurmountable Challenge?

Past and present horizontal gene transfer is a pervasive challenge for the evolutionary reconstruction of early life. The conservative view must be that this process is an insurmountable challenge for reconstruction of LUCA. The thinking goes that it may never be possible to reconstruct the biology of the deepest branch in the tree of

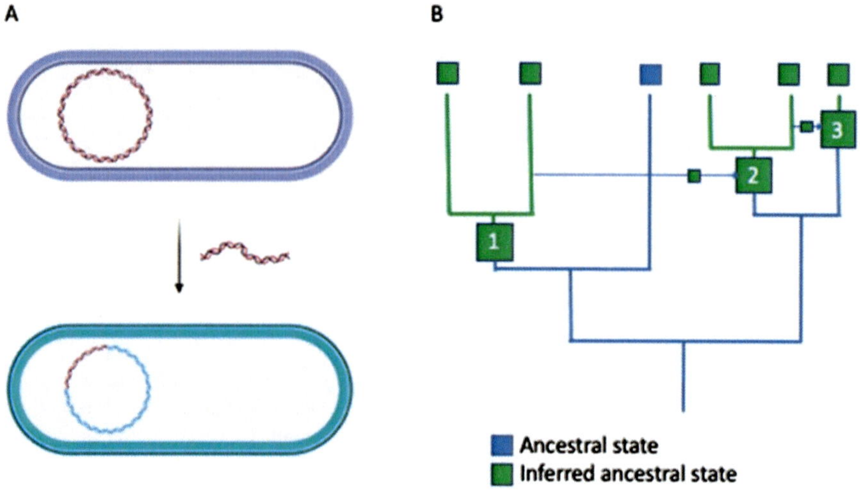

Fig. 9.5 Horizontal gene transfer. (**a**) Idealised illustration of the process of horizontal gene transfer. DNA from the donor cell (purple) is integrated into the genome of the recipient cell (green). This DNA is then be inherited by offspring of the recipient cell (figure created with BioRender.com). (**b**) Horizontal transfer can impact the distribution of a gene across the tree of life. In this case, a young gene arises at point 1, and subsequently spreads to another lineage (Koonin 2003). This is repeated to further spread the gene across the tree (Glansdorff et al. 2008). The common ancestor of the donor and recipient lineages is older than the gene. Using distribution alone to infer ancestry would mean that the ancestral state is incorrectly inferred as the green trait, which is derived and has spread. The actual ancestral state (blue) is obscured by the spread of the derived state

life, and even the tree itself may be challenging to reconstruct as it may be nigh on impossible to demonstrate that a gene has definitively not been subject to horizontal gene transfer at some point in its evolutionary history (Dagan and Martin 2006; Doolittle 1999).

For horizontal gene transfer, there is optimism to be had in that we can detect some of the transfer events that complicate the placement of putatively ancient genes in LUCA. However, it remains challenging to systematically generate large scale reconstructions as putative signals of vertical descent interleaved with ongoing horizontal gene transfer (Weiss et al. 2016) have on closer inspection turned out to not be reliable (Berkemer and McGlynn 2020). Having said that, it is possible to quantify the extent of transfer versus vertical descent (Berkemer and McGlynn 2020), so there is still cause for cautious optimism.

In terms of signal, there is another reason for optimism: protein structure is overall better conserved than sequence (Illergård et al. 2009), and these data can be used successfully to reconstruct evolutionary relationships between proteins where there is clear structural homology but sequence similarities have been overrun by noise (Brindefalk et al. 2013; Lundin et al. 2012; Malik et al. 2020). It may therefore be possible to use such data to augment evolutionary reconstruction from sequence data, which may in turn lead to more reliable evolutionary trees.

9.3 Approaches to LUCA

In this section I will very briefly summarise some of the key factors relevant to reconstructing LUCA. There are a number of ways that researchers have approached the reconstruction of LUCA despite the challenges of signal, horizontal gene transfer and uncertainty around the precise evolutionary relationships between life forms on Earth. The following is a brief overview of some key approaches and what they do and do not tell us.

9.3.1 Tree of Life (TOL) LUCA

In this approach, the aim is to build trees from genes and find those that show the same topology as those that were used to reconstruct the tree of life. In this approach, any genes that show a '3-domains' topology, that is, a branching pattern that is equivalent to the ribosomal tree of life (Woese et al. 1990), can be placed in LUCA. Such reconstructions (Harris et al. 2003) yield a very conservative reconstruction, which limits itself to only those genes that show a particular phylogenetic signal. Thus, only a small number of genes are compatible with LUCA; the rest are unclear because of confounding processes such as HGT or loss of phylogenetic signal.

It turns out that this approach actually only identifies those genes that are suitable for phylogenetic reconstruction of a tree of life from their sequence data. It is worth noting that these genes are regularly rolled out for three domains reconstructions (Gribaldo et al. 2010) because, when concatenated, they can be used to boost evolutionary signal (Ciccarelli et al. 2006; Gribaldo et al. 2010). A second challenge is that it is not clear that the relationships between the three domains adhere to the '3-domain tree' topology. A large number of more recent results have instead found support for a '2-domain tree', which places eukaryotes as a group within the diversity of Archaea (Fig. 9.6). This 2-domain interpretation is currently favoured by the majority of practitioners and has been extensively discussed elsewhere (Gribaldo and Brochier-Armanet 2020; Williams et al. 2020; Raymann et al. 2015; Spang et al. 2015; Zaremba-Niedzwiedzka et al. 2017). The important point for LUCA reconstruction is that, if the 2-domains tree is correct, we only need to worry about whether traits can be placed in the ancestors of both bacteria and archaea; if the eukaryotes are indeed a more recent evolutionary group that emerged from within the Archaea, then, as a derived subgroup of the archaea, they are less important for LUCA reconstruction. However, the key genes that are used for phylogenetic reconstruction of the three domains, depending on the analysis, show either the 2- or the 3-domain topology, so it is not automatically the case that one of the alternative 2-domain topologies expands the number of genes (Gribaldo et al. 2010) available for a ToL LUCA reconstruction.

Lower resolution efforts are possible, where, instead of using sequence-based approaches to reconstruct the history for each gene, one instead uses presence

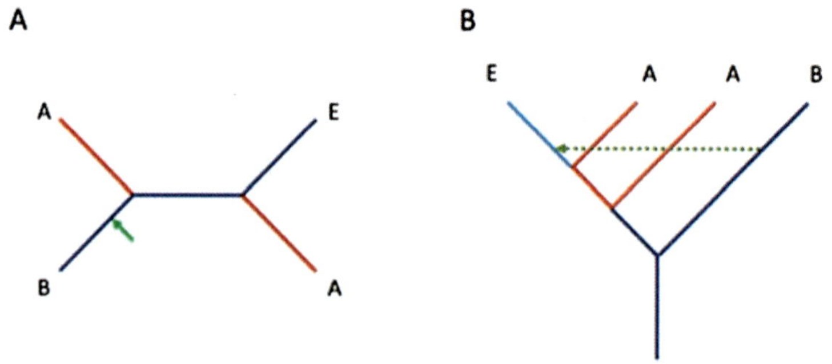

Fig. 9.6 The 2-domain tree of life. (**a**) The 2-domain tree as an unrooted tree. In this tree topology, Archaea (red branches) are present on both sides of the internal branch. This means that, there is no rooting where Archaea form a monophyletic group. Bacterial rooting is indicated (green arrow). (**b**) The 2-domain tree under the bacterial rooting. In this tree, the non-monophyly of the Archaea is clear. A monophyletic group is where the ancestor and all descendents share a common ancestor. Thus, Archaea (red branches) plus Eukarya (light blue branch) together form a monophyletic group, but Archaea alone are paraphyletic. This topology indicates that Eukarya evolved from within the Archaea, so, on parsimony, eukaryote-specific traits are unlikely to trace to LUCA. The implication is that, for reconstructing LUCA, we would be more interested in traits that can be placed in the archaeal or bacterial ancestors. Traits found in bacteria and eukaryotes are likely to be the result of transfer to eukaryotes from the mitochondrial ancestor (a bacterium), or from later gene transfers (dashed arrow)

or absence data (Snel et al. 1999). There are important considerations with this type of approach. First, there is not a requirement for the genes to carry sufficient information for reconstructing a tree carrying all three domains. This approach therefore has the potential to allow us to identify putatively ancient genes based on distribution alone, meaning that we may be able to place genes in LUCA even if they have too little signal for building a reliable tree from them. The second consideration is the assumption being made: the evolutionary history of the gene(s) is assumed to be based on vertical descent, not horizontal gene transfer. As shown in Fig. 9.5, horizontal transfer events may lead to the erroneous placement of young traits in LUCA.

For RNA genes, which might be good candidates for ancient origins based on possible continuity with the RNA world (Poole et al. 1998; Hoeppner et al. 2012), signal is the major challenge as RNA gene conservation decays much faster than protein-coding genes (Lindgreen et al. 2014). Hoeppner et al. (2012) were able to take a character-based approach to assessing the antiquity of noncoding RNA genes, with a conservative approach yielding an uncontroversial placement of a small number of RNAs in LUCA. For RNA genes it has been possible to identify cases where a 3-domains distribution can be accounted for by gene transfer (Hoeppner et al. 2012), and establishing genuine antiquity in the face of HGT likely requires careful case-by-case assessment.

While the number of genes that can be identified using this phylogenetic congruence approach is low, it is noteworthy that methodologically independent reconstructions with RNA and protein-coding genes both recover the same core set of functions, primarily associated with translation and transcription (Harris et al. 2003; Hoeppner et al. 2012).

9.3.2 *Planetary Megaorganism LUCA*

The challenges of reconstructing ancestral states in the face of HGT, has led to a very different model, dubbed the planetary megaorganism (Poole et al. 2015; Caetano-Anollés et al. 2014). This reconstruction aims to reconcile the tree of life with gene transfer. The idea traces to Woese (1998), and appears to be his attempt to reconcile his major life's work—the rRNA gene tree of life—with an appreciation of the process of horizontal gene transfer. It also handily served as a riposte to the numerous alternative trees of life that were emerging in the 1990s as other groups attempted to reconstruct the tree using other ancient genes and got different answers.

In brief, Woese's idea was that gene transfer rates were extremely high early in the evolution of life. Life then started to 'crystallise' around the ribosome, with other processes crystallising later, as gene transfer rates began to drop. There is no direct evidence for such a model (and none against it as such), but it is worth noting that a universal genetic code can in principle evolve under such conditions (Vetsigian et al. 2006).

The corollary of extreme rates of horizontal gene transfer is the 'planetary megaorganism' (Caetano-Anollés et al. 2014) which Woese dubbed the communal ancestor (Woese 1998). This model proposes that early life was one enormous planetary-scale communal system with genes swapping freely. Woese maintained that the rates of horizontal transfer were much higher than vertical descent, meaning that genes could move more readily between cells by gene transfer and inheritance of genes via cell division was not important. The idea is appealing in that HGT is often assumed to have been higher earlier in evolution, though there are no direct data to show this.

The planetary megaorganism model thus takes a conservative view (the gene-based tree of life), and then adds in speculative scenarios for gene swapping. There are two challenges with this model. The first is biophysical: while controlled, regular cell division was plausibly not a feature of early cells, the movement of genes across membranes (i.e. from one cell to another) is no less surmountable. Perhaps regular fusion and fission of cells could account for this, but it is another thing to propose that this was the predominant mechanism, with cellular growth and division being comparatively rare.

More importantly for this model, unconstrained gene swapping turns out to be a major Achilles' heel for any model of early evolution. It has been shown, time and again, that in well-mixed populations, parasites will spread so effectively that they will bring the system down, whereas, where there are limits to transmission, parasites cannot have this effect (Hogeweg and Takeuchi 2003; Szathmáry and Demeter 1987; Boerlijst and Hogeweg 1991; Zintzaras et al. 2002). Unconstrained, extreme rates of transfer in a planetary megaorganism might enable rapid spread of new innovations and testing of new ideas, but it will also yield the perfect conditions for a parasite to bring the entire system down. The consensus from modelling therefore requires that we step back from a planetary megaorganism and embrace more limited gene transfer (Poole 2009).

9.3.3 Non-orthologous Gene Displacement (NOGD) LUCA

Hypothesising the existence of a planetary megaorganism is not the only way to consider the nature of LUCA in the face of HGT. The term nonorthologous gene displacement (NOGD) refers to the fact that a gene may be replaced by an evolutionarily unrelated gene that performs the same function (Koonin et al. 1996) (Fig. 9.7). This may or may not explicitly require horizontal gene transfer—one route is indeed receipt of a gene with identical function from another genome, such as a virus (Forterre 1999). The other however is for an evolutionarily-unrelated gene from the same genome to take over the original role. Such replacements may occur because gene products can have more than one function, creating redundancy. Two critical features of proteins may be responsible for this phenomenon: catalytic promiscuity (O''Brien and Herschlag 1999) and moonlighting (Copley 2012). Moonlighting refers to the fact that a protein may possess more than one function, and our knowledge of the complete set of functions of gene products is still incomplete. A well-understood example of relevance to early evolution is the observation that multiple ribosomal proteins are known to possess other activities (Warner and McIntosh 2009; Wool 1996). Catalytic promiscuity refers to the fact that an enzyme may have the potential to undertake other chemical reactions, albeit at a very low level (O'Brien and Herschlag 1999; Copley 2012). This means that, under the right conditions there is potential for the promiscuous function to be refined by natural selection. This may lead to new functions emerging, but it is also possible that the promiscuous enzyme displaces an existing enzyme. Both moonlighting and catalytic promiscuity may thus lead to the specific genes responsible for a particular function changing during evolution.

Indeed, there is evidence to support NOGD in replication systems. Here, clearly unrelated proteins carry out the exact same function—examples include primases responsible for 'priming' DNA synthesis, and even the replicative DNA polymerases (Leipe et al. 1999; Forterre 2001, 2002; Forterre et al. 2004). What is notable is that some parts of the replication machinery are universal, suggesting that the differences between the replication machineries could be explained by NOGD, with one view being that most of these events are the result of genes moving into cellular systems from viruses, with acquisition perhaps yielding immunity to the virus (Forterre 2002).

For much more recent cases, NOGD has been readily identified. A notable case is the acquisition of mitochondrial DNA polymerase and other parts of the replication machinery from T-odd bacteriophage (Shutt and Gray 2006). This is clearly a

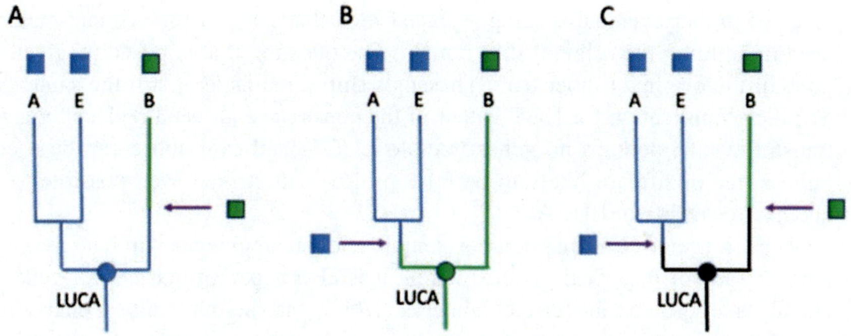

Fig. 9.7 Non-orthologous gene displacement (NOGD). Three possible interpretations of the presence of a biologically equivalent trait where the gene(s) involved have different evolutionary origins. (**a**) In this scenario the blue trait is evolutionarily older and was present in LUCA. The green trait has been acquired more recently and has displaced the ancestral blue trait in the bacterial lineage. As the genes underlying the traits are unrelated in evolution, this displacement is termed non-orthologous. (**b**) The reverse scenario is shown, with the green trait dating to LUCA and the blue trait having more recent origins, again via NOGD. (**c**) In this scenario, neither trait traces to LUCA, with each being acquired independently. In this interpretation, LUCA did not carry the trait, which has evolved twice independently. This third scenario is not actually a case of NOGD. However, as this is a possible interpretation of the observed distribution of the two traits, it is included for completeness. See text for discussion of possible mechanisms of non-orthologous gene displacement

case of NOGD via HGT. Importantly, this provides a mechanistic underpinning to the proposal that genes for parts of the replication machinery were replaced with alternative genes of equivalent function via NOGD from a viral source.

Given the patchwork patterning, with some parts of the modern machinery being traceable to LUCA, and some appearing to be more restricted, the data seem most compatible with DNA replication being a feature of LUCA, but with some parts being subsequently swapped out through NOGD events. However, this is not the only formal interpretation—the challenge with NOGD is that, without clear phylogenetic data, we cannot easily distinguish between competing scenarios (Fig. 9.7). For DNA replication for instance, it is formally possible that either one of the bacterial or the archaeal/eukaryotic systems are ancestral, or that neither is, having evolved later (or even having been acquired from viruses). The latter implies that LUCA had an RNA genome (Leipe et al. 1999), an idea that, based on what is known of modern RNA polymerases, is certainly within the bounds of plausibility (Poole and Logan 2005; Poole 2006).

9.4 The Ship of Theseus and Genes Refractory to HGT

Doolittle (2004) has invoked the ship of Theseus as an analogy for considering the impact of HGT on LUCA reconstruction (Andam et al. 2010). This is a thought experiment where a ship is subjected to repairs over its lifespan, resulting in every piece of timber eventually being replaced such that none of the original materials used in building the original ship remain. The question that is posed is, given that none of the original timber from Theseus's ship remains, is it still the same ship? The clear implication for HGT is that, if the genes have all been replaced via gene transfer events, perhaps no genes trace to LUCA, and even those can be used to build a tree of life are likely to be false positives. Is it then even possible to talk about a tree of life or LUCA?

While it seems clear that that horizontal gene transfer events will heavily impact the accuracy of historical reconstruction, it is also apparent that not all genes are equally likely to be transferred (Jain et al. 1999), and this may in turn enable us to establish which kinds of gene are less likely to be false positives (of the kind seen in Fig. 9.5b) and thus more likely to reveal a conservative signal consistent with vertical descent.

At one end of the spectrum are genes that are likely to be highly transferred owing to the fact that they are only occasionally used. Such periodically-selected functions (Poole et al. 2003) may be easily lost from a genome as they are only sometimes required. Genes for antibiotic resistance are a clear example. In the presence of an antibiotic, a resistance gene is strongly advantageous, so is retained in the genome. However, when the antibiotic is not present in the environment, this gene is not required for survival so may be readily lost—hence it is only 'periodically selected', with selection being dependent on the environment. The flip side of this is that, without horizontal transfer, such genes would go extinct. There are good reasons to expect that, where a phenotype requires multiple genes, there is selection for those genes to be organised in a way that promotes their transfer. The selfish operon model (Lawrence and Roth 1996) argues that, where several genes must act together to form a phenotype, these genes will be organised into a tightly colocated group with a single mechanism for gene expression, which maximises their probability of being physically co-transferred. This in turn maximises their evolutionary success under conditions where they may be frequently lost from an individual owing to periodic selection as they are also more frequently spread via horizontal gene transfer. Periodically-selected genes are well known to be frequently found on mobile genetic elements such as plasmids (Hacker and Carniel 2001), which in turn improves their chance of successfully spreading. These kinds of elements form part of what has been described as the 'accessory genome' which, together with the 'core genome' (those genes common to all members of a species) form the 'pan genome' (Lapierre and Gogarten 2009). Interestingly, in cases where the product of a gene is released into the environment, it both becomes a public good (i.e. a shared resource that cannot be monopolised) and a driver of gene loss (as individuals that can acquire the product rather than synthesising it can freeload off the public good). Where there are multiple public goods, the expected outcome is a mutualistic network as no one lineage produces everything (Fullmer et al. 2015). This in turn suggests that early systems may have been distributed and interdependent (Poole et al. 2015).

At the other end of the spectrum are genes that are refractory to HGT. One obvious reason for a gene being refractory to HGT is that it performs a function that is essential to the basic biology of a cell. However, as demonstrated by

NOGD, such genes may also be replaced by unrelated genes that perform the same function. The complexity hypothesis (Jain et al. 1999) argues that the genes that are most refractory to HGT have an additional feature: they are often parts of multi-component complexes. The basic idea is that, in addition to undertaking some important role, the individual gene products (RNA and protein) that make up a complex cellular structure must also fit together correctly. As species diverge, the components within an individual lineage coevolve such that they are less likely to interact well with components from a functionally-identical, but structurally divergent complex. In other words, the accumulation of evolutionary changes means the individual parts cease to be interchangeable, or, if they are interchangeable, they work suboptimally. This means that, if a horizontal gene transfer event leads to incorporation of a suboptimal piece within a multi-component complex, the recipient is at a disadvantage and is therefore more likely to be lost from the population than to spread.

Unfortunately, those genes that are most refractory to HGT are also largely 'informational' genes that code for complex multi-compnent systems such as the ribosome (Jain et al. 1999; Rivera et al. 1998), so this does not particularly expand our capacity to reconstruct LUCA. The genes that would shed greatest light on the metabolism of LUCA, which Lake and colleagues have dubbed 'operational' genes (Rivera et al. 1998), are among the more readily transferred. Lake and colleagues argued that the reason that these genes may be more likely to be successfully transferred is that the phenotype is associated with a chemical reaction, and the specifics of the enzyme may be less important than the generation of the product from a substrate. This, coupled with the multifunctionality of proteins, in effect means that enzymes may be quite interchangeable over large evolutionary timescales.

9.5 Rescuing LUCA from Theseus?

One important aspect of the ship of Theseus thought experiment is that, while all the individual timbers have been replaced, the ship is still a ship. By analogy, gene transfers where one gene is replaced by another retain the original phenotype; it is just that the specific gene and the chemistry underlying the enzymatic reaction have changed. It may therefore be that it is the phenotype (the functions) rather than the individual genes that is preserved. Perhaps this could provide a silhouette of the nature of LUCA?

This is plausibly the case for even very centrally important molecules, such as thymidylate (dTMP), one of the four building blocks of DNA. There are two functionally equivalent but evolutionarily unrelated thymidylate synthase enzymes (ThyA and ThyX), with broad distribution across the tree of life (Myllykallio et al. 2002). From the perspective of being able to generate this particular building block, it may not be particularly important whether one has a *thyA* or *thyX* gene. There may well be differences between the enzymes, but the patchy distribution suggests a history of gene transfer (Stern et al. 2010), and one possible interpretation is that they are broadly equivalent, with one copy having replaced the other on a number of occasions.

One could therefore make the case that a phenotype (thymidylate synthesis) was present in LUCA (Poole et al. 2001), even though it is unclear if one of the two gene families is older than the other. Thus, one could consider reconstructing LUCA based on gene functions rather than individual genes: the capacity to produce the four DNA building blocks (dA, dC, dG and dT) is conserved, and this phenotype is traceable to LUCA, but the specific gene (*thyA* vs *thyX*) is not.

In the simple analogy of the ship of Theseus, the design remains identical (i.e. production of thymidine for DNA synthesis), so by ignoring the specific history of the individual planks of wood (*thyX* or *thyA* genes), we might still be able to reconstruct an accurate picture of the original ship even though some details are missing (e.g. which gene was originally used). Unfortunately, biological systems present a more complex case since they can evolve—by analogy with the ship, entire structures can be eliminated, and new ones added; what started as a ship may end up being converted into a catamaran.

The question is thus, to what extent do we trust a broader, lower resolution reconstruction? We will next consider a couple of cases where an argument can be made for placement of a trait in LUCA using this logic.

9.6 Are There Other General Traits That We Might Place in LUCA?

9.6.1 Membrane-Associated Proteins Support a Cellular, Membrane-Bound LUCA

Membranes are universal to life, and it seems likely that these would have been a feature of LUCA. However, this has been subject to debate, not least because there is an important difference between the chemistry of phospholipids in archaea versus bacteria and eukaryotes. The phospholipids of all cellular systems consist of a glycerol phosphate moiety linked to hydrocarbon chains. However, the nature of both the glycerol phosphate and the hydrocarbons differ between bacterial (and eukaryotic) phospholipids versus archaeal phospholipids In brief, in bacterial phospholipids glycerol-3-phosphate (G3P) is connected to fatty acid chains via an ester-linkage, while archaea possess ether-linked lipids, where glycerol-1-phosphate (G1P, a stereoisomer of G3P) is ether-linked to isoprenoids. Both G3P and G1P are synthesised by dehydrogenases, but these are completely unrelated, belonging to different enzyme families (Peretó et al. 2004). This 'lipid divide' has been viewed as a major challenge for reconstructing LUCA as it is not strictly possible to say which is older. This led to the suggestion that membranes in archaea and bacteria evolved independently from a non-cellular LUCA that lacked a membrane altogether (Koonin and Martin 2005). Under this model, which takes the common approach that, where there is ambiguity, it is better to hedge bets and place none of these traits in the ancestor. The implication is then that membranes evolved late, after the primary split that gave rise to bacteria and archaea.

Against this backdrop is the observation that a number of membrane-associated proteins, including ATPases, are universal (Jékely 2006). Using these independent indicators, it is possible to conclude, as Jékely has (Jékely 2006) that LUCA did possess a membrane, even though we cannot with certainty state which of the two types of phospholipid is ancestral. It is of course possible (and perhaps likely) that the ancestral state was neither of these specific states. There are several key observations that support the antiquity of the membrane. First, clearly these universal proteins can interact with both archaeal and bacterial-type lipids. Thus, a change of lipid system does not apparently require a redesign of the membrane proteins required for interaction with the environment. Second, there is biophysical evidence that membranes containing both types of lipid are stable (Shimada and Yamagishi 2011). This means that, if one of these forms was ancestral, the advent of the other would not be disastrous—an intermediate membrane state would not likely be lethal. Finally, it appears that the lipid divide may actually not be real. Bacteria and eukaryotes produce ether-linked lipids and, from a metagenomic screen, Villanueva and colleagues (2020) have recently identified bacteria from the Fibrobacteres–Chlorobi–Bacteroidetes (FCB) group superphylum that carry genes for both the bacterial and archaeal pathways. When expressed in *E. coli*, these genes resulted in a mixed membrane (Villanueva et al. 2020). A recent analysis has also concluded that there has been ongoing interdomain transfer of lipid biosynthesis genes, predominantly from archaea to bacteria (Coleman et al. 2019).

While there are challenges with unravelling the ancestral state, it does seem that, on balance, the data favour a membrane-bound LUCA, even though the specific chemical makeup of the membrane is not clear. The key point here is that the addition of other data may help to weigh in on a general answer rather than the specific one created by solely focusing on the separate origins of the dehydrogenase genes central to the production of G1P and G3P-linked lipids.

9.6.2 Positional Conservation of RNA Modifications Places them in LUCA

A second case concerns RNA modification. The most ancient RNA molecules include ribosomal RNAs and transfer RNAs, both of which are decorated by way of extensive posttranscriptional modifications. There is a general consensus that such modifications are ancient and trace to LUCA (Cermakian and Cedergren 1998; Forterre and Grosjean 2009), particularly as a number of modifications are positionally conserved across all three domains of life. One such case are 2'-*O*-methylation and pseudouridylation modifications of ribosomal RNA, including at the peptidyl transferase centre (PTC) of the large subunit rRNA (Ofengand and Bakin 1997; Penny et al. 2009; King et al. 2003). The PTC is responsible for the peptidyl transfer reaction by which amino acids are linked to form peptide bonds during protein synthesis (Beringer and Rodnina 2007). The positional conservation of key modifications strongly suggests that these modifications trace to LUCA: the modifications are universal and the RNAs that are modified are among those which can be placed in LUCA using phylogenetic analyses. Where things get complicated is the mechanisms by which these are these are generated. In the bacteria, specific protein-coding enzymes are responsible for each modification, whereas in archaea and eukarya, small noncoding RNAs called small nucleolar RNAs (snoRNAs), function as guides to direct the modification machinery to the specific positions of modification (Penny et al. 2009; Omer et al. 2000; Gaspin et al. 2000; Gardner et al. 2010; Lafontaine and Tollervey 1998).

Given the primacy of the RNA world, it is plausible that the snoRNA-based system is the more ancient (Penny et al. 2009), but there is no way to demonstrate this definitively. It is not possible to demonstrate common ancestry for archaeal and eukaryotic snoRNAs, on account of the limited sequence conservation between them (Hoeppner et al. 2012; Gaspin et al. 2000; Hoeppner and Poole 2012) despite conservation of function, and despite the fact that the protein families that interact with these RNAs are conserved in eukaryotes and archaea (Gardner et al. 2010). This is a general challenge for noncoding RNA, where conservation signal decays at much shallower phylogenetic distances than for protein-coding genes (Lindgreen et al. 2014).

Returning to the bacterial versus the archaeal/eukaryotic systems, given that these are fundamentally different, the current distribution must surely be the result of some form of NOGD. However, given that some of the specific modifications that these distinct machineries undertake are positionally-conserved, it is reasonable to conclude that the modifications trace to LUCA even if there is uncertainty around the relative antiquity of the two modification machineries.

9.7 Towards a Lo-fi or Jack-of-all-Trades LUCA

These two examples indicate that it is possible to derive a partial picture by separating phenotype and the specifics of individual gene geneaologies. Can these be systematically expanded to derive a larger-scale reconstruction? There are good indications that they can. Two groups have made interesting inroads to this question.

Goldman and colleagues approached this question by building a metaconsensus derived from multiple data types (gene sequences, protein structure, enzyme function) (Goldman et al. 2012, 2013). This shifts the focus away from the specifics of a particular gene or enzyme, and enables placement of general classes of catalytic reaction in LUCA. This is an interesting approach because it in effect pools a number of reactions and enzymes, and asks a more general question about catalytic capacity. The upshot is that the answers are much less likely to depend on the specific evolutionary history of one gene or one specific enzymatic reaction, and therefore, the answer is likely to be robust to the false positive problems raised by HGT. The significance of this work appears to have been underappreciated by the wider community, but the conclusion from their work is clear –a number of broad classes of enzyme capabilities had already evolved by the time of LUCA, including oxidoreductases, transferases, lyases, ligases and hyrolases. This information enables a broad reconstruction of metabolic capacity that contains lipid, carbohydrate, amino acid and nucleotide metabolism (Goldman et al. 2012). Moreover, this approach allows incorporation of additional key information such as cofactor usage, which corroborates the view that nucleotide cofactors were a feature of early biological systems (White 1976; Jeffares et al. 1998). The caveat is that this approach to reconstruction can argue for very general classes of reaction, which may in turn fit with multiple scenarios.

Caetano-Anollés and colleagues (Caetano-Anollés et al. 2007; Caetano-Anollés and Caetano-Anollés 2003) took a not dissimilar approach to Goldman and colleagues. They have applied the same broad approach to a range of datasets using two key parameters as a proxy for the antiquity of traits. Their idea was to use the distribution and 'popularity' of a particular protein family (though they have also examined other datasets, including enzyme EC numbers and even gene ontology data)—the more widely distributed across species and the more widely used within a genome, the more ancient the family is (Caetano-Anollés et al. (2014, 2007), Caetano-Anollés and Caetano-Anollés (2003), Kim and Caetano-Anollés (2010)). This makes some intuitive sense, as old families should be widespread because they were present in the ancestor of many later descendents. There is also more time for older families to be coopted into additional functions, so it seems reasonable to expect that the number of ways a protein family has been deployed is a function of its age. This gels well with the idea that early enzymes were generalists, and processes such as gene duplication enabled the evolution of specialist enzymes and expanding metabolism into new areas using the same basic enzymatic toolkit.

These kinds of reconstruction are frustrating in the sense that specifics of individual genes are relaced by general attributes, but they may well represent a more realistic approach reconstruction—we are used to thinking about enzymes as having a specific reaction, such that, given a gene, we can accurately predict phenotype. However, a more realistic view is that early evolution was full of general capacities (Jensen 1976) (and may still be (Soo et al. 2011)), and early enzymes may thus have been capable of perfoming a range of chemical reactions of the same class. In this view, early enzymes were not as specialised as their modern counterparts, and may be better understood as being generalists—a jack-of-all trades, if you will. Thus, a broad array of activities may have been achieved with only a few genes.

Of course, widespread and widely used protein families could be so distributed because they evolved later and spread successfully through HGT (Fig. 9.5b), but again, because these analyses do not depend on the specific history of a single gene, they may be pretty robust to the false positive problem. It is worth noting that, using different datasets, Caetano-Anollés et al. get quite consistent results.

This 'Lo-fi' picture of LUCA tells us that the broad classes of chemistry that are used by modern organisms were in all likelihood already present in LUCA. A class of reaction being placed in LUCA therefore specifies the metabolic potential of LUCA but is somewhat agnostic on the specifics. This metabolic potential gels well with the idea that the nucleotide cofactors are also very old (White 1976), possibly dating to the RNA world phase of the evolution of life. Somewhat frustratingly, because a fairly narrow set of core catalytic reactions are responsible for most of the diversity of modern metabolism, it is not easy to say much more than that.

9.8 Concluding Remarks: A Path Towards a Trait-Based LUCA?

HGT, NOGD, and loss of evolutionary signal are all major factors that impact our capacity to reconstruct LUCA. This has resulted in a tendency to favour reconstruction of a simple LUCA with some key features such as DNA genomes and membranes being interpreted as evolving post-LUCA. However, it may be possible to reconstruct LUCA on the assumption that broad features belie a common ancestral state, even if the reconstruction is lower-resolution than is achievable for more recent ancestors. This review has noted a few cases where a lower-resolution approach may bear fruit, and in several cases has already done so. While it may be difficult to definitively reconstruct the ancestral state of some features, a second look does suggest that an alternative to the 'simple' or 'minimal' LUCA reconstruction is worth pursuing. To do this requires stepping away from a gene or gene-content based reconstruction to instead focus on common traits. Given the challenges of placing individual genes in LUCA, an approach that focuses on general features may represent an alternative to the model of early divergence followed by massive convergence, with broad traits or phenotypes being conserved, even where specific genes have been transferred or replaced. I have tried to illustrate that, for some scenarios where there is ambiguity because of NOGD, it may nevertheless be possible to place a general feature in LUCA on incomplete signal.

References

Andam, C.P., Williams, D., Gogarten, J.P.: Natural taxonomy in light of horizontal gene transfer. Biol. Philos. **25**(4), 589–602 (2010)

Beringer, M., Rodnina, M.V.: The ribosomal peptidyl transferase. Mol. Cell. **26**(3), 311–321 (2007)

Berkemer, S.J., McGlynn, S.E.: A new analysis of archaea-bacteria domain separation: variable phylogenetic distance and the tempo of early evolution. Mol. Biol. Evol. **37**(8), 2332–2340 (2020)

Boerlijst, M.C., Hogeweg, P.: Spiral wave structure in pre-biotic evolution: Hypercycles stable against parasites. Phys. D. **48**(1), 17–28 (1991)

Boussau, B., Blanquart, S., Necsulea, A., Lartillot, N., Gouy, M.: Parallel adaptations to high temperatures in the Archaean eon. Nature. **456**(7224), 942–945 (2008)
Brindefalk, B., Dessailly, B.H., Yeats, C., Orengo, C., Werner, F., Poole, A.M.: Evolutionary history of the TBP-domain superfamily. Nucleic Acids Res. **41**(5), 2832–2845 (2013)
Caetano-Anollés, G., Caetano-Anollés, D.: An evolutionarily structured universe of protein architecture. Genome Res. **13**(7), 1563–1571 (2003)
Caetano-Anollés, G., Kim, H.S., Mittenthal, J.E.: The origin of modern metabolic networks inferred from phylogenomic analysis of protein architecture. Proc. Natl. Acad. Sci. USA. **104**(22), 9358–9363 (2007)
Caetano-Anollés, G., Mittenthal, J.E., Caetano-Anollés, D., Kim, K.M.: A calibrated chronology of biochemistry reveals a stem line of descent responsible for planetary biodiversity. Front. Genet. **5**, 306 (2014)
Cermakian, N., Cedergren, R.: Modified nucleotides always were: an evolutionary model. In: Grosjean, H., Benne, R. (eds.) Modification and Editing of RNA, pp. 535–541. ASM Press, Washington, DC (1998)
Ciccarelli, F.D., Doerks, T., von Mering, C., Creevey, C.J., Snel, B., Bork, P.: Toward automatic reconstruction of a highly resolved tree of life. Science. **311**(5765), 1283–1287 (2006)
Coleman, G.A., Pancost, R.D., Williams, T.A.: Investigating the origins of membrane phospholipid biosynthesis genes using outgroup-free rooting. Genome Biol. Evol. **11**(3), 883–898 (2019)
Copley, S.D.: Moonlighting is mainstream: paradigm adjustment required. BioEssays. **34**(7), 578–588 (2012)
Dagan, T., Martin, W.: The tree of one percent. Genome Biol. **7**(10), 118 (2006)
Dagan, T., Roettger, M., Bryant, D., Martin, W.: Genome networks root the tree of life between prokaryotic domains. Genome Biol. Evol. **2**, 379–392 (2010)
Doolittle, W.F.: Phylogenetic classification and the universal tree. Science. **284**(5423), 2124–2129 (1999)
Doolittle, W.F.: W. Ford Doolittle. Curr. Biol. **14**(5), R176–R177 (2004)
Forterre, P.: Displacement of cellular proteins by functional analogues from plasmids or viruses could explain puzzling phylogenies of many DNA informational proteins. Mol. Microbiol. **33**(3), 457–465 (1999)
Forterre, P.: Genomics and early cellular evolution. The origin of the DNA world. C. R. Acad. Sci. III. **324**(12), 1067–1076 (2001)
Forterre, P.: The origin of DNA genomes and DNA replication proteins. Curr. Opin. Microbiol. **5**(5), 525–532 (2002)
Forterre, P.: The universal tree of life: an update. Front. Microbiol. **6**, 717 (2015)
Forterre, P., Grosjean, H.: The Interplay Between RNA and DNA Modifications: Back to the RNA World. Molecular Biology Intelligence Unit Landes Bioscience. Springer, Austin, TX (2009)
Forterre, P., Philippe, H.: Where is the root of the universal tree of life? BioEssays. **21**(10), 871–879 (1999)
Forterre, P., Filée, J., Myllykallio, H.: Origin and evolution of DNA and DNA replication machineries. In: de Pouplana, L.R. (ed.) The Genetic Code and the Origin of Life. Landes Bioscience, Georgetown, TX (2004)
Fullmer, M.S., Soucy, S.M., Gogarten, J.P.: The pan-genome as a shared genomic resource: mutual cheating, cooperation and the black queen hypothesis. Front. Microbiol. **6**, 728 (2015)
Galtier, N., Tourasse, N., Gouy, M.: A nonhyperthermophilic common ancestor to extant life forms. Science. **283**(5399), 220–221 (1999)
Gardner, P.P., Bateman, A., Poole, A.M.: SnoPatrol: how many snoRNA genes are there? J. Biol. **9**(1), 4 (2010)
Gaspin, C., Cavaille, J., Erauso, G., Bachellerie, J.P.: Archaeal homologs of eukaryotic methylation guide small nucleolar RNAs: lessons from the Pyrococcus genomes. J. Mol. Biol. **297**(4), 895–906 (2000)
Glansdorff, N., Xu, Y., Labedan, B.: The last universal common ancestor: emergence, constitution and genetic legacy of an elusive forerunner. Biol. Direct. **3**, 29 (2008)

Gogarten, J.P., Kibak, H., Dittrich, P., Taiz, L., Bowman, E.J., Bowman, B.J., et al.: Evolution of the vacuolar H+-ATPase: implications for the origin of eukaryotes. Proc. Natl. Acad. Sci. USA. **86**(17), 6661–6665 (1989)

Goldman, A.D., Baross, J.A., Samudrala, R.: The enzymatic and metabolic capabilities of early life. PLoS One. **7**(9), e39912 (2012)

Goldman, A.D., Bernhard, T.M., Dolzhenko, E., Landweber, L.F.: LUCApedia: a database for the study of ancient life. Nucleic Acids Res. **41**(Database issue), D1079–D1082 (2013)

Gribaldo, S., Brochier-Armanet, C.: Evolutionary relationships between archaea and eukaryotes. Nat. Ecol. Evol. **4**(1), 20–21 (2020)

Gribaldo, S., Poole, A.M., Daubin, V., Forterre, P., Brochier-Armanet, C.: The origin of eukaryotes and their relationship with the archaea: are we at a phylogenomic impasse? Nat. Rev. Microbiol. **8**(10), 743–752 (2010)

Hacker, J., Carniel, E.: Ecological fitness, genomic islands and bacterial pathogenicity. A Darwinian view of the evolution of microbes. EMBO Rep. **2**(5), 376–381 (2001)

Harris, J.K., Kelley, S.T., Spiegelman, G.B., Pace, N.R.: The genetic core of the universal ancestor. Genome Res. **13**, 407–412 (2003)

Hoeppner, M.P., Poole, A.M.: Comparative genomics of eukaryotic small nucleolar RNAs reveals deep evolutionary ancestry amidst ongoing intragenomic mobility. BMC Evol. Biol. **12**, 183 (2012)

Hoeppner, M.P., Gardner, P.P., Poole, A.M.: Comparative analysis of RNA families reveals distinct repertoires for each domain of life. PLoS Comput. Biol. **8**(11), e1002752 (2012)

Hogeweg, P., Takeuchi, N.: Multilevel selection in models of prebiotic evolution: compartments and spatial self-organization. Orig. Life Evol. Biosph. **33**(4–5), 375–403 (2003)

Illergård, K., Ardell, D.H., Elofsson, A.: Structure is three to ten times more conserved than sequence: a study of structural response in protein cores. Proteins. **77**(3), 499–508 (2009)

Iwabe, N., Kuma, K., Hasegawa, M., Osawa, S., Miyata, T.: Evolutionary relationship of archaebacteria, eubacteria, and eukaryotes inferred from phylogenetic trees of duplicated genes. Proc. Natl. Acad. Sci. USA. **86**(23), 9355–9359 (1989)

Jain, R., Rivera, M.C., Lake, J.A.: Horizontal gene transfer among genomes: the complexity hypothesis. Proc. Natl. Acad. Sci. USA. **96**(7), 3801–3806 (1999)

Jeffares, D.C., Poole, A.M., Penny, D.: Relics from the RNA world. J. Mol. Evol. **46**(1), 18–36 (1998)

Jékely, G.: Did the last common ancestor have a biological membrane? Biol. Direct. **1**, 35 (2006)

Jensen, R.A.: Enzyme recruitment in evolution of new function. Annu. Rev. Microbiol. **30**, 409–425 (1976)

Kim, K.M., Caetano-Anollés, G.: Emergence and evolution of modern molecular functions inferred from phylogenomic analysis of ontological data. Mol. Biol. Evol. **27**(7), 1710–1733 (2010)

King, T.H., Liu, B., McCully, R.R., Fournier, M.J.: Ribosome structure and activity are altered in cells lacking snoRNPs that form pseudouridines in the peptidyl transferase center. Mol. Cell. **11**(2), 425–435 (2003)

Koonin, E.V.: Comparative genomics, minimal gene-sets and the last universal common ancestor. Nat. Rev. Microbiol. **1**(2), 127 (2003)

Koonin, E.V., Martin, W.: On the origin of genomes and cells within inorganic compartments. Trends Genet. **21**(12), 647–654 (2005)

Koonin, E.V., Mushegian, A.R., Bork, P.: Non-orthologous gene displacement. Trends Genet. **12**(9), 334–336 (1996)

Koumandou, V.L., Wickstead, B., Ginger, M.L., van der Giezen, M., Dacks, J.B., Field, M.C.: Molecular paleontology and complexity in the last eukaryotic common ancestor. Crit. Rev. Biochem. Mol. Biol. **48**(4), 373–396 (2013)

Kyrpides, N., Overbeek, R., Ouzounis, C.: Universal protein families and the functional content of the last universal common ancestor. J. Mol. Evol. **49**(4), 413–423 (1999)

Lafontaine, D.L., Tollervey, D.: Birth of the snoRNPs: the evolution of the modification-guide snoRNAs. Trends Biochem. Sci. **23**(10), 383–388 (1998)

Lapierre, P., Gogarten, J.P.: Estimating the size of the bacterial pan-genome. Trends Genet. **25**(3), 107–110 (2009)

Lawrence, J.G., Roth, J.R.: Selfish operons: horizontal transfer may drive the evolution of gene clusters. Genetics. **143**(4), 1843–1860 (1996)

Leipe, D.D., Aravind, L., Koonin, E.V.: Did DNA replication evolve twice independently? Nucleic Acids Res. **27**(17), 3389–3401 (1999)

Lindgreen, S., Umu, S.U., Lai, A.S., Eldai, H., Liu, W., McGimpsey, S., et al.: Robust identification of noncoding RNA from transcriptomes requires phylogenetically-informed sampling. PLoS Comput. Biol. **10**(10), e1003907 (2014)

Lundin, D., Gribaldo, S., Torrents, E., Sjöberg, B.M., Poole, A.M.: Ribonucleotide reduction - horizontal transfer of a required function spans all three domains. BMC Evol. Biol. **10**, 383 (2010)

Lundin, D., Poole, A.M., Sjöberg, B.M., Hogbom, M.: Use of structural phylogenetic networks for classification of the ferritin-like superfamily. J. Biol. Chem. **287**(24), 20565–20575 (2012)

Malik, A.J., Poole, A.M., Allison, J.R.: Structural phylogenetics with confidence. Mol. Biol. Evol. **37**(9), 2711–2726 (2020)

Myllykallio, H., Lipowski, G., Leduc, D., Filee, J., Forterre, P., Liebl, U.: An alternative flavin-dependent mechanism for thymidylate synthesis. Science. **297**(5578), 105–107 (2002)

Neumann, N., Lundin, D., Poole, A.M.: Comparative genomic evidence for a complete nuclear pore complex in the last eukaryotic common ancestor. PLoS One. **5**(10), e13241 (2010)

O'Brien, P.J., Herschlag, D.: Catalytic promiscuity and the evolution of new enzymatic activities. Chem. Biol. **6**(4), R91–R105 (1999)

Ofengand, J., Bakin, A.: Mapping to nucleotide resolution of pseudouridine residues in large subunit ribosomal RNAs from representative eukaryotes, prokaryotes, archaebacteria, mitochondria and chloroplasts. J. Mol. Biol. **266**(2), 246–268 (1997)

Omer, A.D., Lowe, T.M., Russell, A.G., Ebhardt, H., Eddy, S.R., Dennis, P.P.: Homologs of small nucleolar RNAs in archaea. Science. **288**(5465), 517–522 (2000)

Penny, D., Poole, A.: The nature of the last universal common ancestor. Curr. Opin. Genet. Dev. **9**(6), 672–677 (1999)

Penny, D., Hoeppner, M.P., Poole, A.M., Jeffares, D.C.: An overview of the introns-first theory. J. Mol. Evol. **69**(5), 527–540 (2009)

Peretó, J., López-García, P., Moreira, D.: Ancestral lipid biosynthesis and early membrane evolution. Trends Biochem. Sci. **29**(9), 469–477 (2004)

Poole, A.M.: Getting from an RNA world to modern cells just got a little easier. BioEssays. **28**(2), 105–108 (2006)

Poole, A.M.: Horizontal gene transfer and the earliest stages of the evolution of life. Res. Microbiol. **160**(7), 473–480 (2009)

Poole, A.M., Logan, D.T.: Modern mRNA proofreading and repair: clues that the last universal common ancestor possessed an RNA genome? Mol. Biol. Evol. **22**(6), 1444–1455 (2005)

Poole, A.M., Jeffares, D.C., Penny, D.: The path from the RNA world. J. Mol. Evol. **46**(1), 1–17 (1998)

Poole, A., Penny, D., Sjöberg, B.M.: Confounded cytosine! Tinkering and the evolution of DNA. Nat. Rev. Mol. Cell Biol. **2**(2), 147–151 (2001)

Poole, A.M., Phillips, M.J., Penny, D.: Prokaryote and eukaryote evolvability. Biosystems. **69**(2–3), 163–185 (2003)

Poole, A.M., Lundin, D., Rytkönen, K.T.: The evolution of early cellular systems viewed through the lens of biological interactions. Front. Microbiol. **6**, 1144 (2015)

Raymann, K., Brochier-Armanet, C., Gribaldo, S.: The two-domain tree of life is linked to a new root for the archaea. Proc. Natl. Acad. Sci. USA. **112**(21), 6670–6675 (2015)

Rivera, M.C., Jain, R., Moore, J.E., Lake, J.A.: Genomic evidence for two functionally distinct gene classes. Proc. Natl. Acad. Sci. USA. **95**(11), 6239–6244 (1998)

Shimada, H., Yamagishi, A.: Stability of heterochiral hybrid membrane made of bacterial sn-G3P lipids and archaeal sn-G1P lipids. Biochemistry. **50**(19), 4114–4120 (2011)

Shutt, T.E., Gray, M.W.: Bacteriophage origins of mitochondrial replication and transcription proteins. Trends Genet. **22**(2), 90–95 (2006)

Skophammer, R.G., Servin, J.A., Herbold, C.W., Lake, J.A.: Evidence for a gram-positive, eubacterial root of the tree of life. Mol. Biol. Evol. **24**(8), 1761–1768 (2007)

Snel, B., Bork, P., Huynen, M.A.: Genome phylogeny based on gene content. Nat. Genet. **21**(1), 108–110 (1999)

Soo, V.W., Hanson-Manful, P., Patrick, W.M.: Artificial gene amplification reveals an abundance of promiscuous resistance determinants in *Escherichia coli*. Proc. Natl. Acad. Sci. USA. **108**(4), 1484–1489 (2011)

Spang, A., Saw, J.H., Jorgensen, S.L., Zaremba-Niedzwiedzka, K., Martijn, J., Lind, A.E., et al.: Complex archaea that bridge the gap between prokaryotes and eukaryotes. Nature. **521**(7551), 173–179 (2015)

Stern, A., Mayrose, I., Penn, O., Shaul, S., Gophna, U., Pupko, T.: An evolutionary analysis of lateral gene transfer in thymidylate synthase enzymes. Syst. Biol. **59**(2), 212–225 (2010)

Szathmáry, E., Demeter, L.: Group selection of early replicators and the origin of life. J. Theor. Biol. **128**(4), 463–486 (1987)

Vetsigian, K., Woese, C., Goldenfeld, N.: Collective evolution and the genetic code. Proc. Natl. Acad. Sci. USA. **103**(28), 10696–10701 (2006)

Villanueva, L., von Meijenfeldt, F.A.B., Westbye, A.B., Yadav, S., Hopmans, E.C., Dutilh, B.E., et al.: Bridging the membrane lipid divide: bacteria of the FCB group superphylum have the potential to synthesize archaeal ether lipids. ISME J. **15**, 168–182 (2020)

Warner, J.R., McIntosh, K.B.: How common are extraribosomal functions of ribosomal proteins? Mol. Cell. **34**(1), 3–11 (2009)

Weiss, M.C., Sousa, F.L., Mrnjavac, N., Neukirchen, S., Roettger, M., Nelson-Sathi, S., et al.: The physiology and habitat of the last universal common ancestor. Nat. Microbiol. **1**(9), 16116 (2016)

White 3rd., H.B.: Coenzymes as fossils of an earlier metabolic state. J. Mol. Evol. **7**(2), 101–104 (1976)

Williams, T.A., Cox, C.J., Foster, P.G., Szollosi, G.J., Embley, T.M.: Phylogenomics provides robust support for a two-domains tree of life. Nat. Ecol. Evol. **4**(1), 138–147 (2020)

Woese, C.: The universal ancestor. Proc. Natl. Acad. Sci. USA. **95**(12), 6854–6859 (1998)

Woese, C.R., Fox, G.E.: Phylogenetic structure of the prokaryotic domain: the primary kingdoms. Proc. Natl. Acad. Sci. USA. **74**(11), 5088–5090 (1977)

Woese, C.R., Kandler, O., Wheelis, M.L.: Towards a natural system of organisms: proposal for the domains archaea, bacteria, and eucarya. Proc. Natl. Acad. Sci. USA. **87**(12), 4576–4579 (1990)

Wool, I.G.: Extraribosomal functions of ribosomal proteins. Trends Biochem. Sci. **21**(5), 164–165 (1996)

Zaremba-Niedzwiedzka, K., Caceres, E.F., Saw, J.H., Backstrom, D., Juzokaite, L., Vancaester, E., et al.: Asgard archaea illuminate the origin of eukaryotic cellular complexity. Nature. **541**(7637), 353–358 (2017)

Zintzaras, E., Santos, M., Szathmáry, E.: "Living" under the challenge of information decay: the stochastic corrector model vs. hypercycles. J. Theor. Biol. **217**(2), 167–181 (2002)

Chapter 10
Earliest Traces of Life as a Window on Life's Origins

Barbara Cavalazzi, Keyron Hickman-Lewis, André Brack, and Sherry L. Cady

Abstract Life is the outcome of a complex network of chemical reactions and molecular interactions that emerged on Earth once primitive chemical automata could self-assemble in such a way that enabled them to self-reproduce and evolve. Yet exactly how, where and when life first appeared on our planet remains unknown. In this chapter, we review the various lines of evidence from fossil and geochemical traces of early life preserved in the geological record, which provide fundamental, albeit often rudimentary, insight into early life. The oldest fossils record the nature of life more than half a billion years after it emerged on Earth and suggest that considerable metabolic diversity had already evolved by this time. Microfossils, microbial mats, stromatolites and organic and inorganic geochemical signatures have been interpreted as compelling evidence for Archean biodiversity. In spite of the chemical enigmas of the earliest life and its limited record, characterization of the various classes of biosignatures indicative of life in its geological context provides guidance as to the earliest co-evolution of the geosphere and biosphere.

B. Cavalazzi
Università di Bologna, Bologna, Italy
e-mail: barbara.cavalazzi@unibo.it

K. Hickman-Lewis
Department of Earth Sciences, The Natural History Museum, London, United Kingdom
e-mail: keyron.hickman-lewis@nhm.ac.uk

A. Brack
Centre de Biophysique Moléculaire, CNRS, Orléans, France
e-mail: andre.brack@cnrs-orleans.fr

S. L. Cady (✉)
Pacific Northwest National Laboratory, EMSL, Richland, WA, USA

Department of Geology, Portland State University, Portland OR, USA
e-mail: cadys@pdx.edu

A. Neubeck, S. McMahon (eds.), *Prebiotic Chemistry and the Origin of Life*,
Advances in Astrobiology and Biogeophysics,
https://doi.org/10.1007/978-3-030-81039-9_10

10.1 Introduction

Life emerged on Earth when parts of the chemical automata self-assembled to generate automata capable of self-reproduction and evolution. Occassional minor errors in the assembly of these molecules would have generated more efficient automata that became the dominant entities. With time, repetition of these processes ultimately led to the emergence of life as we know it. Like Luigi Pirandello in his "*Six Characters in Search of an Author,*" chemists remain in search of the ingredients, the environment, and the most plausible scenarios by which life occurred on Earth around four billion years ago. The lack of a prebiotic geochemical record on Earth eliminates the possibility of identifying the essential evidence required to prove the pathway by which these processes occurred. Likewise, the oldest fossil evidence for life is not necessarily representative of the oldest forms of life on Earth. The fossil evidence for ancient life nonetheless elucidates the variety of settings that hosted early life.

The propensity for life to occupy and colonize surfaces in every favorable niche, which is certain to have happened once it emerged on our planet, adds to the ambiguities faced by chemists who are now closer than ever to completing the puzzle of life's origins. A key step in the puzzle will be reconciling the scenario by which the ingredients for life come together in a sufficiently robust form to have withstood the challenges it would have faced in nature, which would have been considerably different than habitats on Earth today (Martin et al. 2008; Westall and Brack 2018; Westall et al. 2018).

Given that the exact steps, pathways and environmental constraints that led to the emergence of life remain hypothetical, a chemist's playground of opportunities remain constrained primarily by the fundamental laws of science and the limitations imposed by physical/geological settings, the latter of which certainly lies on a spectrum, from having been only slightly different geochemically from their modern analogue environments to being drastically different from any type of environment known to us (Westall et al. 2018; Cavalazzi et al. 2019), and therefore having become obsolete as Earth evolved through time. The vagaries of time are certain to have masked nearly any clue of the events that led to the emergence of life in our geological record. Hence we review the necessary steps required to assemble life as we know it, followed by an overview of the evidence for the earliest life on Earth with insights as to how it is recognized, given the possibility that early life could have emerged in environmental settings that are not significantly different from the types of habitats with which we are familiar, such as hydrothermal settings, which have existed throughout Earth history (Nisbet and Sleep 2001).

10.2 The Record of Early Life

Life as we know it emerged during the Hadean Eon (~4.5–4.0 Ga), on an Earth that was truly a "distant world" in terms of the planet's evolutionary timeline (Sleep 2010). Although various niches on Earth could likely have supported life since the early Hadean (Nisbet and Sleep 2001; Dass et al. 2016; Westall et al. 2018; Sleep 2018), the earliest evidence—direct and indirect—for a biosphere on Earth is archived within the fossil record of prokaryotes in the subsequent Archean Eon (4.0–2.5 Ga). This situation presents a worst-case scenario for deducing the nature of the earliest life, since the oldest evidence for life on Earth records the nature of life more than half a billion years after its origin. In spite of this constraint, by scrutinizing the earliest fossil record it becomes possible to extrapolate back in time to different plausible environments that could have hosted the origins of life. Comparisons of the possible environments that could have hosted the earliest forms of life have recently been reviewed (Westall et al. 2018; Camprubí et al. 2019).

Much of what is known about ancient microorganisms on the early Earth stems from a small number of highly contested purported fossils that were collected from rather limited quantities of rock. Prior to two billion years ago (i.e., the advent of the Paleoproterozoic Turee Creek-, Gunflint- and Francevillian-type biotas), almost all puported microfossil morphologies were remarkably simple (filamentous, vibrioidal, coccoidal and spheroidal,though not lenticular). These are the same morphologies that represent the major cellular morphotypes found on Earth today. Some have argued that many of the most ancient "microfossils" arose from mineralization phenomena, such as the silica-carbonate (witherite) biomorphs, silica-metal oxide vesicles, and iron mineral chemical gardens (Garcia-Ruiz et al. 2003, 2020; Bizzarri et al. 2017; McMahon 2019), Such hypotheses underscore the need to determine whether such objects preserve multiple biosignatures, an attribute of *bona fide* (i.e., carbonaceous) microbial fossils (e.g., Cady 2002).

The earliest rock record is highly fragmentary and limited in extent, as dynamic tectonic processes throughout Earth history have inexorably reworked Hadean and Eoarchean strata. Most compelling evidence for early life (>3 Ga) comes from only two regions on Earth: the Barberton greenstone belt of southern Africa and the greenstone belts of the East Pilbara terrane, Western Australia. These tectonostratigraphic entities present sequences of volcanic and sedimentary rocks, with the latter, primarily in the form of cherts and sandstones, preserving the canonical traces of life. This record provides the evidence closest in time to the origins of life and, hence, naturally guides our expectations both of the early evolution of life and the organic (geo)chemical processes that led to the first life. Decoding the evidence for early life remains, thus, one of the greatest challenges facing modern geobiology.

In the sections that follow, we summarize the evidence for early life, propose areas for future research, and illustrate how the various evidence is used as a means to understand the nature of early life on Earth at microbially relevant spatial and temporal scales.

10.2.1 Indicators of Early Life

Biosignatures, the general term for evidence of biological activity archived in the rock record can be broadly classified into five groups:

- Cellular remains: fossilized (carbonaceous) cells and the extracellular remains of microbial communities;
- Microbialites: bio-sedimentary constructions, including small-to-large scale (millimeter- to meter-scale) structures such as stromatolites, thrombolites, microbially induced sedimentary structures (MISS), and microbially influenced rock fabrics, e.g., biolaminites and mineralized mats and biofilms;
- Biochemical signals of the co-evolving geosphere and biosphere: biomarker molecules, isotope biogeochemistry/fractionations, elemental geochemistry/enrichments (passive or paleo-metallomic), and the (ultra)structural characteristics of carbonaceous and organic materials preserved in rocks;
- Biominerals: biologically induced or mediated minerals;
- Bioalteration: textures in rocks that indicate the former presence of microbial activity and include, but are not limited to, tubular alteration and corrosion textures in lava, ambient microbial inclusion trails, and microbial borings at the surfaces of minerals.

Any effort to make a case for biology in ancient rocks remains challenging for the simple reason that any one of the various types of microbial biosignatures (sans rare ancient *bona fide* microbial fossils (c.f. Cady 2002) and uniquely biological molecular biomarker suites (c.f. Summons et al. 2011) can potentially, in isolation, be accounted for by non-biological means (Brasier et al. 2005, 2015). The substantial burden of proof required to demonstrate the presence of early life in geological materials is thus contingent upon the presence of multiple correlated lines of evidence, each of which must be demonstrated with profound rigor after careful collection from geological deposits that represent an unequivocally habitable setting (Wacey 2009).

10.2.1.1 Cellular Remains

Establishing the biogenicity of ancient purported microfossils has led to widespread and often contentious discourse regarding the "characters" that must be exhibited to conclusively support a biological origin (e.g., Brasier et al. 2002; Schopf et al. 2002). As noted in the seminal work of Cloud (1976), "...a given microstructure can be considered demonstrably biogenic only if some of its representatives display a level of cellular, microstructural, or biogeochemical differentiation comparable with that of living organisms and implying a similarity of function and continuity of evolution between them."

Subsequent refinements and restatements of Cloud's criteria for establishing the biogenicity of ancient evidence for life resulted in new criteria for recognizing ancient microfossils and critical appraisals of proposed ancient microfossil assemblages from specific localities. Buick (1990), for example, challenged previous biological interpretations of non-carbonaceous "microbiota" from the ~3.5 Ga Dresser Formation at North Pole, Pilbara, and Brasier et al. (2002, 2005) and Wacey et al. (2016) demonstrated an abiogenic origin for the microfossil-like suite of objects that, at the time, was considered the oldest convincing evidence for ancient fossil remains (the ~3.46 Ga Apex chert, also Pilbara; Schopf and Packer 1987).

Since each ancient microfossil discovery reveals unique combinations of cellular morphotypes, palaeoenvironmental processes, taphonomic pathways, and modes of preservation, the *traditional approach* to microfossil acceptance, which focussed on descriptive and comparative assessments of the characteristics of objects of putative biogenic origins with objects of known biogenic origin, has more recently been followed with a more holistic *contextual approach*, which considers a combination of regional, local, and fossil-scale biogeochemical characteristics of putative microfossils as a means to demonstrate whether microfossil-like objects are not abiogenic (c.f., Tomescu et al. 2016).

This contextual, holistic approach is used in most recent assessments of microfossil biogenicity and places the burden of proof on falsifying alternative nonbiological (null) hypotheses (Brasier et al. 2002, 2005), which is more resilient to inductive reasoning than the *traditional approach* noted above. A discussion of the oldest examples of key morphotypes of microbial body fossils illustrates why a contextual, rather than traditional, approach is necessary when evaluating the most convincing evidence for ancient life.

Filaments

Filamentous cell-like objects have been identified in multiple geological horizons dated to be between 3.0 and 3.5 Ga, including the 3.48 Ga Dresser Formation (Ueno et al. 2001a, b), the 3.45 Ga Hooggenoeg Cherts (Walsh 1992), the 3.416 Buck Reef Chert (Walsh and Lowe 1985; Walsh 1992), unspecified 3.3–3.4 Ga Kromberg Formation cherts (Muir and Grant 1976), the 3.24 Ga Kangaroo Caves Formation (Rasmussen 2000; Wacey 2009), the 3.2 Ga Dixon Island Formation (Kiyokawa et al. 2006, 2014), the 3.020 Ga Cleaverville Formation (Ueno et al. 2006a), and the 3.0 Ga Farrell Quartzite (Sugitani et al. 2007) (Fig. 10.1). In most cases, ancient filamentous structures are solid or hollow thread-like carbonaceous materials up to several micrometers in diameter and possibly tens or hundreds of micrometers in length, bearing morphological and dimensional similarity to microfossils from younger horizons, such as *Gunflintia* spp. from the 1.88 Ga Gunflint Formation. Exceptions to this rule are the pyritized filaments reported from the Kangaroo Caves Formation, an extensively remineralized volcanogenic massive sulfide deposit.

Once the syngenetic context, both regional and local, of ancient filamentous microfossil-like objects has been demonstrated, high-resolution (sub-micrometric) analytical techniques are required to distinguish their biogenic origin (c.f., Oehler and Cady 2014). For example, one of the traditional criteria for recognizing ancient filaments is the presence of microscopically visible septae that regularly subdivide elongate filaments (Buick 1990). However, a high-resolution study that re-examined

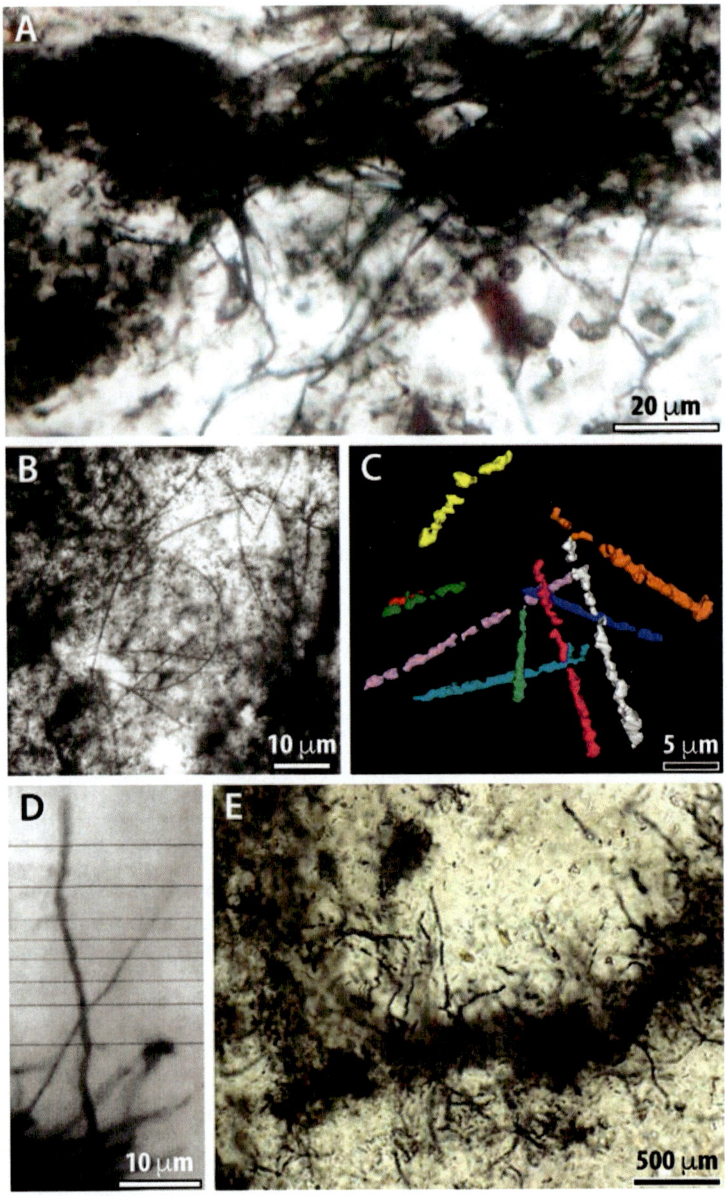

Fig. 10.1 (**a**) Clumps of filaments from the Kromberg Formation; (**b**, **c**) Pyritised filaments from the Kangaroo Caves Formation, (**c**) shows 3D visualisation; (**d**) Carbonaceous filaments from the Dresser Formation; (**e**) Carbonaceous filaments from the Dixon Island Formation. (**a**) Modified from Walsh and Westall (2003); (**b**) Courtesy of Birger Rasmussen; (**c**) Modified from Wacey et al. (2014); (**d**) Modified from Ueno et al. (2001a); (**e**) Courtesy of Shoichi Kiyokawa

the submicroscopic attributes of some iron oxide replaced *Gunflintia* spp. filaments (Maldanis et al. 2020) shows that septae can be lost during mineral replacement of filamentous cells. High-resolution analysis has also proven indispensible in demonstrating the non-biogenic origin of some ancient filamentous objects, as shown when pseudo-septate patterns observed in filament-like objects from the Apex chert were identified as having been produced from the condensation of carbonaceous material around phyllosilicate templates with no biological origin (Wacey et al. 2016).

Spheroids and Coccoids

Ancient spheroidal or coccoidal microfossil-like objects have been identified in the 3.45 Ga Hooggeoneg Formation (Walsh 1992; Glikson et al. 2008), the 3.446 Ga Kitty's Gap Chert (Westall et al. 2006, 2011), the 3.35–3.43 Ga Strelley Pool Chert (Wacey et al. 2011; Sugitani et al. 2015), the 3.26 Ga Fig Tree Group (Knoll and Barghoorn 1977), and the 3.0 Ga Farrell Quartzite (Sugitani et al. 2007, 2009) (Fig. 10.2). Coccoidal objects exhibit substantial size variation. Although in early work, a statistically normal size distribution was considered a requirement for demonstrating the biogenic origin of a coccoid population (e.g., Schopf and Walter 1983), it is now known that a normal distribution of spherical objects can also arise naturally in inorganic particle populations (Buick 1990). Spheroidal microfossil-like objects also exhibit considerable morphological differentiation: within millimeters, hollow and solid forms, both solitary and grouped, have been identified (Sugitani et al. 2007, 2015). Compelling evidence for biogenesis has been found in certain populations of spheroidal microstructures: (i) both Knoll and Barghoorn (1977) and Westall et al. (2006, 2011) identified evidence for cell division; (ii) Walsh (1992) and Sugitani et al. (2007, 2015) showed that coccoidal structures can occur in pairs and groups consistent with biological cell distributions; Walsh (1992) and Sugitani et al. (2007) identified features consistent with the rupturing of cell walls; and Westall et al. (2006) noted deflated coccoid surface textures consistent with cell lysis. In this regard, coccoidal-shaped microfossils are the only ancient fossil morphotype that displays evidence for multiple stages of the cell cycle, though caution is warranted as some of the apparent "life-cycle stages" can also occur during the growth of microfossil-like minerals: silica microspheres and botryoidal oxides exhibit division-like phenomena, and many mineral species show self-organization into arrangements not dissimilar to the pairing and grouping shown by coccoidal microfossil-like objects. Unlike filaments, Early Archean coccoidal microfossils have not, at the time of writing, been subjected to high-resolution microanalysis, which illustrates a potentially impactful future research direction for which the three dimensional tomographic analyses of coccoidal Gunflint-type microfossils (*Huroniospora spp.* and *Eosphaera*) in the works of Wacey et al. (2013), Hickman-Lewis et al. (2017) and Maldanis et al. (2020) provide a benchmark.

Lenticular

A third group of proposed microfossils are described as lenticular (earlier work used the term spindle-like); these are large, thick-walled structures with single or double internal cavities and diaphanous carbonaceous coatings. Lenticular microstructures

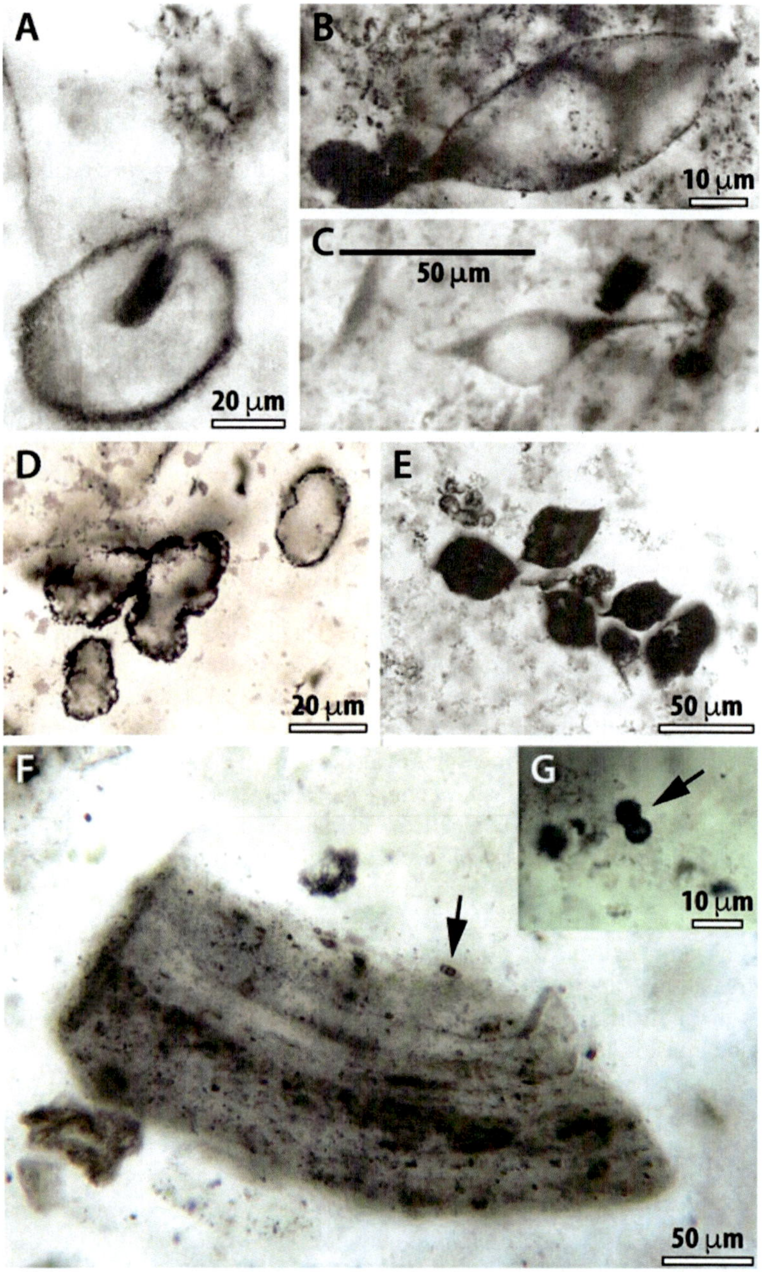

Fig. 10.2 (**a**) Ruptured spheroid from the Kromberg Formation; (**b**, **c**) Lenticular forms from the Kromberg Formation; (**d**) Spheroids from the Strelley Pool Formation; (**e**) Lenticular forms from the Farrell Quartzite; (**f**, **g**) Dividing spheroids from the Sheba Formation (arrows). (**a**, **b**) Modified from Walsh (1992); (**c**) Courtesy of Maud M. Walsh; (**d**) Courtesy of Wacey et al. (2011); (**e**) Courtesy of Kenichiro Sigitani; (**f**, **g**) Courtesy of Andrew H. Knoll

have been identified in the Hooggenoeg Formation (Walsh 1992), the Strelley Pool Chert (Sugitani et al. 2015; Oehler et al. 2017), various Kromberg Formation cherts (Walsh 1992; Oehler et al. 2017; Alleon et al. 2018), and the Farrell Quartzite (Sugitani et al. 2007). They have been ascribed a multitude of biological functions, including the outer sheath of microbial colonies (Walsh 1992) and planktonic organisms (Sugitani et al. 2007, 2015; Oehler et al. 2017; Kozawa et al. 2019) (Fig. 10.2). Their occurrence in pairs and linked chains has been hypothesized to reflect cell division by binary fission (Sugitani et al. 2015); however, no detailed comparison with the few younger microstructures bearing similarity to lenticular microfossils has yet been undertaken and their origin remains open to discussion. One particular challenge with regard to lenticular microstructures is their highly variable morphology; in spite of the large volume of research that supports a biogenic origin for these microstructures, similar populations of objects, some of which are non-carbonaceous, have been found in association with microbial mats of the Middle Marker horizon (Hickman-Lewis et al. 2018).

Analytical Approaches to Archean Microfossils

It is clear that further high-resolution studies of Early Archean microfossils are required to evaluate their biogenicity, demonstrate their metabolic affinities, and explore their evolution through diagenesis. High-resolution approaches have been widely applied to younger Proterozoic microfossils to constrain life processes (e.g., Wacey et al. 2013) and diagenetic changes (e.g., Maldanis et al. 2020). Unfortunately, for the Archean, such studies are hindered by the scarcity of microfossils preserved in rocks of this age. Nonetheless, approaches such as transmission electron microscopy (De Gregorio and Sharp 2006), micro-scale X-ray computed tomography (X-ray μCT; Hickman-Lewis et al. 2017), hydrolysis–pyrolysis (Duda et al. 2018), synchrotron X-ray fluorescence (Sforna et al. 2017) and micro-scale particle-induced X-ray emission (μPIXE; Hickman-Lewis et al. 2020b), have been used to study Archean microfossils and organic materials with promising results. One enduring challenge is that microfossils are rarely preserved within Archean stromatolites and microbial mats (Schopf et al. 2007), and this lack of spatial association leads to reciprocal challenges in proving the biogenicity of both types of traces of life: how can one be assured of the biogenic origins of a stromatolite when no traces of its architects are preserved? Furthermore, many microfossil-oriented studies fail to properly account for the chemistry of the depositional environment at local and regional scales. Taking into account the combination of stratigraphy, sedimentology, and layer-specific geochemistry (both mineralogy and evidence of aqueous alteration) can provide key constraints regarding the conditions of deposition relevant to microbial growth and thus the types of organisms that could plausibly have been present (Kiyokawa et al. 2006; Sugitani et al. 2015; Hickman-Lewis et al. 2020a).

10.2.1.2 Microbialites

Interpretation of the biogenicity of ancient microbialites suffers from similar shortcomings to ancient microfossils, since several macroscopic biological fabrics can be reproduced through abiotic processes (Lowe 1994; McLoughlin et al. 2008; Hickman-Lewis et al. 2018, 2019). Suites of biogenicity criteria have been devised to critically assess the origins of microbial mat fabrics (Hickman-Lewis et al. 2018) and MISS (Noffke 2010), taking into account both the macroscopic and microscopic morphologies of the structures and their relationships with the surrounding stratigraphy. When using these nested microbial–environmental biogenicity criteria, the biogenic origin of fossilized microbial mats, mat relicts, and MISS can be concluded beyond any reasonable doubt. Although the individual constituent parts of microbial mats—filamentous microstructures, carbonaceous aggregates, entrained mineral particles, and laminar macromorphologies, among others—can be accounted for individually by abiotic or hydrodynamic mechanisms, their occurrence in a unified structure cannot. As a consequence of this multi-scale, multi-dimensional approach, fossilized microbial mats, biofilms, and MISS are perhaps the most uncontroversial fossils in deep time and have been identified widely throughout Paleoarchean and Mesoarchean successions, including the Dresser Formation (Noffke et al. 2013), the Middle Marker horizon (Hickman-Lewis et al. 2018, 2019), the Apex Basalt (Hickman-Lewis et al. 2016), the Hooggenoeg Formation (Walsh 1992; Walsh and Lowe 1999) the Buck Reef Chert (Walsh 1992; Walsh and Lowe 1999; Tice and Lowe 2004, 2006a, b; Tice 2009), the Footbridge Chert (Hickman-Lewis et al. 2020a), the Josefsdal Chert (Westall et al. 2006, 2011, 2015), the Mendon Formation (Trower and Lowe 2016), the Moodies Group (Noffke et al. 2006; Heubeck 2009; Homann et al. 2015, 2018) and the Dixon Island Formation (Kiyokawa et al. 2006) (Fig. 10.3).

Fossilized microbial mats and biofilms comprise fine, undulating or crinkly carbonaceous laminations, each level of which represents microbial communities that colonized the sediment surface; non-isopachous laminae (i.e., growth with a low degree of inheritance from underlying laminations) exhibit anastomosis behavior that is incompatible with a sedimentary or diagenetic precursor, but rather evidences trophic or ecophysiological behavior (Hickman-Lewis et al. 2016, 2018). Non-isopachous morphologies on multiple scales are taken to indicate growth according to trophic requirements; for example, photosynthetic microbial mats grow more rapidly at crests, in part due to exposure to sunlight, while permanently submerged benthic mats can form crested or tufted morphologies due to emergence from the laminar bottom flow into hydrodynamically well-mixed, nutrient-rich waters above, or to avoid chemical stresses that result from concentrations of heavy metals arising from EPS (extracellular polymeric substances) detoxification mechanisms. At the sub-millimeter scale, early Archean fossilized microbial mats alternate with pure silica or granular layers and lenses and are laterally discontinuous at micrometer-scales, though broadly continuous at the centimeter-scale (Walsh 1992; Noffke et al. 2013); this denotes their original filament-like morphology, for which limited evidence has been observed in three dimensions (Hickman-Lewis et al. 2017).

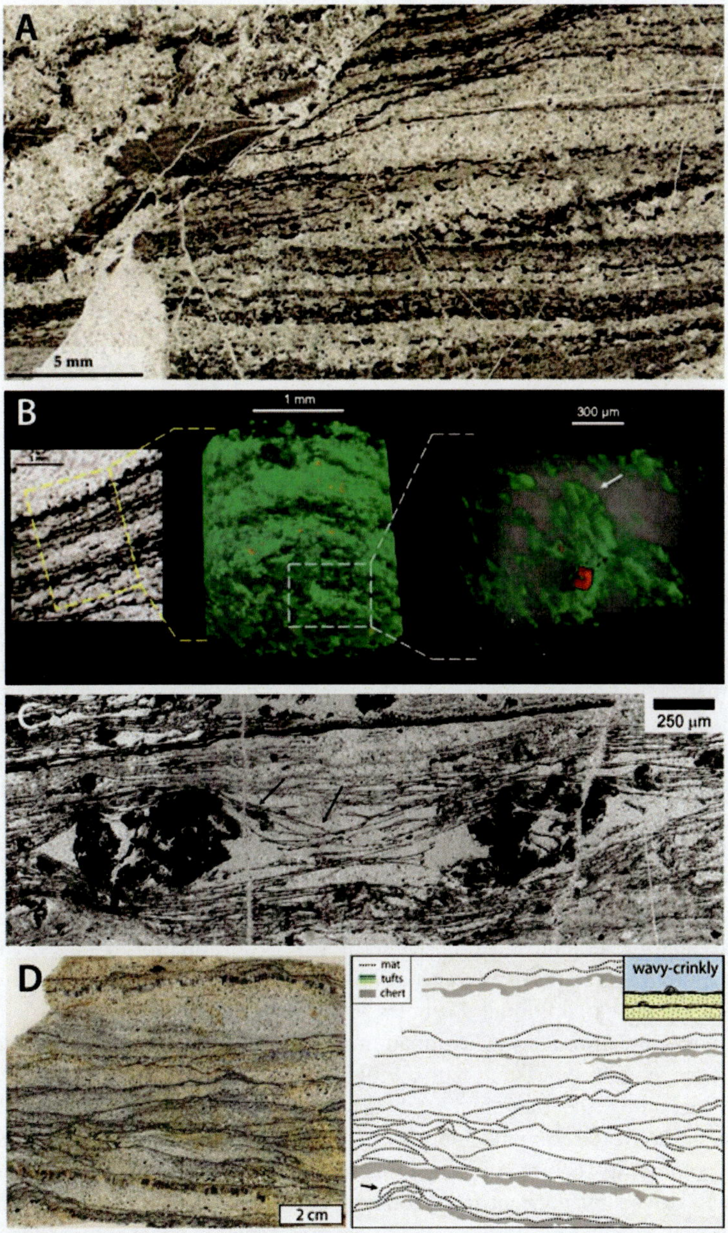

Fig. 10.3 (**a**, **b**) Microbial mats from the Middle Marker horizon, (**b**) shows high-resolution X-ray μCT scan of mats exhibiting micro-tufts (arrow); (**c**) microbial mats from the Buck Reef Chert; (**d**) Microbial mats from the Moodies Group. (**a**, **b**) Credit: Keyron Hickman-Lewis; (**c**) Modified from Tice (2009); (**d**) Modified from Homann et al. (2015)

Key is the entrainment and orientation of sediment particles due to trapping and binding processes by EPS (Noffke 2010; Noffke et al. 2013); preferentially oriented sediment particles are among the key MISS attributes that provide evidence for biological action. The in vivo plasticity of many microbial mats is also denoted by torn and reworked cohesive mat fragments or tangles of filament-like objects that are deposited within the immediately adjacent sediment (Tice and Lowe 2004; Kiyokawa et al. 2006).

Stromatolites are the most iconic microbial fossils and spectacular living examples are directly comparable to ancient forms in many respects, which provides an opportunity to elucidate the nature and role of biological and environmental processes that characterize them (Reid et al. 2003; Riding 2011). The oldest reported stromatolites are found in the Dresser Formation (Walter et al. 1980; Van Kranendonk 2006; Djokic et al. 2017), the Strelley Pool Chert (Hofmann et al. 1999; Allwood et al. 2006, 2009, 2010), and the Mendon Formation (Byerly et al. 1986; Byerly and Palmer 1991). The stromatolites in the Dresser Formation include the domical–conical forms and a diverse assemblage of morphotypes in the Strelley Pool Chert, which include complex conical *Conophyton*-type, encrusting, cuspate, and "egg carton" forms. Geochemical work has shown that the predominance of these forms varies predictably with sedimentary depositional facies (Allwood et al. 2009).

Though controversial, older (~3.7 Ga) stromatolite forms are proposed by Nutman et al. (2016), who argue that ancient laminated conical structures found in the Isua region of Greenland are poorly preserved stromatolites. That work has been challenged by Allwood et al. (2018), whose subsequent analysis of photographs of multiple faces of rock samples and X-ray fluorescence elemental mapping, combined with structural analysis of the outcrop, showed that the structures possessed neither internal lamination nor elemental distributions that could be argued to result from microbial processes. Allwood et al. (2018) found that the putative stromatolites were characterized by ridge-like rather than conical/domical morphologies and argue that the structures are best interpreted as deformation features, with ridge-like morphologies that reflect the applied stress field. In their follow-on study, which addresses additional attributes and geochemical signatures associated with the purported stromatolites, Nutman et al. (2019) claim that the preponderance of evidence remains consistent with a biogenic origin.

Establishing the biogenicity of stromatolites presents formidable challenges since they reflect a complex interplay of multiple processes: biomass growth; microbial community composition; sedimentation rate and character; surface normal growth driven by nutrient diffusion, mineralization, and biophysical constraints; the production and properties of EPS, environmental hydrodynamic processes, and stochastic events (Grotzinger and Knoll 1999; Bosak et al. 2013; Hickman-Lewis et al. 2019). Furthermore, as noted previously, stromatolites rarely include direct evidence and/or expressions of the microorganisms that were involved in their construction. Stromatolite geobiology has thus expanded to include mathematical predictive frameworks founded in the observation that stromatolite morphologies can be partly or fully accounted for by a combination of Kardar-Parisi-Zhang

(KPZ) dynamics, diffusion-limited aggregation, and stochastic growth (Grotzinger and Rothman 1996; Batchelor et al. 2002, 2003; Dupraz et al. 2006; Petroff et al. 2010). The application of such models, however, generates considerable controversy since it remains unclear how best to set the initial conditions of stromatolite growth, incorporate fluctuating palaeoenvironmental parameters, and incorporate stochastic events that may drive the formation of unique morphologies and variants. Many growth models are explicitly deterministic and include no randomness, producing results that depend directly upon the initial settings; this is inconsistent with natural microbialite growth, especially in ecosystem settings where both intrinsic and extrinsic variables fluctuate episodically to produce macroscale and microscale peculiarities (Dupraz et al. 2006; Cuerno et al. 2012). More sophisticated models, while needed, are generally computationally intensive.

Lowe (1994) and Grotzinger and Rothman (1996), among others, argued that simple stromatolite forms, such as domes and cones, can be replicated through abiotic processes. Yet the extraordinary diversity of stromatolite forms with plausibly similar biological genesis—greatest in the Proterozoic, though not insubstantial in the Archean—cannot be explained by abiotic models alone. Indeed, it is difficult even to model conical *Conophyton*-like forms without considering inherent biological attributes (Batchelor et al. 2003; Petroff et al. 2010). It has been suggested more recently that "true" biological signatures in stromatolites may be indicated in macromorphologies that do not conform to certain model expectations (e.g., overhangs; Cuerno et al. 2012) or micromorphologies that can be reproduced through geomicrobiological experimentation (e.g., fabric orientation and density differences within microbial pinnacles; Hickman-Lewis et al. 2019). Observations of modern stromatolites show that such complexity arises from the interaction of the stromatolite-building biota with paleoenvironmental parameters (Reid et al. 2003) (Fig. 10.4).

10.2.1.3 Biogeochemical Signals of the Co-evolving Geosphere and Biosphere

Biomarker Molecules

Chemical fossils include lipid biomarker molecules that can only be derived directly from biological processes (Summons et al. 2011). Regrettably, the biomarker record is plagued with uncertainty due to likely/demonstrable contamination (French et al. 2015) and the inability to consider an individual biomarker compound in isolation as a reliable indicator of Archean metabolism. In the absence of a suite of biomarkers, isotope geochemistry and trace and rare earth element compositions are used to decode aspects of co-evolving ecosystems and environments. Recently, Fourier Transform Infra-Red (FTIR) spectroscopy has been applied to Archean carbonaceous materials as an alternative means of decoding primary aliphatic biochemistry, being sensitive to organic materials but less susceptible to contamination. FTIR investigations illustrate that microfossils and microbial mats comprise organic

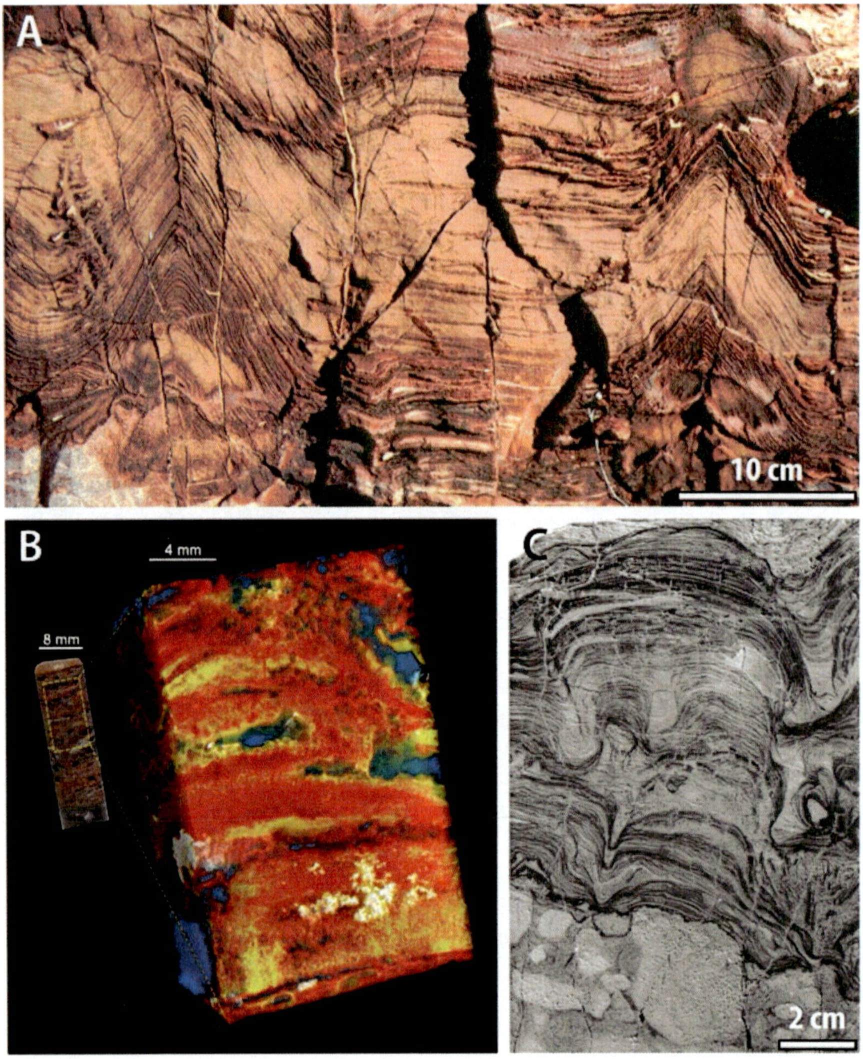

Fig. 10.4 (**a**) Conical stromatolites from the Strelley Pool Formation; (**b**) High-resolution X-ray μCT scan of mineralised stromatolites from the Dresser Formation showing 3D variations in mineralogy and lamination micromorphology (red = iron oxide, yellow = silica, white = baryte, blue = void space); (**c**) Stromatolites from the Mendon Formation. (**a**) Courtesy of S.M. Awramik; (**b**) Credit: Keyron Hickman-Lewis; (**c**) Modified from Byerly and Palmer (1991)

residues from both Bacteria and Archaea based on the exceptional preservation of relict membrane proteins with diverse aliphatic compositions, e.g., the relative abundances of CH_2 and CH_3 moieties (Igisu et al. 2018; Hickman-Lewis et al. 2020c).

Isotope Biogeochemistry

Isotope geochemistry is a powerful approach toward both paleoenvironmental and ecological reconstructions in deep time. Traditional and non-traditional stable isotope approaches using Li, C, N, O, Mg, Si, S, K, V, Cr, Fe, Ni, Cu, Se, Mo, Te, and Tl, among other elements, have been used throughout the Archean geological record to unveil a multitude of biogeochemical and paleoenvironmental parameters (see a comprehensive review by Teng et al. 2017). Most commonly, carbonaceous microfossils are routinely analyzed for their $\delta^{13}C$ composition, since metabolic processes typically yield products with negative $\delta^{13}C$ (Schidlowski 2001). Since the abiotic synthesis of carbonaceous materials can produce similarly wide-ranging $\delta^{13}C$ values (McCollum and Seewald 2006), the argument that negative $\delta^{13}C$ is solely indicative of a biogenic precursor is far from infallible; a more reasonable biogenic signature is a restricted range of $\delta^{13}C$. Coupled isotopic records, such as the $\delta^{13}C$ records of concomitant organic carbon and associated carbonates, can also chart local biogeochemical events (e.g., the pairing of fluctuations in organic and inorganic $\delta^{13}C$ that correspond to terrestrial water influxes into marine environments). Strongly negative C isotope signals ($\delta^{13}C$) associated with Archean organic material, both body fossils and disseminated carbonaceous materials, include $\delta^{13}C$ values of $-28.2‰$ in >3.7 Ga graphite from Labrador, Canada (Tashiro et al. 2017), $-30‰$ to $-19‰$ in >3.7 Ga graphite from Isua (Rosing 1999; Grassineau et al. 2006), as low as $-56‰$ to $-32‰$ in methane inclusions from the Dresser Formation (Ueno et al. 2001a, b, 2006b), and -40 to $-10‰$ in cherts of the Barberton greenstone belt (Grassineau et al. 2006; van Zuilen et al. 2007; Hickman-Lewis et al. 2020b).

An important consideration is the fact that negative C isotope fractionations (between ca. $-40‰$ and $0‰$) are compatible with multiple biological pathways, hence in the absence of strongly grouped $\delta^{13}C$ measurements associated with a single microstructure or a cellular morphotype, it can be difficult or impossible to detect specific metabolisms. Exceptions include a strongly negative $\delta^{13}C$ signal considered incompatible with photosynthesis though consistent with methanogenesis. Less well-constrained distributions of $\delta^{13}C$ associated with different fossil morphotypes could indicate either a mixed non-syntrophic microbial consortium or the influence of non-biological carbon, for example, from FTT processes, which is not expected to conform to a restricted range of values. In such cases, additional evidence is necessitated. As examples, van Zuilen et al. (2007) combined C isotopes with Raman and H/C measurements, Grassineau et al. (2006) used both C and S isotopes from the same horizons, and Hickman-Lewis et al. (2020b) combined C isotopes with Raman, TEM and μPIXE spatially quantified trace element mapping; in all cases, the carbonaceous materials studied were found to be most plausibly of biogenic origin.

Ambiguity in the interpretation of organic $\delta^{13}C$ in ancient microfossils can also be resolved by correlating $\delta^{13}C$ measurements with $\delta^{34}S$, through which the contributions of sulfur-metabolizing pathways can be incorporated into a more comprehensive understanding of the dominant biogeochemical processes (Wacey et al. 2011; Bontognali et al. 2012). Wacey et al. (2011) identified cell-like structures in the Strelley Pool Formation associated with pyrite crystals with $\Delta^{33}S$ values between -1.65 and $+1.43$‰ and $\delta^{34}S$ values from -12 to $+6$‰ (normalized to the Vienna Canyon Diablo Troilite). These values, when considered alongside strongly negative $\delta^{13}C$ (-33 to -46‰) in the same structures, indicate sulfate-reduction and sulfur-disproportionation metabolic pathways. Bontognali et al. (2012), who also analyzed carbonaceous materials from Strelley Pool, measured large $\delta^{34}S$ fractionations at small spatial scales, including values below -15‰, and positive $\Delta^{33}S$ values which, taken together, suggest the sulfurization of organic matter by respiratory sulfur metabolisms alongside elemental sulfur disproportionation. On regional and global scales, the change from dominantly mass-independent to mass-dependent sulfur fractionation during the Archean has been interpreted to reflect a global "chemical shift" from anoxic to oxic environments.

Isotope records are particularly powerful for understanding changing paleoenvironments, the most pronounced expression of which are the major environmental changes associated with episodes of oxygenation, such as the Great Oxygenation Event (GOE), since such shifts reflect and would have impacted the composition and activities of major biological communities. The GOE and numerous "whiffs" of oxygen—for up to several hundred million years prior—have been studied via multiple isotope chemostratigraphy approaches that include, though are not limited to, coupled C-S (Izon et al. 2017), S (Philippot et al. 2018), Fe (Oonk et al. 2016), Cu (Chi Fru et al. 2016), Mo (Duan et al. 2010) and U (Kendall et al. 2013). Although isotope chemostratigraphy has seen limited application to Early Archean rocks, isotope systematics have nonetheless been used to infer the characteristics of habitable environments on the early Earth. Particularly informative isotopic data include a detailed $\delta^{18}C$ record from Paleoarchean–Neoarchean cherts that demonstrate that the early oceans were warm (Knauth 2005; Hren et al. 2009; Tartèse et al. 2017) and $\delta^{30}Si$ isotopes from predominantly South African cherts that indicate shared roles of Si-saturated seawater and hydrothermal fluids in the silicification of sediments (van den Boorn et al. 2007, 2010; Abraham et al. 2011). Deviations of around 2‰ in $\delta^{30}Si$ between sedimentary laminated and massive chemical cherts indicate that, while Si-saturated seawater was the principal agent of silicification in laminated cherts, silicification of massive cherts was due to the mixing of hydrothermal fluids and seawater. Combining O and Si isotopes could place distinct temperature-redox constraints on the earliest habitable environments and facilitate determination of the degree of thermophilic adaptation that would be required by microbial inhabitants. O isotopes indicate that the temperatures of the oceans have gradually decreased over geological time, with temperatures of 40–90 °C being argued for the depositional fluid of Archean cherts, consistent with the hypothesis of thermophilic life occupying the early oceans (Knauth 2005).

Elemental Biogeochemistry

A third branch of biochemical signal includes the various types of elemental enrichments associated with fossil and extracellular fossil remains, a phenomenon that could have been concentrated either passively (Allwood et al. 2010; Hickman-Lewis et al. 2019, 2020b) or actively by life (Fraústo da Silva and Williams

2001; Edwards et al. 2014; Hickman-Lewis et al. 2019, 2020b). Deciphering elemental enrichments in paleoenvironments and fossilized biomass is contingent upon constraining the biogeochemistry of individual elements. Paleoenvironmental reconstruction in ancient geological units primarily utilizes trace and rare earth elements plus yttrium (REE + Y) chemistry, the composition of which exhibits predictable aqueous complexation behavior throughout geological time (Bau and Dulski 1996; Kamber and Webb 2001; Shields and Webb 2004; Gourcerol et al. 2015). Early Archean fossiliferous units typically exhibit normalized REE + Y compositions (i.e., patterns of elemental enrichment and depletion) that indicate influences from seawater and hot hydrothermal fluids (Danielson et al. 1992; Van Kranendonk et al. 2003; Allwood et al. 2010), though most fossiliferous cherts from the Barberton greenstone belt exhibit an additional major component from river-derived inputs (Hickman-Lewis et al. 2020a). In brief, widespread phototrophic microbial communities on the early Earth flourished under disequilibrium conditions at the confluence of marine, terrigenous, and hydrothermal activity (Hickman-Lewis et al. 2020a), whereas ecosystems dominated by chemosynthesis appear to have been concentrated around zones of elevated hydrothermal activity, sustained by the greater magnitude of the positive Eu anomaly (Westall et al. 2015). It has been suggested that some stromatolites exhibit REE + Y enrichment that could function as a biogenicity indicator (Allwood et al. 2010). While this correlation is consistent with known mechanisms of REE-ligand complexation within biomass, particularly in the case of the middle and heavy REEs (Takahashi et al. 2005; Censi et al. 2013), the precise means of REE + Y uptake by living microbial biofilms and bioconstructions remains poorly constrained (Johannesson et al. 2013).

Perhaps a more promising avenue toward an understanding of bio-indicative elemental enrichments are specific elemental distributions that result from uptake by the cell or extracellular polymers. The metallome, first outlined in detail by Williams (2001) and Fraústo da Silva and Williams (2001), was defined to reflect the chemical selection of elements by the cell according to its biochemical requirements. The potential then exists for such microbial communities to preserve "fingerprints" of elemental enrichments that could distinguish them from other communities, as demonstrated by the fact that the elemental compositions of extant organisms vary according to metabolism. For example, methanogens require higher concentrations (or fractional contributions to their metallome) of V and Ni than other groups (Zerkle et al. 2005; Cameron et al. 2012).

The metallome has been considered the inorganic complement to the proteome and genome and, thus, a possible inherent expression of cell biochemistry. The potential for this biosignature to endure geological processing is promising, and it has recently been suggested that diagnostic fragments of the metallome can be preserved even in Earth's oldest fossil record, revealing the metabolisms of the original biomass, for example, enrichments of Fe, Ni, Co, and V in carbonaceous materials, consistent with a putatively lithotrophic consortium cycling methane and nitrogen (Hickman-Lewis et al. 2020b). Although primary elemental concentrations remain associated with biomass through early diagenesis (Edwards et al. 2014), the possibility of their preservation into deep time remains to be stringently tested. In this regard, elemental enrichments in carbonaceous materials that have undergone rapid preservation should be considered when the origin of the carbonaceous material is likely to be biogenic, as such a biosignature can be used to prove biogenicity and identify metabolism in the absence of obvious cellular preservation (Hickman-Lewis et al. 2020b).

10.2.1.4 Biominerals

Many organisms produce mineral phases as a metabolic by-product. Biomineralization, both extracellular and intracellular, is a widespread phenomenon in stromatolite-building communities, for example, cyanobacteria (Benzerara et al. 2006, 2014) and anoxygenic photosynthesizers (Bosak et al. 2007; Bundeleva et al. 2012). The formation of minerals either induced or mediated by microbes, particularly cyanobacteria, is often factored into morphogenetic models of stromatolites as a dominant growth vector (e.g., Grotzinger and Knoll 1999; Petroff et al. 2010). Microscopic pyrite heavily depleted in ^{36}S is commonly associated with microbial sulfate-reducers (Ueno et al. 2008; Wacey et al. 2011; Marin-Carbonne et al. 2018); though pyrite can also be a by-product of the anaerobic oxidation of methane (AOM; Zhang et al. 2014). Among other mechanisms of biomineralization, oriented microcrystals of magnetite formed by magnetotactic bacteria is another potential biosignature that could date back to the Archean, since the gene cluster responsible for the biomineralization of magnetosomes is thought to have originated between 2.7 and 3.6 Ga (Lin et al. 2017). A review of biomineralization and its application to the fossil record is given by Benzerara et al. (2011).

10.2.1.5 Bioalteration

The fifth group of biosignatures is bioalteration, indirect evidence for life manifested as traces of its influence upon the substrate, that is, ichnofossils. In the context of early life, two main groups of bio-induced alteration of the rock are discussed: in vivo microbial corrosion of rocks, usually basalts and hyaloclastites, by endolithic microbes (forming tubules), and diagenetic or post-diagenetic migration of fabric elements due to the alteration (degradation and degassing) of carbonaceous materials (forming ambient inclusion trails). Features that resemble microbial corrosion of basalts have been identified in pillow lavas as old as 3.47 Ga (Furnes et al. 2004, 2007; Banerjee et al. 2006); although their morphology and petrological setting strongly resemble those of modern microbial corrosion textures in ocean crust, subsequent studies have suggested that the fossilized forms may result from post-diagenetic metamorphic processes (Grosch and McLoughlin 2014). Since the formation processes of even the modern examples remain poorly constrained (Staudigel et al. 2015), the debate concerning the origins of such ancient structures remains unresolved. Recently, however, Wacey et al. (2018) identified the co-occurrence of Ti and C in modern microbial corrosion tubules, directly echoing the concentrations of carbonaceous materials in titanite-remineralized tubules from the Archean, which suggests that aspects of the complex formation and mineralization processes outlined for the genesis of these structures in Furnes et al. (2007) are correct.

Ambient inclusion trails are another type of tubular alteration structure, usually with a single microscopic (1–15 μm) pyrite grain at their terminus, that occur adjacent to loci of decaying organic matter. They have been identified throughout the geological record and are attributed to the effects of thermal decomposition of residual organic material that produces CO_2, and potentially CH_4, as propulsive gases that drive the crystal through a microquartz matrix that is simultaneously locally dissolved by organic acids (Wacey et al. 2008a, b, 2016; Olempska and Wacey 2016). The correlation of C and N within ambient inclusion trails has been used to demonstrate a biological role in their formation (Wacey et al. 2008a, b); nonetheless, a range of abiotic means for the formation of ambient inclusion trails have also been demonstrated, for example matrix dissolution due to the influence of clay minerals (Wacey et al. 2016). Such features may thus occur in the absence of biogenic carbonaceous materials.

10.3 Relevance of the Palaeobiological Record for Life's Origin

The ancient geobiological record provides insights into the types of ecosystems early life occupied, how it behaved, and how different populations interacted with one another during that time. Microbial biosignatures preserved in the ancient fossil record have revealed that early life has retained a remarkable degree of morphological conservatism relative to extant life on multiple spatial scales. Cellular morphotypes of early bacteria are similar to those of extant life today, with the exception of lenticular microfossils that are unique to the ancient fossil record. The relatively early diversification of life that colonized available surfaces led to distinct influences of microbial life on sediment fabrics and microbialite structures. The evolution of different metabolic stratigies, evidenced in the isotopic record, enabled early life to occupy a wide range of environments, which for shallow water environments required evolutionary adaptations so photosynthetic microbial populations could shield themselves from the deleterious effects of UV.

Though the exact nature of the earliest form of life is unknown, several lines of evidence suggest it was thermophilic. Hydrothermal systems provide environments where fluids enriched in reduced inorganic chemical species generate proton gradients and redox disequilibria that could support early life metabolisms (Branscomb and Russell 2013). An analysis of metal utilization by microbes throughout early Earth history illustrates that the earliest metabolisms were likely characterized by metal dependencies similar to those of thermophilic extremophiles (Moore et al.

2017). Almost all paleoenvironmental reconstructions of early life habitats invoked some degree of necessity for thermophily, usually as a result of the influence of volcanism and hydrothermal fluids (Danielson et al. 1992; Nisbet and Sleep 2001; Van Kranendonk et al. 2003; Allwood et al. 2010; Sugahara et al. 2010; Hickman-Lewis et al. 2018, 2020a).

The timing of life's origins is unknown and not recorded in the geological record. Though molecular clocks have been developed to extrapolate the timing of key events during the evolution of life, none have yet been calibrated with the fossil evidence preserved in either the Barberton greenstone belt or the numerous greenstone belts of the Zimbabwe craton. As such, the current generation of phylogenetic trees could be improved with a comprehensive interpretation of the evidence preserved in the oldest fossil record.

In summary, and for the purposes of moving closer to unravelling life's origins, "characters" shared by the ancient geological settings that hosted the earliest evidence for life have recorded key aspects of the environmental components of the prebiotic Earth. Water, higher temperatures, (bio)polymer compositions, and the ability to colonize surfaces are examples of these "characters." The progressive complexification of life throughout the Archean and Proterozoic reciprocally informs us of the types of life and settings that are not thought to be relevant to the origins of life, such as oxygenic metabolisms, which certainly post-date anaerobic, anoxygenic cellular machinery. The interested reader is referred to the work of Westall et al. (2018) and Camprubí et al. (2019) for additional discussion of the environmental parameters relevant to the emergence of life on Earth.

Acknowledgements SLC thanks the NASA Astrobiology Institute (NASA CAN7 Award# 16BB06I to the SETI Institute) and the PNNL EMSL user facility (DoE BER program).

References

Abraham, K., Hofmann, A., Foley, S.F., Cardinal, D., Harris, C., Barth, M.G., André, L.: Coupled silicon–oxygen isotope fractionation traces Archaean silicification. Earth Planet. Sci. Lett. **301**, 222–230 (2011)

Alleon, J., Bernard, S., Le Guillou, C., Beyssac, O., Sugitani, K., Robert, F.: Chemical nature of the 3.4 Ga Strelley Pool microfossils. Geochem. Perspect. Lett. **7**, 37–42 (2018)

Allwood, A.C., Walter, M.R., Kamber, B.S., Marshall, C.P., Burch, I.W.: Stromatolite reef from the early Archean era of Australia. Nature. **441**, 713–718 (2006)

Allwood, A.C., Grotzinger, J.P., Knoll, A.H., Burch, I.W., Anderson, M.S., Coleman, M.L., Kanik, I.: Controls on development and diversity of Archean stromatolites. Proc. Natl. Acad. Sci. USA. **106**, 9548–9555 (2009)

Allwood, A.C., Kamber, B.S., Walter, M.R., Burch, I.W., Kanik, I.: Trace element record depositional history of an Early Archean stromatolitic carbonate platform. Chem. Geol. **270**, 148–163 (2010)

Allwood, A.C., Rosing, M.T., Flannery, D.T., Hurowitz, J.A.: Reassessing evidence of life in 3,700-million-year-old rocks of Greenland. Nature **563**(7730), 241–244 (2018)

Banerjee, N.R., Furnes, H., Muehlenbachs, K., Staudigel, H., de Wit, M.: Preservation of ~3.4–3.5 Ga microbial biomarkers in pillow lavas and hyaloclastites from the Barberton Greenstone Belt, South Africa. Earth Planet. Sci. Lett. **241**, 707–722 (2006)
Batchelor, M.T., Burne, R.V., Henry, B.I., Watt, S.D.: Deterministic KPZ model for stromatolite laminae. Phys. A. **282**, 123–136 (2002)
Batchelor, M.T., Burne, R.V., Henry, B.I., Watt, S.D.: Mathematical and image analysis of stromatolite morphogenesis. Math. Geol. **35**, 789–803 (2003)
Bau, M., Dulski, P.: Distribution of yttrium and rare-earth elements in the Penge and Kuruman iron- formations, Transvaal Supergroup, South Africa. Precambrian Res. **79**, 37–55 (1996)
Benzerara, K., Menguy, N., López-García, P., Yoon, T.-H., Kazmierczak, J., Tyliszczak, T., Guyot, F., Brown Jr., G.E.: Nanoscale detection of organic signatures in carbonate microbialites. Proc. Natl. Acad. Sci. USA. **103**(25), 9440–9445 (2006)
Benzerara, K., Miot, J., Morin, G., Ona-Nguema, G., Skouri-Pane, F., Ferard, C.: Significance, mechanisms and environmental implications of microbial biomineralization. Compt. Rendus Geosci. **343**(2–3), 160–167 (2011)
Benzerara, K., Skouri-Panet, F., Li, J., Férard, C., Gugger, M., Laurent, T., Couradeau, E., Ragon, M., Cosmidis, J., Menguy, N., Margaret-Oliver, I., Tavera, R., López-García, P., Moreira, D.: Intracellular Ca-carbonate biomineralization is widespread in cyanobacteria. Proc. Natl. Acad. Sci. USA. **111**(30), 10933–10938 (2014)
Bizzarri, B.M., Botta, L., Pérez-Valverde, M.I., Saladino, R., Di Mauro, E., Garcia-Ruiz, J.M.: Silica metal oxide vesicles catalyze comprehensive prebiotic chemistry. Chem. Eur. J. **24**, 8126–8132 (2017)
Bontognali, T.R.R., Session, A.L., Allwood, A.C., Fischer, W.W., Grotzinger, J.P., Summons, R.E., Eiler, J.M.: Sulfur isotopes of organic matter preserved in 3.45-billion-year-old stromatolites reveal microbial metabolism. Proc. Natl. Acad. Sci. USA. **109**, 15146–15151 (2012)
Bosak, T., Greene, S.E., Newman, D.K.: A likely role for anoxygenic photosynthetic microbes in the formation of ancient stromatolites. Geobiology. **5**(2), 119–126 (2007)
Bosak, T., Knoll, A.H., Petroff, A.P.: The meaning of stromatolites. Annu. Rev. Earth Planet. Sci. **41**, 21–44 (2013)
Branscomb, E., Russell, J.M.: Turnstiles and bifurcators: The disequilibrium converting engines that put metabolism on the road. Biochim. Biophys. Acta **1827**(2), 62–78 (2013)
Brasier, M.D., Green, O.R., Jephcoat, A.P., Kleppe, A.K., Van Kranendonk, M.J., Lindsay, J.F., Steele, A., Grassineau, N.V.: Questioning the evidence for Earth's oldest fossils. Nature. **416**, 78–81 (2002)
Brasier, M.D., Green, O.R., Lindsay, J.F., McLoughlin, N., Steele, A., Stoakes, C.: Critical testing of Earth's oldest putative fossil assemblage from the 3.5 Ga Apex chert, Chinaman Creek Western Australia. Precambrian Res. **140**, 55–102 (2005)
Brasier, M.D., Antcliffe, J., Saunders, M., Wacey, D.: Earth's earliest fossils (3.5-1.9 Ga): changing the picture with new approaches and new discoveries. Proc. Natl. Acad. Sci. **112**, 4859–4864 (2015)
Buick, R.: Microfossil recognition in Archaean rocks: an appraisal of spheroids and filaments from 3500 M.Y. old chert-barite at North Pole, Western Australia. PALAIOS. **5**, 441–459 (1990)
Bundeleva, I.A., Shirokova, L.S., Bénézeth, P., Pokrovsky, O.S., Kompantseva, E.I., Balor, S.: Calcium carbonate precipitation by anoxygenic phototrophic bacteria. Chem. Geol. **291**, 116–131 (2012)
Byerly, G.R., Palmer, M.R.: Tourmaline mineralization in the Barberton greenstone belt, South Africa: early Archean metasomatism by evaporite-derived boron. Contrib. Mineral. Petrol. **107**, 387–402 (1991)
Byerly, G.R., Lowe, D.R., Walsh, M.M.: Stromatolites from the 3300-3500-Myr Swaziland Supergroup, Barberton Mountain Land, South Africa. Nature. **319**, 489–491 (1986)
Cady, S.L.: Formation and preservation of *bona fide* microfossils. In: Signs of Life: A Report Based on the April, 2000 Workshop on Life-Detection Techniques, pp. 149–155. NRC-NAP, Washington, D.C. (2002)

Cameron, V., House CH, Brantley, S.L.: A first analysis of metallome biosignatures of hyperthermophilic archaea. Archaea. (2012). https://doi.org/10.1155/2012/789278

Camprubí, E., de Leeuw, J.W., House CH, Raulin, F., Russell, M.J., Spang, A., Tirumalai, M.R., Westall, F.: The emergence of life. Space Sci. Rev. **215**, 56 (2019)

Cavalazzi, B., Barbieri, R., Gomez, F., Capaccioni, B., Olsson-Francis, K., Pondrelli, M., Rossi, A.P., Hickman-Lewis, K., Agangi, A., Gasparotto, G., Glamoclija, M., Ori, G.G., Rodriguez, N., Hagos M.: The Dallol Geothermal Area, Northern Afar (Ethiopia)–An Exceptional Planetary Field Analog on Earth. Astrobiology **19**, 553–578 (2019)

Censi, P., Cangemi, M., Brusca, L., Madonia, P., Saiano, F., Zuddas, P.: The behavior of rare-earth elements, Zr and Hf during biologically-mediated deposition of silica-stromatolites and carbonate-rich microbial mats. Gondwana Res. **27**, 209–215 (2013)

Chi Fru, E., Rodríguez, N.P., Partin, C.A., Lalone, S.V., Andersson, P., Weiss, D.J., El Albani, A., Rodushkin, I., Konhauser, K.O.: Cu isotopes in marine black shales record the Great Oxidation Event. Proc. Natl. Acad. Sci. USA. **113**(18), 4941–4946 (2016)

Cloud, P.: Beginnings of biospheric evolution and their biogeochemical consequences. Paleobiology. **2**(4), 351–387 (1976)

Cuerno, R., Escudero, C., García-Ruiz, J.-M., Herrero, M.A.: Pattern formation in stromatolites: Insights from mathematical modelling. J. R. Soc. Interface. **9**, 1051–1062 (2012)

Danielson, A., Möller, P., Dulski, P.: The europium anomalies in banded iron formations and the thermal history of the oceanic crust. Chem. Geol. **97**, 89–100 (1992)

Dass, A.V., Hickman-Lewis, K., Brack, A., Kee, T.P., Westall, F.: Stochastic prebiotic chemistry within realistic prebiotic geological systems. Chem. Select. **1**(15), 4906–4926 (2016)

De Gregorio, B.T., Sharp, T.G.: Determining the biogenicity of microfossils in the apex chert, Western Australia, using Transmission Electron Microscopy. Lunar and Planetary Science XXXIV Houston (2003)

Djokic, T., Van Kranendonk, M.J., Campbell, K.A., Walter, M.R., Ward, C.R.: Earliest signs of life on land preserved in ca. 3.5 Ga hot spring deposits. Nat. Commun. **8** (2017). https://doi.org/10.1038/ncomms15263

Duan, Y., Anbar, A.D., Arnold, G.L., Lyons, T.W., Gordon, G.W., Kendall, B.: Molybdenum isotope evidence for mild environmental oxygenation before the great oxidation event. Geochim. Cosmochim. Acta. **74**(23), 6655–6668 (2010)

Duda, J.-P., Thiel, V., Bauersachs, T., Mißbach, H., Reinhardt, M., Schäfer, N., Van Kranendonk, M.J., Reitner, J.: Ideas and perspectives: hydrothermally driven redistribution and sequestration of early Archaean biomass - the hydrothermal pump hypothesis. Biogeosciences **15**, 1535–1548 (2018)

Dupraz, C., Pattisina, R., Verrecchia, E.P.: Translation of energy into morphology: simulation of stromatolite morphospace using a stocastic model. Sedimentary Geology **185**(3-4), 185–203 (2006)

Edwards, N.P., Manning, P.L., Bergmann, U., Larson, P.L., van Dongen, B.E., Sellers, W.I., Webb, S.M., Sokaras, D., Alonso-Mori, R., Ignatyev, K., Barden, H.E., van Veelen, A., Anné, J., Egerton, V.M., Wogelius, R.A.: Leaf metallome preserved over 50 million years. Metallomics. **6**, 774–782 (2014)

Fraústo da Silva, J.J.R., Williams, R.J.P.: The Biological Chemistry of the Elements, p. 600. Oxford University Press, Oxford (2001)

French, K.L., Hallmann, C., Hope, J.M., Schoon, P.L., Zumberge, J.A., Hoshino, Y., Peters, C.A., George, S.C., Love, G.D., Brocks, J.J., Buick, R., Summons, R.E.: Reappraisal of hydrocarbon biomarkers in Archean rocks. Proc. Natl. Acad. Sci. USA. **112**(19), 5915–5920 (2015)

Furnes, H., Banerjee, N.R., Muehlenbachs, K., Staudigel, H., de Wit, M.: Early life recorded in Archean pillow lavas. Science. **304**, 578–581 (2004)

Furnes, H., Banerjee, N.R., Staudigel, H., Muehlenbachs, K., McLoughlin, N., de Wit, M., Van Kranendonk, M.: Comparing petrographic signatures of bioalteration in recent to Mesoarchean pillow lavas: tracing subsurface life in oceanic igneous rocks. Precambrian Res. **158**, 156–176 (2007)

Garcia-Ruiz, J.M., Hyde, S.T., Carnerup, A.M., Christy, A.G., Van Kranendonk, M.J., Welham, N.J.: Self-assembled silica-carbonate structures and detection of ancient microfossils. Science. **302**(5648), 1194–1197 (2003)

Garcia-Ruiz, J.M., van Zuilen, M.A., Bach, W.: Mineral self-organization on a lifeless planet. Phys Life Rev. (2020)

Glikson, M., Duck, L.J., Golding, S.D., Hofmann, A., Bolhar, R., Webb, R., Baiano, J.C.F., Sly, L.I.: Microbial remains in some earliest Earth rocks: comparison with a potential modern analogue. Precambrian Res. **164**, 187–200 (2008)

Gourcerol, B., Thurston, P.C., Kontak, D.J., Côté-Mantha, O.: Interpretations and implications of LA ICP-MS analysis of chert for the origin of geochemical signatures in banded iron formations (BIFs) from the Meadowbank gold deposit, Western Churchill Province, Nunavut. Chem. Geol. **410**, 89–107 (2015)

Grassineau, N.F., Abell, P., Appel, P.W.U., Lowry, D., Nisbet, E.G.: Early life signatures in sulfur and carbon isotopes from Isua, Barberton, Wabigoon (Steep Rock), and Belingwe Greenstone Belts (3.8 to 2.7 Ga). In: Kesler, S.E., Ohmoto, H. (eds.) Evolution of Early Earth's Atmosphere, Hydrosphere, and Biosphere—Constraints from Ore Deposits: GSA Memoir, vol. 198, pp. 33–52. Sage, New York (2006)

Grosch, E.G., McLoughlin, N.: Reassessing the biogenicity of Earth's oldest trace fossil with implications for biosignatures in the search for early life. Proc. Natl. Acad. Sci. USA. **111**, 8380–8385 (2014)

Grotzinger, J.P., Knoll, A.H.: Stromatolites in precambrian carbonates: evolutionary mileposts or environmental dipsticks? Annu. Rev. Earth Planet. Sci. **27**, 313–358 (1999)

Grotzinger, J.P., Rothman, D.R.: An abiotic model for stromatolite morphogenesis. Nature. **383**, 423–425 (1996)

Heubeck, C.: An early ecosystem of Archean tidal microbial mats (Moodies Group, South Africa, ca. 3.2 Ga). Geology. **37**, 931–934 (2009)

Hickman-Lewis, K., Garwood, R.J., Brasier, M.D., Goral, T., Jiang, H., McLoughlin, N., Wacey, D.: Carbonaceous microstructures of the 3.46 Ga stratiform 'Apex chert', Chinaman Creek locality, Pilbara, Western Australia. Precambrian Res. **278**, 161–178 (2016)

Hickman-Lewis, K., Garwood, R.J., Withers, P.J., Wacey, D.: X-ray microtomography as a tool for investigating the petrological context of Precambrian cellular remains. In: Brasier, A.T., McIlroy, D., McLoughlin, N. (eds.) Earth System Evolution and Early Life: A Celebration of the Work of Martin Brasier, vol. 448, pp. 33–56. Geological Society of London Special Publication, London (2017)

Hickman-Lewis, K., Cavalazzi, B., Foucher, F., Westall, F.: Most ancient evidence for life in the Barberton Greenstone Belt: microbial mats and biofabrics of the ~3.47 Ga Middle Marker Horizon. Precambrian Res. **312**, 45–67 (2018)

Hickman-Lewis, K., Gautret, P., Arbaret, L., Sorieul, S., De Wit, R., Foucher, F., Cavalazzi, B., Westall, F.: Mechanistic morphogenesis of organo-sedimentary structures growing under geochemically stressed conditions: Keystone to the interpretation of some Archaean stromatolites? Geosciences. **9**, 359 (2019)

Hickman-Lewis, K., Gourcerol, B., Westall, F., Manzini, D., Cavalazzi, B.: Reconstructing Palaeoarchaean microbial biomes flourishing in the presence of emergent landmasses using trace and rare earth element systematics. Precambrian Res. (2020c). https://doi.org/10.1016/j.precamres.2020.105689

Hickman-Lewis, K., Cavalazzi, B., Sorieul, S., Gautret, P., Foucher, F., Whitehouse, M.J., Jeon, H., Cockell, C.S., Georgelin, T., Westall, F.: Metallomics in deep time and the influence of ocean chemistry on the metabolic landscapes of Earth's earliest ecosystems. Sci. Rep. **10**(1) (2020b). https://doi.org/10.1038/s41598-020-61774-w

Hickman-Lewis, K., Westall, F., Cavalazzi, B.: Diverse communities of bacteria and archaea flourished in Palaeoarchaean (3.5-3.3 Ga) microbial mats. Palaeontology. (2020a). https://doi.org/10.1111/pala.12504

Hofmann, H.J., Grey, K., Hickman, A.H., Thorpe, R.I.: Origin of 3.45 Ga coniform stromatolites in Warrawoona Group, Western Australia. GSA Bull. **111**, 1256–1262 (1999)

Homann, M., Heubeck, C., Airo, A., Tice, M.M.: Morphological adaptations of 3.22 Ga-old tufted microbial mats to Archean coastal habitats (Moodies Group, Barberton Greenstone Belt, South Africa). Precambrian Res. **266**, 47–64 (2015)

Homann, M., Sansjofre, P., van Zuilen, M., Heubeck, C., Gong, J., Killingsworth, B., Foster, I.S., Airo, A., Van Kranendonk, M.J., Ader, M., Lalonde, S.V.: Microbial life and biogeochemical cycling on land 3,220 million years ago. Nat. Geosci. **11**, 665–671 (2018)

Hren, M.T., Tice, M.M., Chamberlain, C.P.: Oxygen and hydrogen isotope evidence for a temperate climate 3.42 billion years ago. Nature. **205**, 205–208 (2009)

Igisu, M., Ueno, Y., Takai, K.: FTIR microspectroscopy of carbonaceous matter in ~3.5 Ga seafloor hydrothermal deposits in the North Pole area, Western Australia. Prog. Earth Planet. Sci. **5** (2018). https://doi.org/10.1186/s40645-018-0242-1

Izon, G., Zerkle, A.L., Williford, K.H., Farquhar, J., Poulton, S.W., Claire, M.W.: Biological regulation of atmospheric chemistry en route to planetary oxygenation. Proc. Natl. Acad. Sci. USA. **114**(13), E2571–E2579 (2017)

Johannesson, K.H., Telfeyan, K., Chevis, D.A., Rosenheim, B.E., Leybourne, M.I.: Rare earth elements in stromatolites–1. Evidence that modern terrestrial stromatolites fractionate rare earth elements during incorporation from ambient waters. In: Dilek, Y., Furnes, H. (eds.) Evolution of Archean Crust and Early Life. Springer, Dordrecht (2013). https://doi.org/10.1007/978-94-007-7615-9_14

Kamber, B.S., Webb, G.E.: The geochemistry of late Archean microbial carbonate: implications for ocean chemistry and continental erosion history. Geochim. Cosmochim. Acta. **65**, 2509–2525 (2001)

Kendall, B., Brennecka, G.A., Weyer, S., Anbar, A.D.: Uranium isotope fractionation suggests oxidative uranium mobilization at 2.50 Ga. Chem. Geol. **362**, 105–114 (2013)

Kiyokawa, S., Ito, T., Ikehara, M., Kitajima: Middle Archean volcano-hydrothermal sequience: Bacterial microfossil-bearing 3.2 Ga Dixon Island Formation, coastal Pilbara terrane, Australia. GSA Bull. **118**(1–2), 3–22 (2006)

Kiyokawa, S., Koge, S., Ito, T., Ikehara, M.: An ocean-floor carbonaceous sedimentary sequence in the 3.2-Ga Dixon Island Formation, coastal Pilbara terrane, Western Australia. Precambrian Res. **255**(1), 124–143 (2014)

Knauth, L.P.: Temperature and salinity history of the Precambrian ocean: implications for the course of microbial evolution. Palaeogeogr. Palaeoclimatol. Palaeoecol. **219**, 53–69 (2005)

Knoll, A.H., Barghoorn, E.S.: Archean microfossils showing cell division from the Swaziland system of South Africa. Science. **198**, 396–398 (1977)

Kozawa, T., Sugitani, K., Oehler, D.Z., House CH, Saito, I., Watanabe, T., Gotoh, T.: Early Archean plaktonic mode of life: implications from fluid dynamics of lenticular microfossils. Geobiology. **17**(2), 113–126 (2019)

Lin, W., Paterson, G.A., Zhu, Q., Wang, Y., Kopylova, E., Li, Y., Knight, R., Bazylinski, D.A., Zhu, R., Kirschvink, J.L., Pan, Y.: Origin of microbial biomineralization and magnetotaxis during the Archean. Proc. Natl. Acad. Sci. USA. **114**(9), 2171–2176 (2017)

Lowe, D.R.: Abiological origin of described stromatolites older than 3.2 Ga. Geology. **22**, 387–390 (1994)

Maldanis, L., Hickman-Lewis, K., Verezhak, M., Gueriau, P., Guizar-Sicairos, M., Jaqueto, P., Trinidade, R.I.F., Rossi, A.L., Berenguer, F., Westall, F., Bertrand, L., Galante, D.: Nanoscale 3D quantitative imaging of 1.88 Ga Gunflint microfossils reveals novel insights into taphonomic and biogenic characters. Sci. Rep. **10**, 8163 (2020)

Marin-Carbonne, J., Remusat, L., Sforna, M.C., Thomazo, C., Cartigny, P., Philippot, P.: Sulfur isotope's signal of nanopyrites enclosed in 2.7 Ga stromatolitic organic remains reveal microbial sufate reduction. Geobiology. **16**(2), 121–138 (2018)

Martin, W., Baross, J., Kelley, D., Russell, M.J.: Hydrothermal vents and the origin of life. Nat. Rev. Microbiol. **6**, 805–814 (2008)

McCollum, T.M., Seewald, J.S.: Carbon isotope composition of organic compounds produced by abiotic synthesis under hydrothermal conditions. Earth Planet. Sci. Lett. **243**, 74–84 (2006)

McLoughlin, N., Wilson, L.A., Brasier, M.D.: Growth of synthetic stromatolites and wrinkle structures in the absence of microbes – implications for the early fossil record. Geobiology. **6**, 95–105 (2008)

McMahon, S.: Earth's earliest and deepest purported fossils may be iron-mineralized chemical gardens. Proc. R. Soc. B. **286**, 20192410 (2019)

Moore, E.K., Jelen, B.I., Giovannelli, D., Raanan, H., Falkowski, P.G.: Metal availability and the expanding network of microbial metabolisms in the Archaean eon. Nat. Geosci. **10**, 629–636 (2017)

Muir, M.D., Grant, P.R.: Micropalaeontological evidence from the Onverwacht Group, South Africa. In: Windley, B.F. (ed.) The Early History of the Earth, pp. 595–608. Wiley/Interscience, London (1976)

Nisbet, E.G., Sleep, N.H.: The habitat and nature of early life. Nature. **409**, 1083–1091 (2001)

Noffke, N.: Microbial Mats in Sandy Deposits from the Archean Era to Today, 175 p. Springer, New York (2010)

Noffke, N., Eriksson, K.A., Hazen, R.M., Simpson, E.L.: A new window into Archean life: microbial mats in Earth's oldest siliciclastic tidal deposits (3.2 Ga Moodies Group, South Africa). Geology. **34**, 253–256 (2006)

Noffke, N., Christian, D., Wacey, D., Hazen, R.M.: Microbially induced sedimentary structures recording an ancient ecosystem in the ca. 3.48 billion-year-old Dresser Formation, Pilbara, Western Australia. Astrobiology. **13**, 1103–1124 (2013)

Nutman, A.P., Bennett, V.C., Friend, C.R., Van Kranendonk, M.J., Chivas, A.R.: Rapid emergence of life shown by discovery of 3,700-million-year-old microbial structures. Nature. **537**, 535–538 (2016)

Nutman, A.P., Bennett, V.C., Friend, C.R.L., Van Kranendonk, M.J., Rothacker, L., Chivas, A.R.: Cross-examining earth's oldest stromatolites: seeing through the effects of heterogeneous deformation, metamorphoism and metasomatism affecting Isua (Greenland) ~3700 Ma sedimentary rocks. Precambrian Res. **331** (2019). https://doi.org/10.1016/j.precamres.2019.105347

Oehler, D., Cady, S.L.: Biogenicity and syngeneity of organic matter in ancient sediments: recent advances in the search for evidence of past life. Challenges. **5**(2), 260–283 (2014)

Oehler, D.Z., Walsh, M.M., Sugitani, K., Liu, M.-C., House, C.H.: Large and robust lenticular microorganisms on the young Earth. Precambrian Res. **296**, 112–119 (2017)

Olempska, E., Wacey, D.: Ambient inclusion trails in Palaeozoic crustaceans (Phosphatocopina and Ostracoda). Palaeogeogr. Palaeoclimatol. Palaeoecol. **441**(4), 949–958 (2016)

Oonk, P., Tsikos, H., Mason, P., Henkel, S., Staubwasser, M., Williams, H.: Species-specific Fe-isotopes on Palaeoproterozoic BIF and their implications. In: 35th International Geological Congress, Cape Town, South Africa, 27 August 2016-4 September 2016 (2016)

Petroff, A.P., Sim, M.S., Maslov, A., Krupenin, M., Rothman, D.H., Bosak, T.: Biophysical basis for the geometry of conical stromatolites. Proc. Natl. Acad. Sci. USA. **107**, 9956–9961 (2010)

Philippot, P., Ávila, J.N., Killingsworth, B.A., Tessalina, S., Baton, F., Caquineau, T., Muller, E., Pecoits, E., Cartigny, P., Lalonde, S.F., Ireland, T.R., Thomazo, C., vanKranendonk, M.J., Busigny, V.: Globally asynchronous sulphur isotope signatls require re-definition of the great oxidation event. Nat. Commun. **9** (2018). https://doi.org/10.1038/s41467-018-04621-x

Rasmussen, B.: Filamentous microfossils in a 3,235-million-year-old volcanogenic massive sulphide deposit. Nature. **405**, 676–679 (2000)

Reid, R.P., James, N.P., Macintyre, I.G., Dupraz, C.P., Burne, R.V.: Shark Bay stromatolites: microfabrics and reinterpretations of origins. Facies. **9**, 243–270 (2003)

Riding, R.: The nature of stromatolites: 3,500 million year history and a century of research. In: Reitner, J., Quéric, N.-V., Arp, G. (eds.) Advances in Stromatolite Geobiology, pp. 29–74. Springer, Berlin, Heidelberg (2011)

Rosing, M.T.: 13C-depleted carbon microparticles in >3700-Ma sea-floor sedimentary rocks from West Greenland. Science. **283**, 674–676 (1999)

Schidlowski, M.: Carbon isotopes as biogeochemical recorders of life over 3.8 Ga of Earth history: evolution of a concept. Precambrian Res. **106**(1–2), 117–134 (2001)

Schopf, J.W., Packer, B.M.: Early Archean (3.3-billion to 3.5-billion-year-old) microfossils from Warrawoona Group, Australia. Science. **237**(4810), 70–73 (1987)

Schopf, J.W., Walter, M.R.: Archean microfossils: new evidence of ancient microbes. In: Schopf, J.W. (ed.) Earth's Earliest Biosphere: Its Origin and Evolution, pp. 214–239. Princeton University Press, Princeton, NJ (1983)

Schopf, J.W., Kudryavtsev, A.B., Agresti, D.G., Wdowiak, T.J., Czaja, A.D.: Laser-Raman imagery of Earth's earliest fossils. Nature. **416**, 73–76 (2002)

Schopf, J.W., Kudryavtsev, A.B., Dzaja, A.D., Tripathi, A.B.: Evidence of archean life: stromatolites and microfossils. Precambrian Res. **158**, 141–155 (2007)

Sforna, M.C., Daye, M., Philippot, P., Somogyi, A., van Zuilen, M.A., Medjoubi, K., Gérard, E., Jamme, F., Dupraz, C., Braissant, O., Glunk, G., Visscher, P.T.: Patterns of metal distribution in hypersaline microbialites during early diagenesis: Implications for the fossil record. Geobiology **15**(2), 259–279 (2017)

Shields, G., Webb, G.: Has the REE composition of seawater changed over geological time? Chem. Geol. **204**, 103–107 (2004)

Sleep, N.H.: The Hadean-Archaean environment. Cold Spring Harb. Perspect. Biol. **2**, a002527 (2010)

Sleep, N.H.: Planetary interior-atmosphere interaction and habitability. In: Deeg, H.J., Belmonte, J.A. (eds.) Handbook of Exoplanets, pp. 1–22. Springer, Berlin (2018)

Staudigel, H., Furnes, H., de Wit, M.: Paleoarchean trace fossils in altered volcanic glass. Proc. Natl. Acad. Sci. USA. **112**, 6892–6897 (2015)

Sugahara, H., Sugitani, K., Mimura, K., Yamashita, F., Yamamoto, K.: A systematic rare-earth elements and yttrium study of Archean cherts at the Mount Goldsworthy greenstone belt in the Pilbara Craton: Implications for the origin of microfossil-bearing black cherts. Precambrian Res. **177**, 73–87 (2010)

Sugitani, K., Grey, K., Allwood, A., Nagaoka, T., Mimura, K., Minami, M., Marshall, C.P., Van Kranendonk, M.J., Walter, M.R.: Diverse microstructures from Archean chert from the Mount Goldsworthy-Mount Grant area, Pilbara Craton, Western Australia: microfossils, dubiofossils, or pseudofossils? Precambrian Res. **158**, 228–262 (2007)

Sugitani, K., Grey, K., Nagaoka, T., Mimura, K., Walter, M.R.: Taxonomy and biogenicity of Archaean spheroidal microfossils (ca. 3.0 Ga) from the Mount Goldsworthy-Mount Grant area in the northeastern Pilbara Craton, Western Australia: Precambrian Res. **173**, 50–59 (2009)

Sugitani, K., Mimura, K., Takeuchi, M., Yamaguchi, T., Suzuki, K., Senda, R., Asahara, Y., Wallis, S., Van Kranendonk, M.J.: A Paleoarchean coastal hydrothermal field inhabited by diverse microbial communities: the Strelley Pool Formation, Pilbara Craton, Western Australia. Geobiology. **13**, 522–545 (2015)

Summons, R.E., Amend, J.P., Bish, D., Buick, R., Cody, G.D., Des Marais, D.J., Dromart, G., Eigenbrode, J.L., Knoll, A.H., Sumner, D.Y.: Preservation of martian organic and environmental records: Final report of the biosignature working group. Astrobiology. **11**(2), 157–181 (2011)

Takahashi, Y., Châtellier, X., Hattori, K.H., Kato, K., Fortin, D.: Adsorption of rare earth elements onto bacterial cell walls and its implication for REE sorption onto natural microbial mat. Chem. Geol. **219**, 53–67 (2005)

Tartèse, R., Chaussidon, M., Gurenko, A., Delarue, F., Robert, F.: Warm Archean oceans reconstructed from oxygen isotope composition of early-life remnants. Geochem. Perspect. Lett. **3**, 55–65 (2017)

Tashiro, T., Ishida, A., Hori, M., Igisu, M., Koike, M., Mejean, P., Takahata, N., Sano, Y., Komiya, T.: Early trace of life from 3.95 Ga sedimentary rocks in Labrador, Canada. Nature. **549**, 516–518 (2017)

Teng, F.-Z., Dauphas, N., Watkins, J.M.: Non-traditional stable isotopes: retrospective and prospective. Rev. Mineral. Geochem. **82**, 1–26 (2017)

Tice, M.M.: Environmental controls on photosynthetic microbial mat distribution on and morphogenesis on a 3.42 Ga clastic-starved platform. Astrobiology. **9**, 989–1000 (2009)

Tice, M.M., Lowe, D.R.: Photosynthetic microbial mats in the 3416-Myr-old ocean. Nature. **431**, 549–552 (2004)

Tice, M.M., Lowe, D.R.: Hydrogen-based carbon fixation in the earliest known photosynthetic organisms. Geology. **34**, 37–40 (2006a)

Tice, M.M., Lowe, D.R.: The origin of carbonaceous matter in pre-3.0 Ga greenstone terrains: a review and new evidence from the 3.42 Ga Buck Reef Chert. Earth Sci. Rev. **76**, 259–300 (2006b)

Tomescu, A.M.F., Klymiuk, A.A., Matsunaga, K.K.S., Bippus, A.C., Shelton, G.W.K.: Microbes and the fossil record. In: Hurst, C.J. (ed.) Advances in Environmental Microbiology. Their World: A Diversity of Microbial Environments, pp. 69–169. Springer, Heidelberg (2016)

Trower, E.J., Lowe, D.R.: Sedimentology of the ~3.3 Ga upper Mendon Formation, Barberton Greenstone Belt, South Africa. Precambrian Res. **281**, 473–494 (2016)

Ueno, Y., Isozaki, Y., Yurimoto, H., Maruyama, S.: Carbon isotopic signatures of individual Archean microfossils (?) from Western Australia. Int. Geol. Rev. **43**(3), 196–212 (2001a)

Ueno, Y., Maruyama, S., Isozaki, Y., Yurimoto, H.: Early Archean (ca. 3.5 Ga) microfossils and 13C- depleted carbonaceous matter in the North Pole area, Western Australia: field occurrence and geochemistry. In: Nakashima, S., Maruyama, S., Brack, A., Windley, B.F. (eds.) Geochemistry and the Origin of Life, pp. 201–236. Universal Academy Press, Tokyo (2001b)

Ueno, Y., Isozaki, Y., McNamara, K.J.: Coccoid-like microstructures in a 3.0 Ga chert from Western Australia. Int. Geol. Rev. **48**(1), 78–88 (2006a)

Ueno, Y., Yamada, K., Yoshida, N., Maruyama, S., Isozaki, Y.: Evidence from fluid inclusions for microbial methanogenesis in the early Archaean era. Nature. **440**, 516–519 (2006b)

Ueno, Y., Ono, S., Rumble, D., Maruyama, S.: Quadruple sulfur isotope analysis of ca. 3.5 Ga Dresser Formation: new evidence for microbial sulfate reduction in the early Archaen. Geochim. Cosmochim. Acta. **72**(23), 5675–5691 (2008)

van den Boorn, S.H.J.M., van Bergen, M., Nijman, W., Vroon, P.: Dual role of seawater and hydrothermal fluids in Early Archean chert formation: evidence from silicon isotopes. Geology. **35**, 939–942 (2007)

van den Boorn, S.H.J.M., van Bergen, M.J., Vroon, P.Z., de Vries, S.T., Nijman, W.: Silicon isotope and trace element constraints on the origin of ~3.5 Ga cherts: implications for early Archaean marine environments. Geochim. Cosmochim. Acta. **74**, 1077–1103 (2010)

Van Kranendonk, M.J.: Volcanic degassing, hydrothermal circulation and the flourishing of early life on Earth: new evidence from the Warrawoona Group, Pilbara Craton, Western Australia. Earth Sci. Rev. **74**, 197–240 (2006)

Van Kranendonk, M.J., Webb, G.E., Kamber, B.S.: Geological and trace element evidence for a marine sedimentary environment of deposition and biogenicity of 3.45 Ga stromatolitic carbonates in the Pilbara Craton, and support for a reducing Archaean ocean. Geobiology. **1**(2), 91–108 (2003)

Van Zuilen, M.A., Chaussidon, M., Rollion-Bard, C., Marty, B.: Carbonaceous cherts of the Barberton Greenstone Belt, South Africa: Isotopic, chemical and structural characteristics of individual microstructures. Geochim. Cosmochim. Acta. **71**(3), 655–669 (2007)

Wacey, D.: ~3,240 Ma, Kangaroo Caves formation, East Pilbara, Western Australia. In: Wacey, D. (ed.) Early Life on Earth, pp. 221–227. Springer, Dordrecht (2009)

Wacey, D., Kilburn, M., Stoakes, C., Aggleton, H., Brasier, M.: Ambient inclusion trails: their recognition, age, range, and applicability to early life on Earth. In: Dikel, Y., Furnes, H., Muehlenbachs, K. (eds.) Links Between Geological Processes, Microbial Activities and Evolution of Life, vol. 4, pp. 113–134. Springer, Dordrecht (2008b)

Wacey, D., Kilburn, M.R., McLoughlin, N., Parnell, J., Stoakes, C.A., Grovenor, C.R.M., Brasier, M.D.: Use of NanoSIMS in the search for early life on Earth: ambient inclusion trails in a c. 3400 Ma sandstone. J. Geol. Soc. **165**, 43–53 (2008a)

Wacey, D., Kilburn, M.R., Saunders, M., Cliff, J., Brasier, M.D.: Microfossils of sulphur-metabolizing cells in 3.4-billion-year-old rocks of Western Australia. Nat. Geosci. **4**, 698–702 (2011)

Wacey, D., McLouglin, N., Kilburn, M.R., Saunders, M., Cliff, J.B., Kong, C., Barley, M.E., Brasier, M.D.: Nanoscale analysis of pyritized microfossils reveals differential heterotrophic consumption in the ~1.9-Ga Gunflint chert. Proc. Natl. Acad. Sci. USA. **110**, 8020–8024 (2013)

Wacey, D., Saunders, M., Cliff, J., Kilburn, M.R., Kong, C., Barley, M.E., Brasier, M.D.: Geochemistry and nano-structure of a putative ~3240 million-year-old black smoker biota, Sulphur Springs Group, Western Australia. Precambrian Res. **249**, 1–12 (2014)

Wacey, D., Saunders, M., Kong, C., Brasier, A.T., Brasier, M.D.: 3.46 Ga Apex chert 'microfossils' reinterpreted as mineral artefacts produced during phyllosilicate exfoliation. Gondwana Res. **36**, 296–313 (2016)

Wacey, D., Saunders, M., Kong, C.: Remarkably preserved tephra from the 3430 Ma Strelley Pool Formation, Western Australia: Implications for the interpretation of Precambrian microfossils. Earth Planet. Sci. Lett. **487**, 33–43 (2018)

Walsh, M.M.: Microfossils and possible microfossils from the early Archean Onverwacht group, Barberton Mountain Land, South Africa. Precambrian Res. **54**, 271–293 (1992)

Walsh, M.M., Lowe, D.R.: Filamentous microfossils from the 3,500-Myr-old Onverwacht Group, Barberton Mountain Land, South Africa. Nature. **314**, 530–532 (1985)

Walsh, M.M., Lowe, D.R.: Modes of accumulation of carbonaceous matter in the early Archaean: a petrographic and geochemical study of the carbonaceous cherts of the Swaziland Supergroup. In: Lowe, D.R., Byerly, G.R. (eds.) Geologic Evolution of the Barberton Greenstone Belt, South Africa, vol. 329, pp. 115–132. Geological Society of America, Boulder (1999)

Walsh, M.M., Westall, F.: Archean biofilms preserved in the Swaziland Supergroup, South Africa. In: Krumbein, W.E., Paterson, D.M., Zavarzin, G.A. (eds.) Fossil and Recent Biofilms, pp. 307–316. Kluwer, Dordrecht (2003)

Walter, M.R., Buick, R., Dunlop, J.: Stromatolites, 3400–3500 Myr old from the North Pole area, Western Australia. Nature. **284**, 443–445 (1980)

Westall, F., Brack, A.: The importance of water for life. Space Sci. Rev. **214**, 50 (2018)

Westall, F., de Vries, S.T., Nijman, W., Rouchon, V., Orberger, B., Pearson, V., Watson, J., Verchovsky, A., Wright, I., Rouzaud, J.-N., Marchesini, D., Anne, S.: The 3.466Ga Kitty's Gap chert, an Early Archean microbial ecosystem. In: Reimold, W.U., Gibson, R.L. (eds.) Processes on the Early Earth, vol. 405, pp. 105–131. Geological Society of America, Boulder (2006)

Westall, F., Foucher, F., Cavalazzi, B., de Vries, S., Nijman, W., Pearson, V., Watson, J., Verchovsky, A., Wright, I., Rouzaud, J.-N., Marchesini, D., Anne, S.: Volcaniclastic habitats for early life on Earth and Mars: a case study from c.3.5 Ga-old rocks from the Pilbara, Australia. Planet. Space Sci. **59**, 1093–1106 (2011)

Westall, F., Campbell, K.A., Bréhéret, F.G., Foucher, F., Gautret, P., Hubert, A., Sorieul, S., Grassineau, N., Guido, D.M.: Complex microbe-sediment systems are ancient (3.33 Ga) and flourished in a hydrothermal context. Geology. **43**, 615–618 (2015)

Westall, F., Hickman-Lewis, K., Hinman, N., Gautret, P., Campbell, K.A., Bréhéret, J.G., Foucher, F., Hubert, A., Sorieul, S., Dass, A.V., Kee, T.P., Georgelin, T., Brack, A.: A hydrothermal-sedimentary context for the origin of life. Astrobiology. **18**(3), 259–293 (2018)

Williams, R.J.P.: Chemical selection of elements by cells. Coord. Chem. Rev. **216–217**, 583–595 (2001)

Zerkle, A.L., House, C.H., Brantley, S.L.: Biogeochemical signatures through time as inferred from whole microbial genomes. Am. J. Sci. **305**, 467–502 (2005)

Zhang, M., Konishi, H., Xu, H., Sun, X., Lu, H., Wu, D., Wu, N.: Morphology and formation mechanism of pyrite induced by the anaerobic oxidation of methane from the continental slope of the NE South China Sea. J. Asian Earth Sci. **92**, 293–301 (2014)

Chapter 11
Origin and Early Evolution of the Eukaryotes: Perspectives from the Fossil Record

Heda Agić

Abstract The emergence of a complex, eukaryotic cell is one of the major steps in the evolution of life on Earth. Eukaryotic organisms include a range of macroscopic life such as animals, plants, fungi, as well as a plethora of microscopic single-celled or colonial protists. The first evidence for eukaryotic life appears in the geologic record around 1650 million years ago (Ma), as organic-walled microfossils—cellular vesicles often preserved as carbonaceous compressions in siliciclastic rocks. Early eukaryotes were predominantly single-celled and minute for about a billion years, until the onset of macroscopic multicellularity in algae and animals in the Ediacaran Period (635–538 Ma). Here I review the earliest evidence of eukaryotic life, including a range of Proterozoic organic-walled microfossils and problematica. These fossils contain a suite of morphological and geochemical characters that offer clues about their palaeobiology. Complex microfossil morphology like spines can be considered a proxy for the appearance of a cytoskeleton, very early in eukaryote history, in the late Paleoproterozoic. Multidisciplinary studies on fossil features such as cell morphology, cell wall ultrastructure and its chemical composition, ancient forms of multicellularity, as well as understanding the environments these microorganisms inhabited, enable the use of the fossil record to inform the timing and mode of eukaryogenesis.

11.1 Introduction

One of the major irreversible events in the evolution of life on Earth was the origin and the subsequent evolution of eukaryotes. Most of the visible life surrounding us in the present is eukaryotic: animals, plants, fungi, though there is also a vast diversity of microscopic forms (collectively called protists) which constitute a great proportion of eukaryotes, and play a crucial role in global ecosystem functions (for

H. Agić (✉)
Department of Earth Science, University of California, Santa Barbara, Santa Barbara, CA, USA
e-mail: hagic@geol.ucsb.edu

A. Neubeck, S. McMahon (eds.), *Prebiotic Chemistry and the Origin of Life*,
Advances in Astrobiology and Biogeophysics,
https://doi.org/10.1007/978-3-030-81039-9_11

example, eukaryotic plankton are the planet's major primary producers in marine ecosystems today; Stockner 1988). In a traditional view, Eukarya constitute one of the three domains of life, alongside prokaryotic organisms: Bacteria and Archaea (Pace 2009). All eukaryotes notably stand out from prokaryotes in possessing membrane-bound intracellular structures such as a nucleus and other organelles, and they include a variety of both single-celled and multicellular organisms (e.g. Bhattacharya et al. 2007; Alberts et al. 2014; Dacks et al. 2016; Eme and Ettema 2018; Adl et al. 2019). Most of the genetic information in a eukaryotic cell is stored in its nucleus, a core organelle encased in a nuclear envelope which allows it to control its contents. The nucleus intrigued the early microscopists since it was first named by Robert Brown ("little nut"; Pederson 2011). This unique feature contributed the name for eukaryotic organisms, which derives from the Greek word κάρυον (*karyon*), meaning a nut or kernel. In contrast to simpler cells of Bacteria and Archaea, eukaryotes have a suite of specialized subunits in addition to a nucleus, such as a complex cytoskeleton, endomembrane system, as well as mitochondria and plastids (e.g. chloroplasts in photosynthetic organisms) that perform specialized functions, allowing the cell versatility in form and modes of life (Alberts et al. 2014; López-García et al. 2017). Möbius (1884) dubbed these "organelles" ("little organs"), as a cell's main functional units akin to the organs of multicellular organisms. Some organelles originated from ancient endosymbiotic events whereby a larger cell of an archaeon-like ancestor engulfed a proteobacterium in the case of a mitochondrion (now employed by the cell in respiration), or a cyanobacterium in case of a chloroplast (now used in photosynthesis) (Sagan 1967; Bhattacharya et al. 2007; López-García et al. 2017).

The origin of the eukaryotic cell and sequence of the endosymbiotic events is a complex, multi-step process that has long presented a mystery (e.g. Martin et al. 2015), because there are no living intermediate organisms and no fossils with preserved intracellular features that would reveal the timing of eukaryogenesis and the order in which the endosymbiotic events occurred. This problem is mainly tackled by modern cell biology and molecular phylogeny (e.g. Koonin 2010; López-García et al. 2017; Eme and Ettema, 2018; Burki et al. 2020), but palaeontology is starting to play a bigger role (Knoll et al. 2006; Butterfield 2015a; Porter 2020). Phylogenetic analyses often calibrated by temporally and phylogenetically well-constrained fossils (molecular clocks) are used to narrow down the timing of the origin of eukaryotes and their early diversification (e.g. Betts et al. 2018; Burki et al. 2020). Although our understanding of relationships between groups of modern eukaryotes is becoming more refined (Adl et al. 2019; Burki et al. 2020), the timing of eukaryotic origins, and especially of the appearance of crown-group eukaryotes remains elusive. Phylogenetic studies incorporating divergence times of major modern eukaryote groups predict the origin of stem-group eukaryotes broadly from 1.8 to 1 Ga (Eme et al. 2014), and a recent study combining genomic and fossil data estimates the origin of crown-group eukaryotes 1.842–1.21 Ga with the appearance of the last eukaryotic common ancestor (LECA) around 1.86–1.68 Ga (Betts et al. 2018). This places the origin of eukaryotes well after the Great Oxidation Event around 2.4 Ga (Holland 2006), which is broadly consistent with the microfossil record (Fig. 11.1).

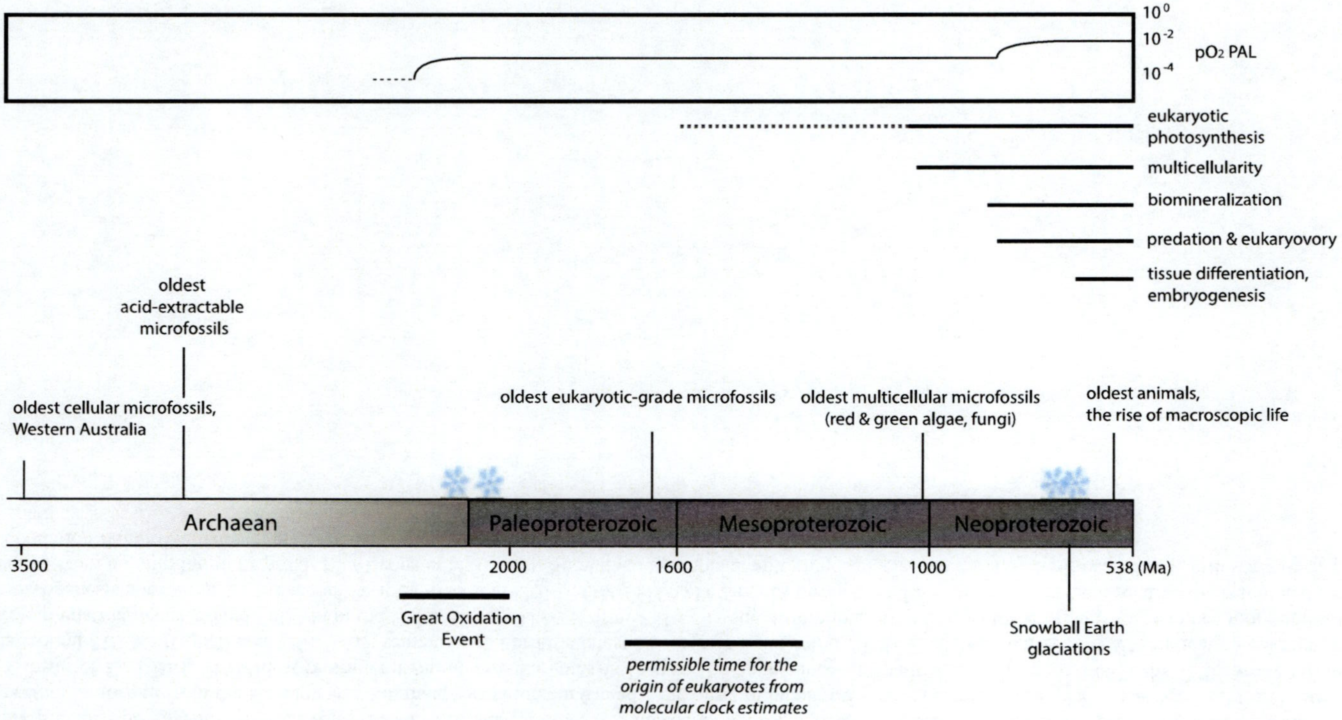

Fig. 11.1 Timeline of major milestones in the history of life as seen from the fossil record in relation to major environmental events in the Precambrian. Appearance of major evolutionary innovations within eukaryotes is shown above the timeline. The oldest microfossil cells and sulphur metabolizers occur in 3.4 Ga Strelley Pool Formation in Australia (Wacey et al. 2011) and the oldest microfossils recovered via palynological acid-maceration are not much younger, from the 3.2 Ga Moodies Group in South Africa (Javaux et al. 2010). The oldest microfossils with eukaryotic characteristics known to date appear

in the 1.65 Ga Changcheng Group in China (Miao et al. 2019), within a permissible time interval for the origin of eukaryotes estimated by phylogenetic molecular clock analyses (Eme et al. 2014; Betts et al. 2018). Multicellularity appears by around 1 Ga, coinciding with the radiation of likely crown-group eukaryotic lineages: red and green algae, and fungi (Butterfield 2000; Loron et al. 2019; Tang et al. 2020) but multicellularity only becomes ubiquitous and macroscopic in the Ediacaran, following the Snowball Earth glaciations (Erwin et al. 2011; Ye et al. 2015; Hoffman et al. 2017). Some Mesoproterozoic taxa were interpreted as photosynthetic eukaryotes (Moczydłowska et al. 2011; Butterfield 2015b), and based the oldest demonstrable photosynthesizing eukaryote, a red alga *Bangiomorpha* from the Hunting and Angmaat formations in Canada, the onset of eukaryotic photosynthesis took place by at least 1.2 Ga (Butterfield 2000; Gibson et al. 2018). The oldest biomineralization in eukaryotes is seen in the appearance of apatitic scale microfossils in the Fifteenmile Group in Canada around 810 Ma (Cohen et al. 2017), but remains rare until the latest Neoproterozoic (Wood 2018). Predation marks on OWM indicating eukaryote-on-eukaryote predation occur around 780 Ma in the Chuar Group, USA (Porter 2016). Tissue differentiation in multicellular organisms becomes more ubiquitous and embryogenesis appears in the Ediacaran Period (Xiao et al. 2014). Oxygen level estimates through the Precambrian in percent relative to present day levels, are adapted from a compilation of proxies by Lyons et al. (2014). Snowflake symbols denote major glacial intervals in the Precambrian, following/during the Great Oxidation Event (Gumsley et al. 2017), and the Cryogenian Snowball Earth low-latitude glaciations (Hoffman et al. 2017)

Rocks of the Phanerozoic Eon (538 million years ago (Ma) to present), a time of "visible life", tell a story of eukaryotic evolutionary success manifested in a rich fossil record of skeletons and soft-bodied remains of plants and animals that increased our understanding of the evolution of aquatic ecosystems, the transition of macroscopic life onto land, and eventually air. Darwin (1859) and many other naturalists were long puzzled by the apparent lack of fossils in older strata that comprise a significant portion of Earth's history. It wasn't until relatively late in the history of geology that older unequivocal fossils were discovered and recognized as explicitly Precambrian in age, in the c. 1.88 billion years old (Ga) Gunflint chert (Tyler and Barghoorn 1954), and this kickstarted a new field of palaeontology incorporating other geoscience disciplines. In addition to a rich record of microbial life since at least 3.4 Ga (Wacey et al. 2011), the field of Precambrian palaeontology exposed a history of eukaryotes far more ancient than multicellular life, dating back to the end of the Paleoproterozoic Era. Complex cells evolved halfway through the planet's history, but eukaryotic organisms remained minute until much later (Fig. 11.1; e.g. Payne et al. 2009; Butterfield 2009; Cohen and Macdonald 2015).

The bulk of the fossil record in the Proterozoic Eon (2500 to 538 Ma) is composed of organic-walled microfossils (OWM; e.g. Knoll et al. 2006; Agić and Cohen 2021). These are usually soft-bodied remains of prokaryotes and single-celled eukaryotes, often preserved in fine-grained sediments as compressed carbonaceous vesicles. Proterozoic OWM include plausible remains of the oldest eukaryotes known to date (see Miao et al. 2019). These microfossils are structurally complex and large (>20 μm), and as such were traditionally considered distinct from the cells of Bacteria and Archaea which rarely exceed a cell diameter of several μm (Knoll et al. 2006). Despite the morphological complexity of some OWM, the use of this microfossil group in studying eukaryogenesis and early eukaryotic phylogeny has been limited, due to few diagnostic characters and the lack of preservable intracellular structures such as organelles that define Eukarya (e.g. the nucleus and mitochondria). Instead, many of these defining components of a eukaryotic cell are considered to have been a part of the last eukaryotic common ancestor (LECA), which is often viewed as a naked amoeba of low preservation potential and presumably pre-dates the first microfossils (e.g. Koonin 2010; Butterfield 2015a). However, recent developments in phylogenomics and microbiology revealed the closest relatives between eukaryotes and their archaeal precursors, the Asgard archaea (Spang et al. 2015), are intriguingly complex. These organisms possess an actin cytoskeleton which provides structure to the cell and allows it to mould its shape (Akıl et al. 2020), or even produce spiny extensions similar to the cell morphology of living and fossil eukaryotes (Imachi et al. 2019). These finds question whether the long-assumed eukaryotic characters among ancient OWM are truly restricted to eukaryotes, and cast uncertainty on the timing of LECA.

Here I explore how the fossil record in deep time informs us about the origin of eukaryotes and their early evolution, in light of new palaeontological and biological discoveries.

11.2 Early Eukaryotes and Where to Find Them

11.2.1 A Key to Identifying Ancient Eukaryotes in the Fossil Record

Criteria for recognizing ancient eukaryotes in the fossil record were traditionally based on morphological and size differences between living eukaryotes and their smaller bacterial and archaeal relatives (e.g. Knoll et al. 2006). Many of the diagnostic features of a eukaryotic cell are largely a reflection of its function and as such reveal information about the palaeobiology of early eukaryotes. However, some criteria have lost their potency when recent microbiological discoveries revealed a remarkable variety and complexity among prokaryotes, including some Asgard archaea that produce cellular extensions similarly to eukaryotes (Imachi et al. 2019). The already equivocal identification of early eukaryotic fossils became an even more formidable challenge. Specific microfossil characters inferred as eukaryotic are hereby closely examined in light of recent finds.

From their origin until the evolution of macroscopic life like plants and animals, eukaryotes are largely represented in the fossil record by organic walled microfossils (Butterfield 2015a; Agić and Cohen 2021). These are most commonly single-celled fossils ranging from a few micrometres to one millimetre in size, composed of a sac of organic tissue called a vesicle. Some vesicles display surface ornamentation (e.g. processes or flanges), distinct wall surface patterns, envelopes, or opening structures (Figs. 11.2 and 11.3). OWM are usually extracted from shales and siltstones via palynological acid maceration in highly corrosive acids (e.g. Grey 1999). Their fossil record is very long, extending from late Palaeoproterozoic to recent. OWM are a polyphyletic fossil group, i.e. they represent a variety of unrelated microorganisms including prokaryotic cell aggregates and filaments and complex cells of early eukaryotes in the Proterozoic, as well as various organisms with modern relatives in the Phanerozoic (e.g. Colbath and Grenfell 1995). Traditionally, eukaryotic OWM were called acritarchs (meaning "of uncertain origin") and are widely used in Palaeozoic biostratigraphy and palaeoenvironmental reconstructions. Due to a lack of diagnostic characters, the specific biological affinities of these microfossils largely remain a mystery, apart from a few exceptions subsequently removed from Acritarcha (e.g. Moczydłowska and Willman 2009; Loron et al. 2019; Tang et al. 2020). The OWM record allows comprehensive studies on prokaryotic and eukaryotic communities in a given environment, and this review focuses on OWM representing eukaryotic cells. But how are they recognized as eukaryotes?

Conventionally, several criteria were used to identify a eukaryotic cell in the fossil record (Javaux et al. 2004; Knoll et al. 2006), although not without flaws. These main diagnostic features are: (1) large size; (2) surface elements such as processes (also called spines) or protrusions, and collectively termed ornamentation; (3) openings in the cell wall, called the excystment structure; (4) resilient carbonaceous vesicle wall with the ability to withstand acid maceration, and (5) complex wall ultrastructure. All of these criteria also need to be considered in light

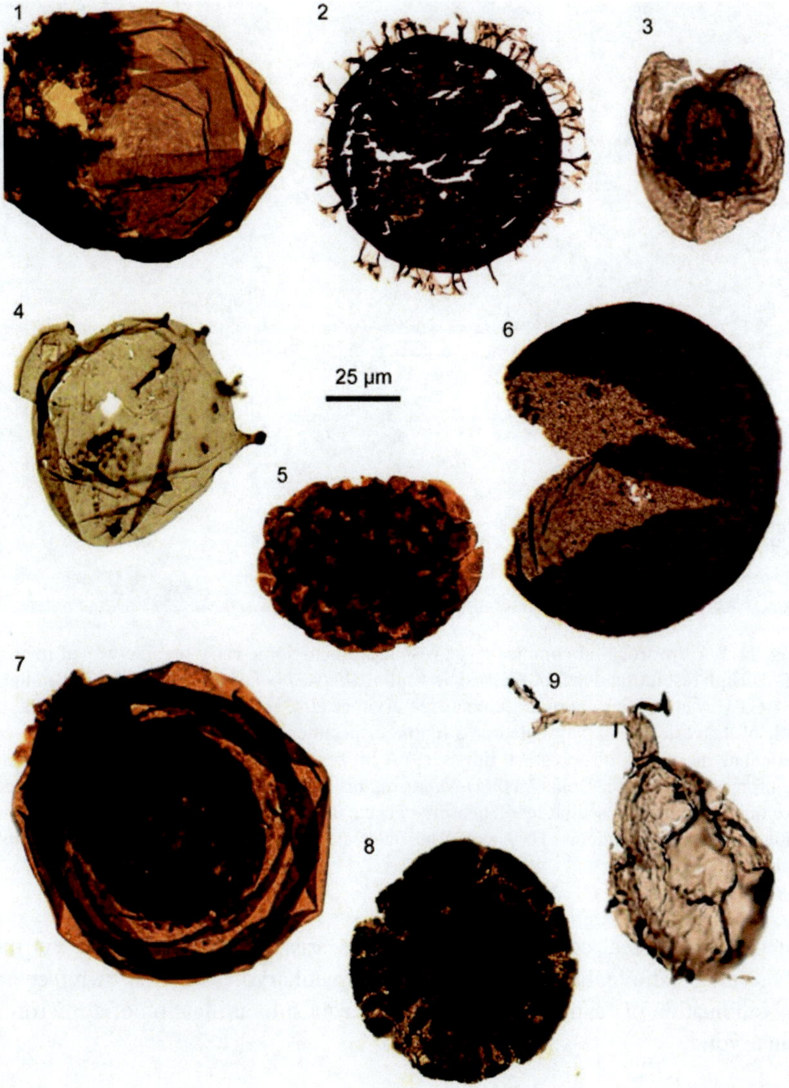

Fig. 11.2 Mesoproterozoic organic-walled microfossils representing potential early eukaryotes. (1) *Valeria lophostriata* from the 1600–1400 Ma Ruyang Group, northern China (see Agić et al. 2017). (2) One of the oldest process-bearing OWM, *Shuiyousphaeridium macroreticulatum* from the Ruyang Group. (3) Envelope-bearing *Pterospermopsimorpha insolita* from the Ruyang Group. (4) Process-bearing *Tappania plana* from the Ruyang Group. (5) *Squamosphaera colonialica* bearing bulbous protrusions, from the c. 1048 Ma lower Bylot Supergroup in Nunavut, Canada. (6) Sphaeromorph with complex vesicle wall *Dictyosphaera macroreticulata* from the Ruyang Group. (7) A large leiosphaerid with a large internal body from the Bylot Supergroup. (8) *Caelatimurus foveolatus* bearing a distinct sculptured wall surface, from the Bylot Supergrup. (9) A candidate for the oldest fossil fungus, *Ourasphaira giraldae* from the 1232–892 Ma Shaler Supergroup in Canada (see Loron et al. 2019). Image courtesy of C. Loron

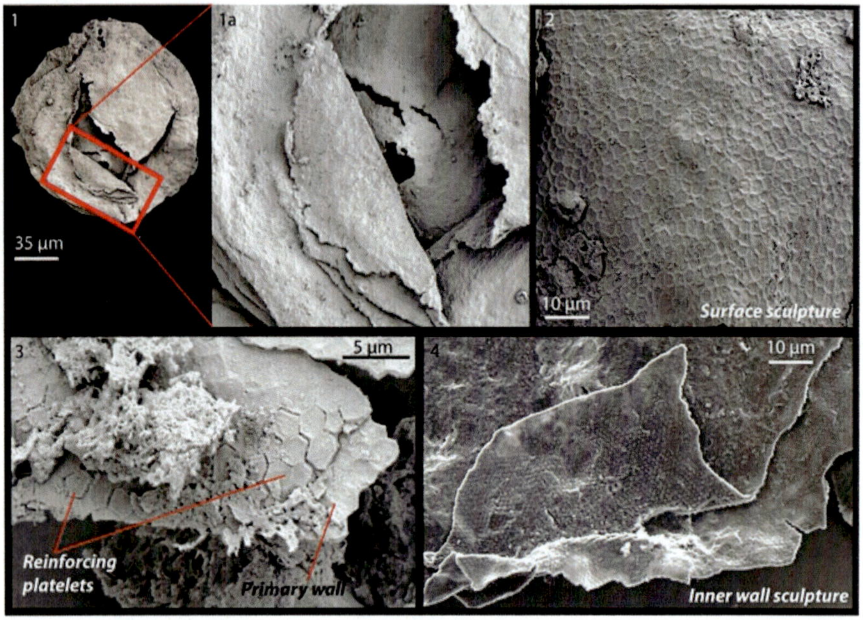

Fig. 11.3 Ultrastructural complexity of vesicle walls of some early organic-walled microfossils. (1–3) High resolution detail of the vesicle wall of *Dictyosphaera macroreticulata* from the c. 1.6–1.4 Ga Ruyang Group, northern China (see Agić et al. 2015). (1a) Morphology on the internal side of the vesicle wall is revealed on a ruptured specimen. (2) The external wall surface contains reticulate meshwork of polygonal ridges. (3) A broken specimens shown a possible primary wall reinforced by polygonal platelets. (4) *Nunatsiaquus* sp. from the c. 1048 Ma Bylot Supergroup, Arctic Canada bears sculpture of densely-packed, low-relief, circular protrusions on the inner side of the vesicle (credit: H. Agić). The outer surface of the microfossil lacks sculpture or ornamentation

of their geological context, to avoid biases arising from diagenesis. As most of these diagnostic features are not restricted to eukaryotes on their own (see below), a combination of features is required for a robust identification of a microfossil as eukaryotic.

11.2.1.1 Size

The majority of Proterozoic OWM (including many of the oldest OWM) fit the size range of a modern eukaryotic cell, from 10 to 100+ µm (Huntley et al. 2006; Alberts et al. 2014). On the other hand, cells of some extant prokaryotes like sulphur oxidising bacteria (e.g. *Thiomargarita namibiensis*) can grow up to 700 µm (Schulz et al. 1999), and eukaryotic picoplankton can be as tiny as bacteria. For instance, the smallest modern eukaryote, a prasinophyte alga *Ostreococcus taurii* is only 0.2 to 2 µm in diameter (Courties et al. 1994). Therefore, size on its own is not sufficient to narrow down a microorganism's affinity. Sizes of Proterozoic microfossils are commonly 10–20 μm (e.g. Schopf 1992; Huntley et al. 2006) which overlap with those of some prokaryotes. Cell size among archaea ranges from 0.1 to 15 μm, and the cell of a process-bearing archaeon *Prometheoarchaeum* (Lokiarchaeota) is around 0.5 μm in diameter (sans protrusions; Imachi et al. 2019). Large cell size is still somewhat informative because the modern eukaryotic cells are generally larger than prokaryotic (Alberts et al. 2014) minus some exceptions, but for a confident interpretation of eukaryotic affinity, it should come alongside other diagnostic features like those indicating cellular complexity.

11.2.1.2 Shapeshifting

Proterozoic OWM have a simple anatomy, shared with a myriad extant microscopic and usually unicellular eukaryotes (protists), which presents a problem in identifying robust characters for cladistic analysis to investigate their evolutionary relationships. But although some modern protists are very simple themselves, and their exterior doesn't fully distinguish them from prokaryotic cells or sheaths, many form complex cellular structures at least during a part of their existence (e.g. Corliss 2001; Schaap and Schilde 2018; Adl et al. 2019). To produce complex cellular morphology, a microorganism requires intricate intracellular machinery that most prokaryotes generally lack (e.g. McFadden et al. 1986; Alberts et al. 2014).

Some OWM bear processes/spines that are similar to those in extant eukaryotes (e.g. spiny extensions on the cells of dinoflagellates or ciliates). Processes may have had more than one function. Based on the physiology of modern protists, we can infer that the processes increase a cell's surface area, which has been suggested to increase buoyancy (Hoops et al. 1999). This would help the microorganism to remain suspended either in the photic zone to harvest sunlight, or in a desired part of the water column following nutrients. The ornamentation in modern protists also improves the efficiency of nutrient uptake by increasing the cell's surface/volume ratio for osmosis (Pahlow et al. 1997). In addition to their function during a metabolically active phase of a microorganism's life, processes in some eukaryotic cells originate as a part of cyst-formation during reproduction or during development of a resting stage, and are thus only present during a part of a life cycle, for example in modern dinoflagellates (Kokinos and Anderson 1995), zygnematalean algae (Calderón and Tavera 2020), or crustacean resting egg stages (Belmonte et al. 1997). Cell ornamentation may also aid in burial prevention during a sedimentary disturbance when the organism is encysted. For example, the spines on cysts of the resting stages of modern plankton are used to keep the mineral particles from the sediment at a distance from the cyst wall itself, to prevent damage caused by sediment scraping or rupturing the cyst (Belmonte et al. 1997). The resting egg stage of copepod *Diaptomnus* occurs up to 24 cm depth in the sediment and remains viable for over 300 years (Hairston et al. 1995). The ability to reduce sediment disturbance with ornamentation could have been beneficial for the resting cysts of early eukaryotes by allowing them to dwell deep in the sediment and avoid environmental perturbance.

The process-bearing morphology requires the presence of a cytoskeleton which would enable the cell to quickly change its shape. This ability however, is more ancient than eukaryotes, being also present in the closest archaean relatives of eukaryotes (Ettema et al. 2011; Imachi et al. 2019; Akıl et al. 2020). Thus, the presence of simple processes is not sufficient to infer eukaryotic affinity, and can be considered a feature of a 'protoeukaryote', pre-dating LECA. However, it should be noted that many fossil and extant protists have complex processes that can have multiple order branching or possess elaborate terminations which could aid in buoyancy or attachment. Some processes of Mesoproterozoic taxa are hair-like fibres encased in an outer envelope like in *Gigantosphaeridium* (Agić et al. 2015), some branch or flare outwards like in *Shuiyousphaeridium* (Xiao et al. 1997; Agić et al. 2015), and some are thick extensions with bulbous tips like in *Tappania*, randomly distributed across cell's surface (Nagovitsin 2009; Adam et al. 2017). Most processes in Proterozoic OWM are even in morphology (homomorphic) and distributed regularly on the surface of a cell, with the exception of heteromorphic processes (several types on one taxon) in *Tappania* (Fig. 2.4) and *Trachyhistrychosphaera* (Butterfield et al. 1994).

Research on the morphologic plasticity in the Asgard archaea is in progress (e.g. Akıl et al. 2020). Presently, it is unknown if they would have been able to produce complex extensions that could branch or support an outer membrane, and the only example of archaeal processes includes irregularly distributed ones in *Prometheoarchaeum syntrophicum* (Imachi et al. 2019). Branching extensions appear to be derived, commonly found in fungi (see Loron et al. 2019), and the envelope-supporting processes are associated with cyst formation (see Moczydłowska et al. 2011; Agić et al. 2015), so these can be viewed as eukaryotic features.

11.2.1.3 Cyst Production: Openings, Sturdy Walls, and Intracellular Complexity

Many eukaryotic clades produce resting or reproductive cysts with variable patterns of ornamentation and composition (Corliss 2001; Schaap and Schilde 2018). A distinguishing feature of a cyst is an opening through which the organism emerges upon the return of favourable environmental conditions, or it releases its gametes in the case of a reproductive cyst (e.g. Kremp et al. 2009). Proterozoic OWM usually have two ancient opening types. A median-split opening is a simple partial rupture of the vesicle, common in the oldest OWM. Paleoproterozoic taxon *Schizofusa* has an epityche opening (Miao et al. 2019), which is a slit with a curved outline that allows a part of the vesicle wall to fold outwards like a flap. A regular and circular opening with a defined outline (pylome) is much rarer in the fossil record, e.g. in *Leiosphaeridia kulgunica* (Beghin et al. 2017). Some pylomes additionally bear a plug-like operculum, though this character is rare and appears rather late, e.g. in the Tonian taxon *Kaibabia* (Porter and Riedman

2016). Simple openings could be formed simply by rupture of the cell wall from internal pressure, without a control mechanism. Considering this, even the basic morphological plasticity allowed by an archaeal cytoskeleton could presumably form a simple split. Therefore, a median split-opening is less demonstrative of a specific affinity. On the other hand, circular/pylome and especially lidded openings (i.e. bearing an operculum) are rarer and research on modern unicellular algae indicates that such structures are more challenging to produce. Extant lid-forming algae like *Acetabularia* possess a specialized organelle called the lid-forming apparatus (Neuhaus-Url and Schweiger 1984) which controls cyst formation and opening though several steps. The analogues of this organelle have appeared by the Ordovician Period (Agić et al. 2016), or possibly earlier. Considering this, and due to its specialization and required cellular complexity, a pylome is a strong character of a cyst-forming eukaryote.

Extant protists, including some ciliates, dinoflagellates, algae, and amoebae, often have morphologically and biochemically complex cell or cyst walls. Complex wall ultrastructure is present in many modern green algae, whereby the wall is additionally reinforced, appearing as a trilaminar sheet structure (TLS). The same feature was observed in a few Precambrian OWM taxa mainly from younger, Ediacaran-Cambrian rocks (Moczydłowska and Willman 2009), and potentially even in a simple sphaeromorph from one of the oldest OWM assemblages (c. 1.65 Ga; Peng et al. 2009). Elaborate and metabolically expensive, wall formation during encystment in modern protists is usually conducted by a range of organelles like the cytoskeleton, the endoplasmatic reticulum, and the Golgi apparatus (e.g. McFadden et al. 1986; Corliss 2001). Cyst walls displaying complex wall ultrastructure or composite construction require complex intracellular machinery for their construction, so this morphology serves as an indirect indicator for the presence a cytoskeleton and an endomembrane system (e.g. Javaux et al. 2001, 2004; Agić et al. 2015).

In extant phytoplankton, a slew of chemical and ultrastructural changes take place during cyst formation and lead to increased wall thickness (Hagen et al. 2002; Suda et al. 2004). Compounds that reinforce the cyst wall may also increase its preservation potential. Examples include dinosporin in dinoflagellates (Bogus et al. 2012), sporopollenin in pollen and spores (Brooks and Shaw 1978), algaenan in some green algae (de Leeuw et al. 2006; Kodner et al. 2009), and chitin in some fungi and cilliates (Morozov and Likhoshway 2016). Analyses of the chemical composition of the wall of some Proterozoic OWM showed that they are also made of complex biopolymers (Marshall et al. 2005). OWM are soft-bodied and have a recalcitrant organic-walled vesicle. They are robust enough to resist digestion in highly corrosive acids (hydrofluoric), which, to some extent, allows for their extraction from a rock matrix. The origin of this resistant, non-hydrolysable property is unclear though. Some modern microeukaryotes have robust cell walls build in part by non-hydrolysable aliphatic macromolecules like algaenan (Kodner et al. 2009) or long-chain polysaccharides like chitin (Morozov and Likhoshway 2016). Some of these structural macromolecules end up as components of kerogen and are recognised as contributors to the refractory organic matter in sedimentary rocks (Hedges

1995), and may thus aid in OWM preservation. However, organic molecules are prone to alteration through geologic history due to decay and diagenesis. Organic cuticles of ancient fossils tend to undergo additional polymerization which can cause further resistance (Briggs and Summons 2014). For this reason, as well as due to external influences on the preservation of soft-bodied fossil material (Anderson et al. 2020), the acetolysis-resistance of OWM does not necessarily always reflect the original composition of a microoganism's cell wall, and chemical composition of ancient microfossils is of limited use for phylogenetic studies. Considering the generally low preservation potential of most prokaryotes, a recalcitrant vesicle was traditionally assumed as an indicator of eukaryotic affinity (Knoll et al. 2006; Moczydłowska et al. 2011). It is less informative, however, about a more specific phylogenetic placement. Pathways for the synthesis of specific compounds tend to be widespread across Eukarya (e.g. chitin, Morozov and Likhoshway 2016). When phylogenetic distribution of a specific compound is versatile or not fully explored and mapped among the majority of eukaryotes, the compound loses some of its diagnostic value.

Bacterial microfossils can also survive the acid maceration process, but it should be noted that this record largely consists of bacteria with sturdy filamentous sheaths, with few exceptions such as cell aggregate *Ostiana* (Butterfield et al. 1994) or enigmatic large cells of likely prokaryotic affinity from the Archean (c. 3.2 Ga) rocks (Javaux et al. 2010). Single cells of bacteria of low preservation potential are more commonly fossilised by permineralisation in cherts or phosphorites (e.g. Golubic and Seong-Joo 1999). Although not all spherical acid-extractable Precambrian microfossils should be considered eukaryotic, combined with other indicators of eukaryotic affinity like large size and complex cellular morphology it may additionally help in distinguishing between prokaryotic and eukaryotic affinity.

11.2.1.4 Life Cycle and Reproduction

Prokaryotes usually reproduce via binary fission whereby the genetic material of the cell is copied and split as the parent cell divides (Alberts et al. 2014), so features like budding, reproductive openings, or protrusions from the vesicle are more likely to have been produced by eukaryotes.

Cyst formation is a common reproductive and/or survival strategy among free-living protists (e.g. Corliss 2001), and modern yeasts reproduce by budding (Herskowitz 1988). Although some cyanobacteria produce an endocyst (e.g. *Azotobacter*; Sadoff 1975), those are rather simple and do not have a well-defined opening. The appearance of cellular microfossils with openings in the fossil record, at the end of the Paleoproterozoic (Miao et al. 2019), coincides with the oldest estimate for the origin of eukaryotes suggested by molecular clocks (Eme et al. 2014). Reproductive features among ancient OWM include a cyst-like vesicle, openings that could have been reproductive, and large internal bodies (Fig. 2.7). Based on living and fossil dinoflagellates, Kokinos and Anderson (1995) suggested that encysted stages in the reproductive cycle of some species were more durable and more frequently produced than asexual temporary cysts, and were thus more likely to contribute to the fossil record. A bimodal life cycle has been inferred for Paleo-Mesoproterozoic OWM *Dictyosphaera* and *Shuiyousphaeridium*, where each taxon represents a part of a reproductive cycle of a single organism, based on comparison to extant protists (Agić et al. 2015).

11.2.1.5 Preservation

Lastly, geological context is important when examining ancient fossils. Microfossils, especially organically preserved ones, are prone to alteration on long time scales (e.g. Briggs and Summons 2014), and as such are often composed of derivatives of the initial biomolecules used in cell wall construction. These derivatives from labile original material (e.g. lipids, proteins) are often additionally polymerized and/or grafted during diagenesis (de Leeuw et al. 2006; Briggs and Summons 2014). Although this may aid in microfossil preservation by making the cells more resilient, it masks information about their original composition to some degree. Organic matter in the sediments is also prone to change due to metamorphism, and this degree of organic geochemical alteration and thermal maturity can be assessed by Raman microspectroscopy and imagery (e.g. Schopf et al. 2005; Marshall et al. 2005). Javaux and Lepot (2018) advised that the studies on early fossil record must show that OWM are of the same age as the host rock (syngenetic), biogenic (i.e. not morphologies mimicking microfossils that are formed through abiotic processes), and not contaminants.

11.3 Early Eukaryotic Body Fossils

How prevalent are these eukaryotic characters among Proterozoic fossils and how old are fossil eukaryotes? Molecular clocks suggest emergence of stem-group eukaryotes in the Proterozoic Eon (c. 1.8–1.0 Ga, Eme et al. 2014). This is a time after the Great Oxidation Event around 2.4–2.1 Ga, when biologically produced molecular oxygen began to accumulate in Earth's oceans and eventually in the atmosphere (e.g. Holland 2006; Lyons et al. 2014; Gumsley et al. 2017). This phenomenon is regarded as a consequence of oxygenic photosynthesis performed by recently evolved cyanobacteria (Fischer et al. 2016), causing massive change to Earth's biogeochemical cycling. The oxygenation was followed by a severe decline in primary productivity (Hodgskiss et al. 2019) as well as a perturbation to the global carbon cycle (the Lomagundi-Jatuli positive carbon isotope excursion; Gumsley et al. 2017). The fossil record during most of the Paleoproterozoic is relatively scarce, and dominated by bacteria, usually preserved in chert (Schirrmeister et al. 2016; Javaux and Lepot 2018; Hodgskiss et al. 2019). Fossils of even putative eukaryotes appear for the first time in much younger rocks, so their emergence likely isn't directly linked to the immediate aftermath of the Great Oxidation Event (contra e.g. Gross and Bhattacharya 2010).

11.3.1 Macroscopic Compressions

One of the oldest fossils occasionally interpreted in the literature as a eukaryote, *Grypania spiralis*, appears after this interval, around 1.87 Ga (Han and Runnegar 1992). *Grypania* is a tubular, ribbon-like fossil, preserved as a compression on bedding planes, often in bundles. One to a few mm long and visible to the naked eye, *Grypania* was proposed to be an alga because of its large size and structural rigidity and the reassessment by Knoll et al. (2006) suggested that *Grypania* was a single organism and not an amalgamation of prokaryotic filaments. However, large size is not restricted to eukaryotes (e.g. there are macroscopic bacterial colonies *Nostoc*; Dodds and Castenholz 1988) and bacteria are also capable of producing sturdy sheaths (e.g. Schirrmeister et al. 2016). *Grypania* is a very long-ranging taxon, occurring sporadically in Proterozoic strata until the Ediacaran Period (635–541 Ma; Wang et al. 2016), although the lack of diagnostic characters and discrepancies in size between *Grypania* specimens from different units cast doubt on whether or not the fossils included in this taxon actually represent the same organism. With the absence of concrete eukaryotic characters to distinguish it from a filamentous bacterium, a eukaryote is not the most likely affinity for *Grypania*. Narrowing down the timing of the origin of eukaryotes and understanding their early evolution requires a reliable fossil record, so caution is necessary when interpreting potential fossil eukaryotes in such ancient sediments. In this conservative approach, the default interpretation should be that a fossil is likely a prokaryote, unless it possesses demonstrable eukaryote-grade characters.

11.3.2 The Oldest OWM Assemblages

A significant shift in the Proterozoic fossil record from assemblages predominantly in chert to those in shales is seen around 1.67 Ga, coinciding with the appearance of the oldest widely accepted evidence for eukaryotic life (e.g. Prasad et al. 2005; Agić et al. 2017; Miao et al. 2019). These are organic-walled microfossils; large (>70 μm) and morphologically complex vesicles, most commonly preserved in shales and siltstones. Such early eukaryotic assemblages are relatively rare compared to younger time intervals, in part due to a dearth of sediments of the right age and metamorphism and thermal alteration hindering their preservation (e.g. Schiffbauer et al. 2012). Nevertheless, OWM occur in strata on several palaeocontinents, notably the 1.67–1.65 Ga Changcheng Group in China (Miao et al. 2019) and 1.63–1.6 Ga Semri Group in India (Prasad et al. 2005). Late Paleoproterozoic OWM include large, round and elongate cells with median-split openings—presumable cyst structures, as well as ornamented and envelope-bearing microfossils. This broadly fits with molecular clock estimates for the origin of Eukarya (Eme et al. 2014). The Changcheng OWM possess few diagnostic characters, but include candidates for early eukaryotes, e.g. *Schizofusa* which, in addition to large size, has a large epityche opening. Taxa with characters like a reticulate wall, encasing envelopes (*Pterospermopsimorpha*), surface pattern or ridges arranged in concentric circles (*Valeria*), or tail-like extensions (*Germinosphaera*), possess morphology that is generally more complex than commonly produced by bacteria, and were thus also identified as eukaryotes (Miao et al. 2019).

Valeria lophostriata is one of the oldest microfossils with a sculptured wall. Its spherical vesicle is marked by a unique pattern of ridges arranged as concentric circles (Fig. 2.1). Analyses utilizing transmitted electron microscopy (TEM) on younger *Valeria* specimens revealed a homogenous vesicle wall (Javaux et al. 2004) and with more electron-tenuous ridges on the surface (Pang et al. 2015). Biomechanical spherical pressure modelling further suggested that the sculpture of concentric circles in *Valeria* could have enabled an easier split of the two cell hemispheres, thus working as a mechanism for "biologically programmed excystment" (Pang et al. 2015, p. 251), and this could have assisted the organism in reproduction, or in emergence from its resting cyst upon the return of favourable conditions in its environment. With occurrences in Australia, China, and India (Javaux et al. 2004; Prasad et al. 2005; Miao et al. 2019), *Valeria* was widely dispersed by the end of the Paleoproterozoic. Interestingly, although it is often a rare component within an OWM assemblage (e.g. Agić et al. 2017), it is a lingering presence in Proterozoic rocks with nearly a billion years long stratigraphic range (around 1650–740 Ma). Perhaps its morphologically unique vesicle performed an important function in its reproduction or survival in adverse condition, for example in cyst-release (cf. Pang et al. 2015), that led to this higher evolutionary success compared to other Mesoproterozoic taxa. It is not unreasonable that a relatively simple morphology like the concentric ridges could have been produced by archaeal cytoskeleton. However, *Valeria* possessed a combination of large size, a unique cell morphology currently unknown in modern prokaryotes, and a potentially complex and regulated excystment structure, which makes it a suitable candidate for one of the oldest eukaryotic microfossils.

11.3.3 OWM Indicative of Cytoskeletal Complexity

Shortly following their initial appearance, eukaryotic OWM diversified and became globally distributed. Historically, the search for the oldest eukaryotes was galvanized by discoveries of spinose (process-bearing, acanthomorphic) microfossils *Tappania* and *Shuiyousphaeridium* in lower Mesoproterozoic rocks (Xiao et al. 1997); prior to that, morphologically complex microfossils were mainly known from the latest Proterozoic and Phanerozoic strata. Microfossils of complex cells are now known from numerous sedimentary successions of terminal Paleoproterozoic and early Mesoproterozoic age: c. 1.65–1.4 Ga Ruyang Group on the North China Craton (Agić et al. 2017), c. 1.6–1.4 Ga Vindhyan Supergroup in India (Prasad et al. 2005), c. 1.5–1.4 Ga Roper Group of the Northern Territory, Australia (Javaux and Knoll 2017), c. 1.45 Ga Belt Supergroup in Montana, USA (Adam et al. 2017), c. 1.4 Kaltasy Formation on East European Platform, Russia (Sperling et al. 2014) and c. 1.4–1.2 Ga Kamo Group in Siberia (Nagovitsin 2009). Although age constraints on many Mesoproterozoic units are not firm, the fossil record divulges that diverse eukaryotes inhabited shelfal marine environments of multiple palaeocontinents by at least 1.4 Ga.

Examples of ancient microfossils with inferred eukaryotic characteristics include Paleo-Mesoproterozoic taxa *Dictyosphaera* and *Shuiyousphaeridium*. Both taxa are large (reaching up to 300 μm in diameter) and distinguished by a composite vesicle wall with a polygonal pattern on the external surface and reinforced by platelets on the internal side (Figs. 2.6, 3.1, 3.2, and 3.3; Javaux et al. 2004; Agić et al. 2015; Miao et al. 2019). *Shuiyousphaeridium* (Fig. 2.2) stands out as the oldest process-bearing taxon. It appears in strata below the earliest occurrence of another spinose taxon *Tappania* in the Ruyang Group (Agić et al. 2017), and it is one of few such taxa in the Paleo-Mesoproterozoic. It bears funnel-like, widening processes that occasionally prop up a thin, organic membrane surrounding the microfossil (Xiao et al. 1997; Agić et al. 2015, 2017). TEM analysis further revealed a multi-layered vesicle wall (Javaux et al. 2004). The platelets reinforce a primary cell wall on the vesicle's interior of both *Dictyosphaera* and *Shuiyousphaeridium* (Figs. 3.1, 3.2, and 3.3). A similar pattern of wall construction in modern eukaryotes (e.g. unicellular algae *Nephroselmis* and *Scherffelia*) is mediated by the organelles endoplasmatic reticulum and the Golgi apparatus (McFadden et al. 1986), which secrete organic granules that become subsequently attached to the inner cell membrane during zygotic cyst formation (Suda et al. 2004). The composite wall construction in both *Dictyosphaera* and *Shuiyousphaeridium* would have likely required intracellular complexity, so these taxa may be viewed as indirect evidence for the presence of organelles involved in cell wall construction and protein and lipid transport like the Golgi apparatus—a part of the endomembrane system.

Striking similarity between the wall construction of both taxa hints at their conspecificity, whereby each taxon potentially represents a morphotype in a life cycle of a single biological species based on comparisons to modern eukaryotes (Xiao et al. 1997; Agić et al. 2015). This reconstruction implies the presence of a bimodal life cycle employed by many extant protists, by the earliest Mesoproterozoic. A rupture in the vesicle is common in both taxa and a pylome possibly appears in *Shuiyousphaeridium*. Together, all of these morphologic characters support a eukaryotic affinity for both *Dictyosphaera* and *Shuiyousphaeridium*. Their likely function as a reproductive or a resting cyst is shared with modern cyst-forming protists, especially unicellular green algae, but their precise affinity remains unknown. The Ruyang OWM assemblage also contains one of the largest pre-Ediacaran microfossils, around 600 μm in diameter: *Gigantosphaeridium*. A thick and opaque vesicle is surrounded by a membrane that is supported by dense, fibrelike processes. An opening was not found in *Gigantosphaeridium*, but its sturdy nature suggests a protective role, for instance a cyst (Agić et al. 2015).

Germinosphaera is a common taxon throughout the Proterozoic and it consists of a spherical cell with a single process (e.g. Butterfield et al. 1994). In some complete specimens, the process can branch. Although this branching is likely a eukaryotic feature, most documented *Germinosphaera* specimens have a truncated process (see Miao et al. 2019). As a single cellular extension could potentially be produced by the cellular machinery possessed by an Asgard archaeon or a protoeukaryote (based on Imachi et al. 2019), perhaps only fully preserved *Germinosphaera* with a bifurcating process should be included within Eukarya.

Process-bearing microfossil *Tappania* was also globally distributed by c. 1.4 Ga (Fig. 2.4). A vesicle of 30–80 μm in size is ornamented by heteromorphic processes unevenly distributed on the vesicle surface, and it occasionally bears a wide, bulbous protrusion (Prasad et al. 2005; Adam et al. 2017, Agić et al. 2017). Its processes sometimes bifurcate or end in a bulbous tip, and vary greatly in length and thickness. Such variable morphology and irregularly distributed extensions prompted an interpretation where *Tappania* was an actively growing vegetative cell or a germinating cell instead of an inactive cyst (Javaux et al. 2001). This is supported by the lack of a typical excystment opening in *Tappania*, but another possibility is that the neck-like protrusion acted as a release. Substantial disparity of specimens and complexity of processes suggest that the cell was actively changing it shape during growth, which would have demanded the presence of a cytoskeleton. *Tappania* is most commonly found in distal shelf deposits (Javaux and Knoll 2017).

Among the standout Mesoproterozoic taxa is a budding microfossil *Blastanosphaira kokkoda*, discovered in the Roper Group in Australia (Javaux and Knoll 2017). The vesicle varies from spheroidal to oval in shape with faint folding of the wall. It forms budding structures at various stages of development: starting with small protrusions ("buds") emerging from the vesicle wall, to larger ones >10s μm long, and eventually to two or several joined vesicles compressed at margins. Scars present in some specimens are likely caused by detaching buds or cells that became disjointed. Varying size of the buds and potential detachment scars suggest that the organism exercised its reproduction by budding, similar to the reproductive strategy of extant protists. Buds at various stages of development also imply that the organism underwent active growth, likely as a vegetative cell. Budding similar to *Blastanosphaira* is employed by modern yeasts like *Saccharomyces* (Herskowitz 1988). However, such structures are also produced by some cyanobacteria during reproduction (Waterbury and Stanier 1977), so budding on its own is not an explicit eukaryotic character. Together though, large size, faint ornamentation, and budding suggest *Blastanosphaira* was likely a eukaryote. Budding microfossils are rather rare through the Proterozoic. *Blastanosphaira* is not a unique example, but similar taxa occur much later, around the Mesoproterozoic-Neoproterozoic transition. *Aimia*, *Eosaccharomyces*, and *Gemmuloides* from the 1.0 Ga Lakhanda Formation in Siberia (up to 100 μm in size) bear numerous, small, spherical buds or occasionally, in case of *Gemmuloides*, oval protrusions emerging from the vesicle

(Samuelsson and Butterfield 2001). In addition to yeasts, modern aggregating slime moulds employ a similar strategy during reproduction (Bonner 1971).

Envelope-bearing OWM are abundant throughout the Proterozoic. Although their morphology is simple and they do not possess a suite of explicitly eukaryotic characters, these microfossils were traditionally identified as planktonic eukaryotes owing to their size and similarity to cysts of extant prasinophyte green algae (Moczydłowska et al. 2011). *Simia* and *Pterospermopsimorpha* (Fig. 2.3) are common in most Mesoproterozoic and Neoproterozoic fossiliferous units (e.g. Schopf 1992; Butterfield et al. 1994; Riedman and Porter 2016; Agić et al. 2017; Beghin et al. 2017). *Pterospermopsimorpha* is a sphere-within-sphere, with an inner cell surrounded by a spherical cell of varying width. The narrow envelope in *Simia*, on the other hand, closely hugs the central cell, appearing as extension in the equatorial plane. Inferring eukaryotic affinity for envelope-bearing OWM in deep time merits caution, because their simple morphology is not as definitive as in taxa that exhibit a combination of complex characters. Nevertheless, *Simia*'s envelope with an equatorial extension is distinct from envelopes or sheaths in living or fossil prokaryotes (cf. Sadoff 1975; Schirrmeister et al. 2016). In conjunction with a vesicle size up to 200 μm and a thick-walled central cell, *Simia* was probably a kind of an encysting eukaryote or even an early alga (Moczydłowska et al. 2011), functioning equivalently to a prasinophyte phycoma.

Aside from various cell extensions, Proterozoic microfossils are also unique in a variety of their vesicle surface sculpture, which can be decorated by grooves, pits, warts, or extensions. Similar to *Valeria*, late Mesoproterozoic OWM are distinguished by range of different patterning. *Squamosphaera* (Fig. 2.5) is common in late Mesoproterozoic and early Neoproterozoic units (Tang et al. 2015), and its vesicle contains bulbous protrusions. *Nunatsiaquus* (Fig. 3.4) from the c. 1232–892 Ma Shaler Supergroup in Canada possess an inner wall riddled with bulbous structures (Loron et al. 2019). The cell surface of *Caelatimurus* from the c. 1048 Ma Bylot Supergroup in Canada consists of grooves and pits, giving the microfossil a fluffy appearance (Fig. 2.8). Utility and benefits of versatile vesicle sculpture to a microorganism are not well understood. Similar morphology of grooves and pits in modern pollen exines supports the structural integrity of a cell, so its role was probably protective.

11.3.4 Multicellularity and Cellular Differentiation

The oldest unambiguous evidence for eukaryotic multicellularity appears at the end of the Mesoproterozoic Era, around 1 Ga. One of its earliest examples is *Bangiomorpha pubecens*, interpreted as a red alga (Butterfield 2000). The fossil has a differentiated basal holdfast structure that probably served as a substrate attachment. Some specimens have multiseriate filaments and differentiated reproductive cells (gametes) with radial symmetry, which suggests sexual reproduction. Included in *Bangiomorpha* are several ontogenetic stages, from single cells, to double-celled and 4–8-celled stages. These features are virtually indistinguishable from living bangiophycean red algae, and no modern bacteria or green algae are known to show a radial division of cells. *Bangiomorpha* is exceptionally preserved in early diagenetic chert of the c. 1200 Ma Hunting Formation and c. 1048 Ma Angmaat Formation in Arctic Canada (Butterfield 2000; Gibson et al. 2018).

Shortly after, multicellular green algae also appear. *Proterocladus antiquus*, from the c. 1000 Ma Nanfen Formation in North China is a thallus-like fossil with differentiated cells and asymmetrical branches, and it also contains a holdfast structure (Tang et al. 2020). It was interpreted as a siphonocladalean green alga based on similarities to the living *Cladophora*. Holdfasts are indicators of an attached lifestyle, so both *Bangiomorpha* and *Proterocladus* were likely erect, epibenthic organisms, and potentially reflect early ecosystem tiering on a microscopic scale (Butterfield 2000). *Proterocladus* also occurs in younger strata, c. 800 Ma Svanbergfjellet Formation on Svalbard, alongside other multicellular microfossils, including another possible chlorophyte *Palaeastrum* (Butterfield et al. 1994; Porter and Riedman 2016). The cells of *Palaeastrum* are connected to one another by robust attachment discs similar to multicellular coenobia in modern chlorococcalean green algae, which makes them stand out from other common contemporary cell aggregates of likely prokaryotic affinities. Microfossils showing cell-to-cell adhesion (e.g. *Arctacellularia)* become more widespread by the early Tonian, occurring in the c. 1 Ga Lakhanda Formation in Siberia (Hermann 1990) and Mbuji-Mayi Supergroup in DR Congo (Baludikay et al. 2016), but parenchymatous multicellular tissue is to date unknown until the Ediacaran (e.g. Xiao et al. 2014).

Ourasphaira giraldae (Fig. 2.9), potentially the oldest fossil fungus, was found in the c. 1232–892 Ma Shaler Supergroup (Loron et al. 2019). It consists of a large vesicle with a branching extension which sometimes develops second and third order branches, separated by septa. Chitosan and α-chitin, known to be used by modern fungi in construction of their cell walls, were detected in *Ourasphaira* via Fourier transform infrared microspectroscopy (FTIR). However, compounds like chitin can be synthesised by different groups across the eukaryotic tree, e.g. ciliates and choanoflagellates (Morozov and Likhoshway 2016), so chemical composition on its own is not sufficient to infer eukaryotic affinity. *Ourasphaira*'s morphology of branching, hyphae-like cellular extensions is also similar to fungi and the fossil contains a bilayered wall like some modern fungal cells (Loron et al. 2019). Separately, these features could be convergent, but in combination they hint at fungal affinity. Most of the Proterozoic record of putative fungi to date (Butterfield 2005) has been questioned (see discussion in Loron et al. 2019), and there are currently no other unambiguous fungal fossils until the appearance of terrestrial fungi at the Silurian-Devonian transition (Smith 2016; Strullu-Derrien et al. 2019).

Conspicuously, in addition to other multicellular microfossils appearing in the rocks of similar ages, *Ourasphaira* corroborates the origin of multicellularity and a radiation of crown-group eukaryotic lineages around one billion years ago. Moreover, if it is indeed an early fungus, its age profoundly pushes back the timing of the divergence of crown-group opisthokonts (a eukaryotic group which includes all animals and fungi and the closest relatives of each; cf. Torruella et al. 2015).

The Neoproterozoic fossil record is rich in large and multicellular algal compressions (Ye et al. 2015). *Bangiomorpha* and *Proterocladus* were both interpreted as photosynthesizers, red and green algae respectively. Recent precise age constraints on the rock unit containing *Bangiomorpha* helped assess the time for the origin of eukaryotic photosynthesis. Cross-calibrated molecular clock analyses using the new ages of *Bangiomorpha* to test the divergence of red and green algae, provided models for the timing of the primary plastid endosymbiosis around 1422–1246 Ma (Gibson et al. 2018).

11.3.5 Early Eukaryotic Diversity

A compilation and statistical analyses of the current record of eukaryotic fossils from Proterozoic strata showed that eukaryotic diversity in the Mesoproterozoic is relatively low and constant through time (Cohen and Macdonald 2015), though some newly reported OWM assemblages have nearly as high diversity as Neoproterozoic units (e.g. Agić et al. 2017; Beghin et al. 2017; Loron et al. 2019). A marked increase in eukaryotic within-assemblage diversity occurs in the late Tonian, starting around 800 Ma (coinciding with the Bitter Springs positive carbon isotope excursion and a pulse in oxygenation; cf. Wörndle et al. 2019) and ending before the Cryogenian low-latitude glaciations (Cohen and Macdonald 2015). The increase in diversity at this time also includes the emergence of novel microfossil groups in addition to OWM, notably microorganisms capable of biomineralization, i.e. producing biomineralized components like tests and scales, that first appeared in the Tonian, starting around 810 Ma (Porter and Knoll 2000; Cohen et al. 2017).

The Tonian diversification and increased abundance of eukaryotes is further reflected in a change in the marine zinc cycle and the organic Zn/C ratio in the oceans around 800 Ma (Isson et al. 2018). Eukaryotic cells utilise Zn to a higher degree than prokaryotes (e.g. Quigg et al. 2011), so a shift towards a more efficient sequestration of organic-derived Zn at this time suggests a rise in the abundance of eukaryotes in the Tonian oceans (Isson et al. 2018), as well as their greater ecological importance. And increase in biomarkers more commonly produced by eukaryotes (steranes, derivatives of sterols) also seemingly supports a shift towards an ecological dominance of eukaryotic primary producers at this time (Brocks et al. 2017).

This increase in diversity and ecological relevance of eukaryotes has been attributed to a few potential drivers: (1) a higher availability of nutrients in the Neoproterozoic oceans (Reinhard et al. 2020) from the increased weathering following the break-up of the palaeocontinent Rodinia, (2) a prior nitrate limitation in euxinic oceans that were widespread until the late Mesoproterozoic (Stüeken 2013), and (3) oxygenation of the atmosphere and the oceans (Lyons et al. 2014). The latter may have benefitted the early eukaryotes by supplying oxygen for a higher energy demand of a eukaryotic cell (Lane 2011), or a better bioavailability of some essential nutrients like phosphorus. However, diverse early OWM occur in strata representing environments with both oxic and ferruginous bottom waters, meaning that some early eukaryotes may have also inhabited anoxic ferruginous environments (Porter et al. 2018), so the low oxygen concentrations in the oceans may not have presented a problem. Further insights into the palaeoecology of early eukaryotes from the fossil record, such as understanding their habitats, will help resolve between these potential drivers of eukaryotic radiation.

11.4 Molecular Fossils (Biomarkers)

In addition to body fossils, another important biosignature in Precambrian rocks are biomarkers or molecular fossils – molecular remains of specific organic compounds usually from an organism's cell wall. A group of compounds called sterols play an important role in physiology, e.g. in cellular membrane construction, and are predominantly produced by eukaryotes and were thought to have probably been present in LECA (Desmond and Gribaldo 2009). Sterol-producing genes are not present in archaea (Gold et al. 2017). The sterol biomarkers, i.e. their fossilized remains, are steranes. Unlike the antiquity of the eukaryotic body fossil record, steranes appear in the mid Tonian Period and their absence from older rocks has been interpreted to reflect a low abundance of eukaryotes in earlier seas (Brocks et al. 2017). As oxygen is required for sterol synthesis, the absence of steranes from older rocks either suggests that there were no eukaryotes in earlier times contrary to the evidence from the body-fossil record, or that early eukaryotes were not able to produce sterols due to low oxygen conditions (see Brocks et al. 2017; Porter et al. 2018). Alternatively, eukaryotes may not have evolved the sterol-synthesis pathways until later in their evolutionary history, potentially through horizontal gene transfer between crown group eukaryotes (Porter 2020), or even from bacteria because some modern ones are also able to produce sterols (Wei et al. 2016).

Biomarker data from mid-Neoproterozoic rocks show a shift in the contribution of eukaryotic and bacterial lipids to sedimentary organic matter shown by the sterane/hopane ratio, as well an increased abundance of eukaryote-characteristic steranes cholestane, ergostane and stigmastane, which suggests a change from principally prokaryotic to eukaryotic primary production (Brocks et al. 2017; Hoshino et al. 2017). Eukaryotes could have been making a contribution to primary productivity earlier, but without sterol production, this is not reflected in the rock record. However, additional geochemical evidence also points to an increased impact of eukaryotes onto biogeochemical cycles in the Tonian Period (Isson et al. 2018), suggesting that this ecological shift is a real signal.

11.5 How Can the Fossil Record Inform Us About the Sequence of Events in Eukaryotic Evolution?

Using the diagnostic eukaryotic characters in Proterozoic fossils outlined above, a timeline of the evolution of main eukaryotic features can be mapped. Figure 11.4 shows occurrence of key eukaryotic features in the fossil record from some well-studied geologic units. Such an overview has limits; the list of fossiliferous units is not exhaustive and some of them lack decent age-constraints. Furthermore, the fossil record only provides a minimum age constraint on the appearance of these key features. Caution is needed with this approach because the absence of evidence is not the evidence of absence, and a lot of critical information may be missing from the fossil record due to the low preservation potential of some eukaryotic characters, a lack of lithologies conducive to preservation of cells, a lack of rocks representing environments which early eukaryotic cells may have inhabited, and lastly, a general dearth of fossiliferous Proterozoic units compared to the Phanerozoic—more remains to be explored. However, an examination of the literature of Proterozoic palaeobiology allows an informed view into the minimum age of key events in eukaryotic evolution and helps us to develop and test hypotheses about it.

The cell's ability to change its shape is an ancient feature, seen in taxa of the oldest OWM assemblages. This would have required a cytoskeleton, present both in eukaryotes and their closest relatives, the Asgard archaea. Archaea lack many homologs of cytoskeleton-related genes in eukaryotes, so the early eukaryotes probably had a more complex cytoskeleton which was then inherited and quickly modified (cf. Eme and Ettema 2018). And a comparative increase in morphological complexity is reflected in the fossil record. The 1.65 Ga Changcheng assemblage contains one taxon marked by a single extension from the cell—*Germinosphaera* (Miao et al. 2019). Slightly younger assemblages (1.6–1.4 Ga) contain several OWM taxa with more intricate processes (e.g. Adam et al. 2017; Agić et al. 2017), including branching and membrane-supporting processes, and those that are interpreted as actively growing (e.g. in *Tappania*; Javaux et al. 2001). From this, the evolution of a more complex, eukaryotic cytoskeleton able to modify a cell's shape can be inferred at around 1.6 Ga to by at least 1.4 Ga.

Cyst production is a common strategy used by many protists either in reproduction or as a protective response to environmental stresses (Schaap and Schilde 2018), and features of a cyst are another common phenomenon observed among the oldest eukaryotic microfossils. Indicative of a cyst are characters like excystment openings, processes, complex and reinforced walls. All of these are present in the oldest OWM assemblages (Peng et al. 2009; Miao et al. 2019), suggesting the importance of cyst formation for the early eukaryotes. However, a true controlled excystment opening (a circular pylome) or a pylome with an operculum appears slightly later in the Mesoproterozoic (Agić et al. 2017; Beghin et al. 2017) and are rare until the late Neoproterozoic and Phanerozoic (see Agić et al. 2016; Porter and Riedman 2016). This means that although the earliest fossilized eukaryotes were able to produce robust cyst-like cells, modes of excystment were refined at a later point in their evolution.

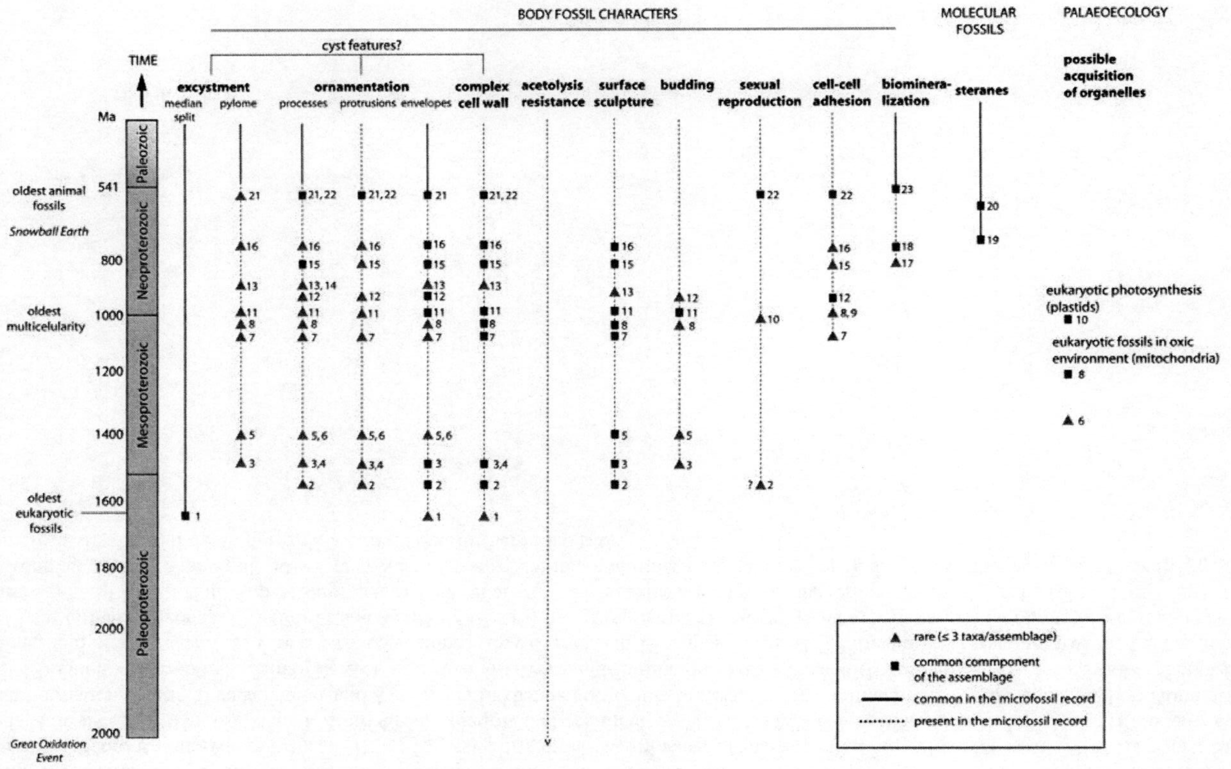

Fig. 11.4 Schematic overview of the occurrence of early eukaryotic characters through time, observed though the fossil record of organic-walled microfossils. Each symbol represents occurrence of a key feature in one or more well-documented assemblages at a point in time, marked by a numbered reference listed

below. This list is not exhaustive and it should be noted that some of these fossiliferous strata are poorly age-constrained, so the illustrated overview only represents a rough timeline of appearances of specific characters in the fossil record of early eukaryotes. (1) Changcheng Group, China (Miao et al. 2019). (2) Ruyang Group, China (Agić et al. 2015, 2017). (3) Roper Group, Australia (Javaux and Knoll 2017). (4) Greyson Formation, Montana, USA (Adam et al. 2017). (5) Kamo Group, Siberia, Russia (Nagovitsin 2009). (6) Kaltasy Formation, central Russia (Sperling et al. 2014; Sergeev et al. 2016. (7) Shaler Supergroup, Arctic Canada (Loron et al. 2019). (8) Atar/El Mreïti Group, Mauritania (Beghin et al. 2017). (9) Mbuji-Mayi Supergroup, DR Congo (Baludikay et al. 2016). (10) Angmaat Formation, Arctic Canada (Butterfield 2000). (11) Lakhanda Formation, Siberia, Russia (Hermann 1990). (12) Liulaobei Formation, China (Tang et al. 2013). (13) Gouhou Formation, China (Tang et al. 2015). (14) Wynniatt Formation, Arctic Canada (Butterfield 2005). (15) Svanbergfjellet Formation, Svalbard (Butterfield et al. 1994). (16) Chuar Group, Arizona, USA (Porter and Riedman 2016). (17) Fifteenmile Formation, Yukon, Canada (Cohen et al. 2017). (18) Chuar Group, Arizona, USA (Porter and Knoll 2000). (19) See compilation of biomarker data from various Neoproterozoic units in Brocks et al. 2017; Hoshino et al. 2017. (20) Huqf Supergroup, Oman (Love et al. 2009). (21) Ediacaran units of the Officer Basin, Australia (Grey 2005). (22) Doushantuo Formation, Weng'an, China (Xiao et al. 2014). (23) Nama Group, Namibia and other units (Wood 2018). The listed characters and their frequency in early eukaryotic microfossils through the Proterozoic are discussed further in the text

Some features of a cyst seen in early OWM can inform about intracellular complexity. In addition to processes indicating the presence of a cytoskeleton, elaborate wall ultrastructure and a composite wall reinforced by organic platelets in *Dictyosphaera* and *Shuiyousphaeridium* implies the presence of an internal membrane system including the endoplasmatic reticulum and the Golgi apparatus (Agić et al. 2015). These organelles are involved in protein/lipid trafficking and membrane construction, and are not directly related to endosymbiotic events. In extant protists, they produce scales that are integrated into the cell membrane (McFadden et al. 1986) and are crucial for the construction of distinctive cyst walls or envelopes (Corliss 2001). The fossil record shows that these features were present early in the evolutionary history of a eukaryotic cell, consistent with the view that a primitive endomembrane system would have been present in a protoeukaryote or even the last archaeal ancestor of eukaryotes (e.g. Eme and Ettema 2018; Grant et al. 2018).

Acetolysis resistance and organic preservation of a microfossil partially suggest its innate recalcitrance, and was long thought to be a feature of a eukaryotic cyst (e.g. Grey 2005; Moczydłowska et al. 2011). While most OWM interpreted as eukaryotes are indeed acetolysis-resistant, this property is also seen in much older, Archean microfossils (Javaux et al. 2010) and bacterial sheaths. Additionally, it doesn't necessarily reflect the original chemical composition of the organism as biomolecules tend to transform on geological timescales (Briggs and Summons 2014) and other, external factors also contribute to organic matter preservation in shales (e.g. Anderson et al. 2020; Woltz et al. 2021). Unlike most other eukaryotic characters that appear en masse around 1.65 Ga, acetolysis-resistance is considerably older, and occurs through most of the Precambrian and Phanerozoic microfossil record (Fig. 11.4), i.e. in every instance where an OWM assemblage has been recovered from a rock via acid maceration. As such, it is not the most useful character for identifying early eukaryotes, and if used it should be accessory to other identifying features.

Cysts may be resting or reproductive, but this is difficult to recognize in fossils (though see Kokinos and Anderson 1995). The former function is more commonly used in the literature (e.g. Javaux and Knoll 2017), yet in most cyst-like microfossils there is too little evidence to recognize its specific function. Reproductive cysts can be formed during either asexual or sexual reproductive phases (Hagen et al. 2002; Adl et al. 2019). Independently, phylogenomic studies predict the origin of sexual reproduction very early on in the evolutionary history of eukaryotes (Goodenough and Heitman 2014), so presence of a zygote-like reproductive cyst would not be unexpected among early cyst-like OWM (see Agić et al. 2015). On the other hand, assuming that all Paleo-Mesoproterozoic cyst-like unicellular microfossils represent resting stages and not reproductive cysts, some of the oldest evidence for a reproduction strategy, via budding similar to modern yeasts, is seen around 1.5 Ga in *Blastanosphaira* (Javaux and Knoll 2017).

The oldest unambiguous evidence for sexual reproduction, manifested in differentiated gametes, is seen around 1.04 Ga in one of the oldest multicellular algae—*Bangiomorpha* (Butterfield 2000). This late appearance of sexual reproduction in the fossil record suggests that it perhaps evolved just before the radiation of crown group eukaryotes and enabled them greater evolutionary success. The sexual nature of LECA and even a protoeukaryote is generally accepted because all living eukaryotic lineages contain regulatory genes exclusively involved in meiosis (Goodenough and Heitman 2014). It follows that at least some early eukaryotes reproduced sexually. Gross and Bhattacharya (2010) further proposed that because emergence of sex and eukaryogenesis were probably simultaneous and synergic, both could be explained by a common selective pressure, for instance rising oxygen levels. Early oxic environments may have been toxic to Archean/Paleoproterozoic microbes through the accumulation of reactive oxygen species (ROS) that damage DNA, and this would create pressure for efficient repair of ROS- or UV-damaged genetic material through meiosis (Gross and Bhattacharya 2010; Speijer 2016). However, the fossil record holds no evidence of sexual reproduction until later, around 1 Ga and the time of the radiation of presumable crown group lineages like red algae. The reason for this could partially be that it is difficult to confidently interpret the evidence for sexual reproduction in unicellular fossils and in the absence of differentiated cell types (and multicellularity) or well-defined life cycles, but it is also possible that sexual reproduction arose later in eukaryotic evolution. If the latter is correct, it also coincides with the progressive increase in oxygen levels in the oceans and the atmosphere (cf. Lyons et al. 2014), and still allows the possibility that sexual reproduction arose in response to oxygenation (cf. Gross and Bhattacharya 2010). Future detailed examination of the Mesoproterozoic fossil record of reproductive structures will cast more light on the evolutionary history of sexual reproduction in eukaryotes; an older record of cell differentiation and complex life cycles would support its greater antiquity (following Goodenough and Heitman 2014).

The radiation of multicellular crown-group lineages probably occurred prior to 1 Ga. The oldest fossils widely accepted as algae belong to rhodophytes and chlorophytes, and appear around 1 Ga (Butterfield 2000; Tang et al. 2020). This suggests the acquisition of plastids by at least 1.24 Ga (see Gibson et al. 2018) consistent with some molecular clock estimates (Eme et al. 2014), though some older eukaryotic taxa may have also been photosynthetic (Moczydłowska et al. 2011; Butterfield 2015b). Although most Paleo-Mesoproterozoic OWM of unresolved affinities are not demonstrably photosynthetic, Butterfield (2015b) pointed out that due to limited metabolic capabilities of crown-group eukaryotes and generally poor preservation potential of most heterotrophic protists, photosynthesis should not be unexpected in OWM, especially if they show vegetative growth.

According to the shift in the ratio of predominantly bacteria- to eukaryotes-produced (hopane/sterane) biomarkers, eukaryotic photosynthesizers likely only become prominent contributors to primary production in the mid-Neoproterozoic (Brocks et al. 2017). Butterfield (2015b) suggested that eukaryotes acquired their plastids by the Paleoproterozoic, but remained outcompeted by prokaryotic photosynthesizers (cyanobacteria) until the late Neoproterozoic and the appearance of suspension-feeding animals.

The appearance of mitochondria, another crucial organelle acquired during eukaryogenesis is difficult to recognize (see Porter 2020). It is generally assumed early eukaryotes (whether members of stem or crown groups) would have been capable of aerobic respiration and thus possessed mitochondria. This 'powerhouse of the cell' is important in sustaining larger eukaryotic genomes (Lane 2011) and was perhaps required in large and complex species from the earliest OWM assemblages in order for a cell to be able to produce recalcitrant and metabolically expensive structures. However, many extant anaerobic protists living in low-oxygen or even sulfidic conditions are also complex and large (e.g. Edgcomb et al. 2011) so aerobic metabolism is not a requirement of complex cells (see Gold et al. 2017). Porter (2020) suggested that the presence of mitochondria should not be assumed in complex early OWM and proposed that it was acquired in the late Mesoproterozoic. Some Proterozoic eukaryotic microfossils occur in sediments deposited under both oxic and anoxic ferruginous bottom water conditions (Porter et al. 2018), and the eukaryotic microfossils in the c. 1.65 Ga Changcheng Group occur in sediments recording ferruginous conditions (Li et al. 2015). Early eukaryotes may have been restricted to oxygenated surface waters, but it is also possible some inhabited parts of the anoxic water column, similar to living anaerobic protists from a variety of clades (e.g. foraminifera, ciliates, cercozoans; Bernhard 1993; Edgcomb et al. 2011; Sergeeva et al. 2012). Eukaryotes evolved through a time of much lower oxygen concentrations in the oceans and atmosphere than at present (Lyons et al. 2014; Li et al. 2015) and their first common ancestor was likely an anaerobe similarly to their closest relatives (Spang et al. 2015; Dacks et al. 2016). A later (c. 1 Ga) acquisition of a mitochondrion and/or transition to mainly aerobic lifestyle is also partially supported by the late appearance of sterane biomarkers in the rock record, which are indicators of sterol synthesis—a process in cell membrane construction that requires oxygen (see Porter et al. 2018). A widespread presence of eukaryotic microfossils in fully oxic environments can be taken as an indicator for the acquisition of a mitochondrion (Porter 2020).

Because all crown-group eukaryotes have mitochondria, this organelle is assumed to have been present in LECA, so presumably mitochondrion must have been incorporated before the split of the crown groups. Multiple evidence shows the appearance of fossils from the crown-group lineages (red and green algae, and fungi; Butterfield 2000; Loron et al. 2019; Tang et al. 2020) by at least around 1 Ga. Although the timing of the acquisition of the mitochondria in total-group eukaryotes is a bit unclear, the fossil record indicates this most likely occurred before 1 Ga at the latest (accepting the crown group affinity of the early multicellular eukaryotes mentioned above).

11.6 Takeaways

The Proterozoic record of organic-walled microfossils provides a direct window into early eukaryotic evolution. The oldest known eukaryotic fossils, appear in 1650 million-year-old siliciclastic rocks as carbonaceous cellular vesicles. They remain mostly single-celled and minute for over a billion years, until the rise of macroscopic multicellularity in algae and animals in the Neoproterozoic Era. These fossils provide clues about major steps in the evolution of early eukaryotes on a microscopic level, and these cellular innovations may be tracked through time to some extent (Fig. 11.4).

The cytoskeleton and its ability to quickly mould a cell's shape arose very early in eukaryotic evolution, seen in microfossils with cellular extensions and forms of ornamentation. Recent work on the close archaeal relatives of eukaryotes indicates that this ability was likely passed on from the Asgard archaea, which possessed actin-regulatory genes used in cell shaping (e.g. membrane remodelling and formation of processes; Imachi et al. 2019; Akıl et al. 2020). The ability to form a cyst, common among living protists, also appeared early, seen in microfossils with opening structures, complex cell walls, and processes (Fig. 11.4). Also ancient is a multivariant life cycle. By the early Mesoproterozoic, microfossils exhibit budding and germination similar to reproduction in modern yeasts, as well as potentially separate life cycle stages that represent reproductive or resting stages (Agić et al. 2015; Javaux and Knoll, 2017). The biomarker record suggests that the ability to synthesise sterols appeared rather late (cf. Brocks et al. 2017).

Although some Paleo-Mesoproterozoic fossils have been suggested as members of distinct eukaryotic clades (e.g. green algae; Moczydłowska et al. 2011), unambiguous eukaryotic fossils that also exhibit independently evolved multicellularity and cell differentiation, interpreted as photosynthesizers (Butterfield 2000; Tang et al. 2020), appear at the very end of the Mesoproterozoic Era, nearly half a billion years after the first eukaryotes. This makes the timing of the acquisition of plastids unclear, but evidence from the fossil record informs us that this endosymbiotic event must have taken place by at least c. 1.2–1 Ga (Gibson et al. 2018). The mitochondrion is another organelle of crucial function in modern eukaryotic cells that could have also been acquired late (Porter 2020). Eukaryotic microfossils are found in sediments deposited under oxic and anoxic ferruginous bottom waters throughout Mesoproterozoic and early Neoproterozoic, so perhaps aerobic respiration (through mitochondria) could have been a later event in their evolution. The delayed appearance of some of these distinguishing eukaryotic features led to the view that crown group eukaryotes may have diversified a long time after the group's origin (Porter 2020). This is consistent with a late Mesoproterozoic and Neoproterozoic diversifications of eukaryotic forms and the appearance of more fossil groups with features similar to modern clades, and the evidence for crown-group clades around 1 Ga suggests a mitochondrion was acquired by this time. The unambiguous evidence for sexual reproduction also appears late in the fossil record, by 1 Ga, seen in differentiated cells of a red alga (Butterfield 2000).

New unicellular forms and multicellularity appear around the Mesoproterozoic-Neoproterozoic transition, c. 1 Ga. The appearance of multicellular taxa with features shared with modern bangiophycean red algae and possibly cladophoran green algae (Butterfield 2000; Tang et al. 2020) indicate a divergence of major eukaryotic lineages and crown groups by 1 Ga, consistent with molecular clock analyses (Eme et al. 2014; Betts et al. 2018). Diversity assessments of Proterozoic microfossils have shown an increase in eukaryotic diversity by the middle Neoproterozoic Era (Cohen and Macdonald 2015). An ecological shift from bacterially-dominated primary production to eukaryotic algae becoming the main contributors to productivity is also seen at this time (Brocks et al. 2017).

The fossil record, with its caveats, presents us a view of eukaryotic evolution punctuated by major milestones. The long time between each of these, i.e. from the first eukaryotic fossil cells around 1.65 Ga to the first multicellular eukaryotes around 1 Ga, and subsequently to the onset of macroscopic life in the Ediacaran Period (Fig. 11.1), and the potential drivers for these developments, are still not well understood. Porter (2020) proposed that this delay could reflect the late emergence of the eukaryotic crown group. On the other hand, Butterfield (2015a) argued that although crown-group eukaryotes were probably established early on, they did not become ecologically significant until the evolution of animals—a unique clade within Eukarya with particular propensity to drive evolutionary and environmental change.

Future multidisciplinary studies on Proterozoic eukaryotes and their palaeoecology have the potential to narrow down the environmental conditions in which complex cells appeared and diversified, and ultrastructural studies on OWM will shed further light on the construction of complex cells and eukaryogenesis.

Acknowledgements The editors Sean McMahon and Anna Neubeck are thanked for their kind invitation to contribute to this book. Constructive discussions about prokaryotic and protistan evolution and ecology with David Chapman and Susannah Porter (UC Santa Barbara) greatly benefitted the manuscript. Corentin Loron (U. Liège) kindly provided additional fossil images. I am grateful to Ross Anderson, Ben Slater, and Sebastian Willman for providing valuable reviews which improved this chapter. This work was supported by NSF grant EAR-1855092.

References

Adam, Z.R., Skidmore, M.L., Mogk, D.W., Butterfield, N.J.: A Laurentian record of the earliest fossil eukaryotes. Geology. **45**, 387–390 (2017)

Adl, S.M., Bass, D., Lane, E.C., Lukeš, J., Schoch, C.L., Smirnov, A., Agatha, S., Berney, C., Brown, M.W., Burki, F., Cárdenas, P.: Revisions to the classification, nomenclature, and diversity of eukaryotes. J. Euk. Microbiol. **66**, 4–119 (2019)

Agić, H., Cohen, P.A.: Non-pollen palynomorphs in deep time: unravelling the evolution of early eukaryotes, vol. 511, pp. 321–342. Geological Society, London, Special Publications (2021)

Agić, H., Moczydłowska, M., Yin, L.: Affinity, life cycle, and intracellular complexity of organic-walled microfossils from the Mesoproterozoic of Shanxi. China. J. Paleontol. **89**, 28–50 (2015)

Agić, H., Moczydłowska, M., Canfield, D.E.: Reproductive cyst and operculum formation in the Cambrian-Ordovician galeate-plexus microfossils. GFF. **138**, 278–294 (2016)

Agić, H., Moczydłowska, M., Yin, L.: Diversity of organic-walled microfossils from the early Mesoproterozoic Ruyang group, North China craton: a window into the early eukaryote evolution. Precambrian Res. **297**, 101–130 (2017)

Akıl, C., Kitaoku, Y., Tran, L.T., Liebl, D., Choe, H., Muengsaen, D., Suginta, W., Schulte, A., Robinson, R.C.: Mythical origins of the actin cytoskeleton. Curr. Opin. Cell Biol. **68**, 55–63 (2020)

Alberts, B., Johnson, A., Lewis, J., Morgan, D., Raff, M., Keith Roberts, P.W.: Molecular Biology of the Cell, 6th edn, p. 1464. W. W. Norton & Company, Boca Raton (2014)

Anderson, R.P., Tosca, N.J., Cinque, G., Frogley, M.D., Lekkas, I., Akey, A., Hughes, G.M., Bergmann, K.D., Knoll, A.H., Briggs, D.E.: Aluminosilicate haloes preserve complex life approximately 800 million years ago. Interface Focus. **10**, 20200011 (2020)

Baludikay, B.K., Storme, J.Y., François, C., Baudet, D., Javaux, E.J.: A diverse and exquisitely preserved organic-walled microfossil assemblage from the Meso–Neoproterozoic Mbuji-Mayi Supergroup (Democratic Republic of Congo) and implications for Proterozoic biostratigraphy. Precambrian Res. **281**, 166–184 (2016)

Beghin, J., Storme, J.Y., Blanpied, C., Gueneli, N., Brocks, J.J., Poulton, S.W., Javaux, E.J.: Microfossils from the late Mesoproterozoic–early Neoproterozoic Atar/El Mreïti group, Taoudeni Basin, Mauritania, northwestern Africa. Precambrian Res. **291**, 63–82 (2017)

Belmonte, G., Miglietta, A., Rubino, F., Boero, F.: Morphological convergence of resting stages of planktonic organisms: a review. Hydrobiologia. **355**, 159–165 (1997)

Bernhard, J.M.: Experimental and field evidence of Antarctic foraminiferal tolerance to anoxia and hydrogen sulfide. Mar. Micropal. **20**, 203–213 (1993)

Betts, H.C., Puttick, M.N., Clark, J.W., Williams, T.A., Donoghue, P.C., Pisani, D.: Integrated genomic and fossil evidence illuminates life's early evolution and eukaryote origin. Nat. Ecol. Evol. **2**, 1556–1562 (2018)

Bhattacharya, D., Archibald, J.M., Weber, A.P., Reyes-Prieto, A.: How do endosymbionts become organelles? Understanding early events in plastid evolution. BioEssays. **29**, 1239–1246 (2007)

Bogus, K., Harding, I.C., King, A., Charles, A.J., Zonneveld, K.A., Versteegh, G.J.: The composition and diversity of dinosporin in species of the *Apectodinium* complex (Dinoflagellata). Rev. Palaeobotany Palynol. **183**, 21–31 (2012)

Bonner, J.T.: Aggregation and differentiation in the cellular slime molds. Ann. Rev. Microbiol. **25**, 75–92 (1971)

Briggs, D.E., Summons, R.E.: Ancient biomolecules: their origins, fossilization, and role in revealing the history of life. BioEssays. **36**, 482–490 (2014)

Brocks, J.J., Jarrett, A.J., Sirantoine, E., Hallmann, C., Hoshino, Y., Liyanage: The rise of algae in Cryogenian oceans and the emergence of animals. Nature. **548**, 578–581 (2017)

Brooks, J., Shaw, G.: Sporopollenin: a review of its chemistry, palaeochemistry and geochemistry. Grana. **17**, 91–97 (1978)

Burki, F., Roger, A.J., Brown, M.W., Simpson, A.G.: The new tree of eukaryotes. Trends Ecol. Evol. **35**, 43–55 (2020)

Butterfield, N.J.: *Bangiomorpha pubescens* n. gen., n. sp.: implications for the evolution of sex, multicellularity, and the Mesoproterozoic/Neoproterozoic radiation of eukaryotes. Paleobiology. **26**, 386–404 (2000)

Butterfield, N.J.: Probable proterozoic fungi. Paleobiology. **31**, 65–182 (2005)

Butterfield, N.J.: Modes of pre-Ediacaran multicellularity. Precambrian Research. **173**, 201–211 (2009)

Butterfield, N.J.: Early evolution of the Eukaryota. Palaeontology. **58**, 5–17 (2015a)

Butterfield, N.J.: Proterozoic photosynthesis—a critical review. Palaeontology. **58**, 953–972 (2015b)

Butterfield, N.J., Knoll, A.H., Swett, K.: Paleobiology of the Neoproterozoic Svanbergfjellet formation, Spitsbergen. Fossils Strata. **34**, 1–84 (1994)

Calderón, E., Tavera, R.: New observations on the sexual and asexual reproductive stages of *Staurastrum gracile* (Desmidiaceae, Zygnematophyceae). Phycologia. **59**, 409–421 (2020)

Cohen, P.A., Macdonald, F.A.: The Proterozoic record of eukaryotes. Paleobiology. **41**, 610–632 (2015)

Cohen, P.A., Strauss, J.V., Rooney, A.D., Sharma, M., Tosca, N.: Controlled hydroxyapatite biomineralization in an ~810 million-year-old unicellular eukaryote. Sci. Adv. **3**, e1700095 (2017)

Colbath, G.K., Grenfell, H.R.: Review of biological affinities of Paleozoic acid-resistant, organic-walled eukaryotic algal microfossils (including "acritarchs"). Rev. Palaeobot. Palynol. **86**, 287–314 (1995)

Corliss, J.O.: Protozoan cysts and spores. eLS. **1**, 1–8 (2001)

Courties, C., Vaquer, A., Troussellier, A., Lautier, J., Chrétiennot-Dinet, M.J., Neveux, J., Machado, C., Claustre: Smallest eukaryotic organism. Nature. **370**, 255–255 (1994)

Dacks, J.B., Field, M.C., Buick, R., Eme, L., Gribaldo, S., Roger, A.J., Brochier-Armanet, C., Devos, D.P.: The changing view of eukaryogenesis–fossils, cells, lineages and how they all come together. J. Cell Sci. **129**, 3695–3703 (2016)

Darwin, C.: On the Origin of Species by Means of Natural Selection, or the Preservation of Favoured Races in the Struggle for Life, p. 502. John Murray, London (1859)

de Leeuw, J.W., Versteegh, G.J., van Bergen, P.F.: Biomacromolecules of algae and plants and their fossil analogues. Plant Ecol. **182**, 209–233 (2006)

Desmond, E., Gribaldo, S.: Phylogenomics of sterol synthesis: insights into the origin, evolution, and diversity of a key eukaryotic feature. GBE. **1**, 364–381 (2009)

Dodds, W.K., Castenholz, R.W.: Effects of grazing and light on the growth of *Nostoc pruniforme* (cyanobacteria). Br. Phycol. J. **23**, 219–227 (1988)

Edgcomb, V.P., Leadbetter, E.R., Bourland, W., Beaudoin, D., Bernhard, J.: Structured multiple endosymbiosis of bacteria and archaea in a ciliate from marine sulfidic sediments: a survival mechanism in low oxygen, sulfidic sediments? Front. Microbiol. **2**, 1344–1356 (2011)

Eme, L., Ettema, T.J.: The eukaryotic ancestor shapes up. Nature. **562**, 352–353 (2018)

Eme, L., Sharp, S.C., Brown, M.W., Roger, A.J.: On the age of eukaryotes: evaluating evidence from fossils and molecular clocks. CSH Perspect. Biol. **6**, a016139 (2014)

Erwin, D.H., Laflamme, M., Tweedt, S.M., Sperling, E.A., Pisani, D., Peterson, K.J.: The Cambrian conundrum: early divergence and later ecological success in the early history of animals. Science. **334**, 1091–1097 (2011)

Ettema, T.J., Lindås, A.C., Bernander, R.: An actin-based cytoskeleton in archaea. Mol. Microbiol. **80**, 1052–1061 (2011)

Fischer, W.W., Hemp, J., Valentine, J.S.: How did life survive Earth's great oxygenation? Curr. Opin. Chem. Biol. **31**, 166–178 (2016)

Gibson, T.M., Shih, P.M., Cumming, V.M., Fischer, W.W., Crockford, P.W., Hodgskiss, M.S., Wörndle, S., Creaser, R.A., Rainbird, R.H., Skulski, T.M., Halverson, G.P.: Precise age of *Bangiomorpha pubescens* dates the origin of eukaryotic photosynthesis. Geology. **46**, 135–138 (2018)

Gold, D.A., Caron, A., Fournier, G.P., Summons, R.E.: Paleoproterozoic sterol biosynthesis and the rise of oxygen. Nature. **543**, 420–423 (2017)

Golubic, S., Seong-Joo, L.: Early cyanobacterial fossil record: preservation, palaeoenvironments and identification. Eur. J. Phycol. **34**, 339–348 (1999)

Goodenough, U., Heitman: Origins of eukaryotic sexual reproduction. CSH Perspect. Biol. **6**, a016154 (2014)

Grant, C.R., Wan, J., Komeili, A.: Organelle formation in Bacteria and archaea. Annu. Rev. Cell Dev. Biol. **34**, 217–238 (2018)

Grey, K.: A modified palynological preparation technique for the extraction of large Neoproterozoic acanthomorph acritarchs and other acid insoluble microfossils, vol. 10, pp. 1–23. Geological Survey of Western Australia Record, Nottingham (1999)

Grey, K.: Ediacaran palynology of Australia. Mem. Assoc. Austral. Palaeontol. **31**, 1–439 (2005)

Gross, J., Bhattacharya, D.: Uniting sex and eukaryote origins in an emerging oxygenic world. Biol. Dir. **5**, 53 (2010)

Gumsley, A.P., Chamberlain, K.R., Bleeker, W., Söderlund, U., de Kock, M.O., Larsson, E.R., Bekker, A.: Timing and tempo of the great oxidation event. PNAS. **114**, 1811–1816 (2017)

Hagen, C., Siegmund, S., Braune, W.: Ultrastructural and chemical changes in the cell wall of *Haematococcus pluvialis* (Volvocales, Chlorophyta) during aplanospore formation. Eur. J. Phycol. **37** 217–226 (2002)

Hairston Jr., N.G., Van Brunt, R.A., Kearns, C.M., Engstrom, D.R.: Age and survivorship of diapausing eggs in a sediment egg bank. Ecology. **76**, 1706–1711 (1995)

Han, T.M., Runnegar, B.: Megascopic eukaryotic algae from the 2.1-billion-year-old Negaunee iron-formation, Michigan. Science. **257**, 232–235 (1992)

Hedges, J.I.: Sedimentary organic matter preservation: an assessment and speculative synthesis. Mar. Chem. **49**, 81–115 (1995)
Hermann, T.N.: Organic world billion years ago. Nauka, Leningrad (1990)
Herskowitz, I.: Life cycle of the budding yeast *Saccharomyces cerevisiae*. Microbiol. Rev. **52**, 536 (1988)
Hodgskiss, M.S., Dagnaud, O.M., Frost, J.L., Halverson, G.P., Schmitz, M.D., Swanson-Hysell, N.L., Sperling, E.A.: New insights on the Orosirian carbon cycle, early cyanobacteria, and the assembly of Laurentia from the Paleoproterozoic Belcher Group. Earth Planet. Sci. Lett. **520**, 141–152 (2019)
Hoffman, P.F., Abbot, D.S., Ashkenazy, Y., Benn, D.I., Brocks, J.J., Cohen, P.A., Cox, G.M., Creveling, J.R., Donnadieu, Y., Erwin, D.H., Fairchild, I.J.: Snowball Earth climate dynamics and cryogenian geology-geobiology. Sci. Adv. **3**, e1600983 (2017)
Holland, H.D.: The oxygenation of the atmosphere and oceans. Philos. Trans. R. Soc. B. **361**, 903–915 (2006)
Hoops, H.J., Brighton, M.C., Stickles, S.M., Clement, P.R.: A test of two possible mechanisms for phototactic steering in Volvox carteri (Chlorophyceae). J Phycol. **35**(3), 539–547 (1999)
Hoshino, Y., Poshibaeva, A., Meredith, W., Snape, C., Poshibaev, V., Versteegh, G.J., Kuznetsov, N., Leider, A., van Maldegem, L., Neumann, M., Naeher, S., Moczydłowska, M., Brocks, J.J., Jarret, A.J.M., Tang, Q., Xiao, S., McKirdy, D., Kumar, S., Alvaro, J.J., Sansjofre, P., Hallmann, C.: Cryogenian evolution of stigmasteroid biosynthesis. Sci. Adv. **3**, e1700887 (2017)
Huntley, J.W., Xiao, S., Kowalewski, M.: 1.3 billion years of acritarch history: an empirical morphospace approach. Precambrian Res. **144**, 52–68 (2006)
Imachi, H., Nobu, M.K., Nakahara, N., Morono, Y., Ogawara, M., Takaki, Y., Takano, Y., Uematsu, K., IkutaT, I.M., Matsui, Y.: Isolation of an archaeon at the prokaryote–eukaryote interface. Nature. **577**, 519–525 (2019)
Isson, T.T., Love, G.D., Dupont, C.L., Reinhard, C.T., Zumberge, A.J., Asael, D., Gueguen, B., McCrow, J., Gill, B.C., Owens, J., Rainbird, R.H.: Tracking the rise of eukaryotes to ecological dominance with zinc isotopes. Geobiology. **16**, 341–352 (2018)
Javaux, E.J., Knoll, A.H.: Micropaleontology of the lower Mesoproterozoic Roper Group, Australia, and implications for early eukaryotic evolution. J. Paleontol. **91**, 199–229 (2017)
Javaux, E.J., Lepot, K.: The Paleoproterozoic fossil record: implications for the evolution of the biosphere during Earth's middle-age. Earth Sci. Rev. **176**, 68–86 (2018)
Javaux, E.J., Knoll, A.H., Walter, M.R.: Morphological and ecological complexity in early eukaryotic ecosystems. Nature. **412**, 66–69 (2001)
Javaux, E.J., Knoll, A.H., Walter, M.R.: TEM evidence for eukaryotic diversity in mid-Proterozoic oceans. Geobiology. **2**, 121–132 (2004)
Javaux, E.J., Marshall, C.P., Bekker, A.: Organic-walled microfossils in 3.2-billion-year-old shallow-marine siliciclastic deposits. Nature. **463**, 934–938 (2010)
Knoll, A.H., Javaux, E.J., Hewitt, D., Cohen, P.: Eukaryotic organisms in Proterozoic oceans. Philos. Trans. R. Soc. B. **361**, 1023–1038 (2006)
Kodner, R.B., Summons, R.E., Knoll, A.H.: Phylogenetic investigation of the aliphatic, non-hydrolyzable biopolymer algaenan, with a focus on green algae. Org. Geochem. **40**, 854–862 (2009)
Kokinos, J. P., Anderson, D. M.: Morphological development of resting cysts in cultures of the marine dinoflagellate *Lingulodinium polyedrum* (= *L. machaerophorum*). Palynology. **19**, 143–166 (1995)
Koonin, E.V.: The origin and early evolution of eukaryotes in the light of phylogenomics. Genome. Biol. **11**, 209 (2010)
Kremp, A., Rengefors, K., Montresorc, M.: Species specific encystment patterns in three Baltic cold-water dinoflagellates: the role of multiple cues in resting cyst formation. Limnol. Oceanogr. **54**, 1125–1138 (2009)
Lane, N.: Energetics and genetics across the prokaryote-eukaryote divide. Biol. Dir. **6**, 1–31 (2011)
Li, C., Planavsky, N.J., Love, G.D., Reinhard, C.T., Hardisty, D., Feng, L., Bates, S.M., Huang, J., Zhang, Q., Chu, X., Lyons, T.W.: Marine redox conditions in the middle Proterozoic Ocean

and isotopic constraints on authigenic carbonate formation: insights from the Chuanlinggou formation, Yanshan Basin, North China. Geochim. Cosmochim. Acta. **150**, 90–105 (2015)

López-García, P., Eme, L., Moreira, D.: Symbiosis in eukaryotic evolution. J. Theor. Biol. **434**, 20–33 (2017)

Loron, C.C., François, C., Rainbird, R.H., Turner, E.C., Borensztajn, S., Javaux, E.J.: Early fungi from the Proterozoic era in Arctic Canada. Nature. **570**, 232–235 (2019)

Love, G.D., Grosjean, E., Stalvies, C., Fike, D.A., Grotzinger, J.P., Bradley, A.S., Kelly, A.E., Bhatia, M., Meredith, W., Snape, C.E., Bowring, S.A.: Fossil steroids record the appearance of Demospongiae during the Cryogenian period. Nature. **457**, 718–721 (2009)

Lyons, T.W., Reinhard, C.T., Planavsky, N.J.: The rise of oxygen in Earth's early ocean and atmosphere. Nature. **506**, 307–315 (2014)

Marshall, C.P., Javaux, E.J., Knoll, A.H., Walter, M.: Combined micro-Fourier transform infrared (FTIR) spectroscopy and micro-Raman spectroscopy of Proterozoic acritarchs: a new approach to palaeobiology. Precambrian Res. **138**, 208–224 (2005)

Martin, W.F., Garg, S., Zimorski, V.: Endosymbiotic theories for eukaryote origin. Philos. Trans. R. Soc. B. **370**, 20140330 (2015)

McFadden, G.I., Preisig, H.R., Melkonian, M.: Golgi apparatus activity and membrane flow during scale biogenesis in the green flagellate *Scherffelia dubia* (Prasinophyceae). II: Cell wall secretion and assembly. Protoplasma. **131**, 174–184 (1986)

Miao, L., Moczydłowska, M., Zhu, S., Zhu, M.: New record of organic-walled, morphologically distinct microfossils from the late Paleoproterozoic Changcheng Group in the Yanshan Range, North China. Precambrian Res. **321**, 172–198 (2019)

Möbius, K.: Das Sterben der einzelligen und der vielzelligen Tiere. Vergleichend betrachtet. Biologisches Centralblatt. **4**, 389–392 (1884)

Moczydłowska, M., Willman, S.: Ultrastructure of cell walls in ancient microfossils as a proxy to their biological affinities. Precambrian Res. **173**, 27–38 (2009)

Moczydłowska, M., Landing, E.D., Zang, W., Palacios, T.: Proterozoic phytoplankton and timing of chlorophyte algae origins. Palaeontology. **54**, 721–733 (2011)

Morozov, A.A., Likhoshway, Y.V.: Evolutionary history of the chitin synthases of eukaryotes. Glycobiology. **26**, 635–639 (2016)

Nagovitsin, K.: *Tappania*-bearing association of the Siberian platform: biodiversity, stratigraphic position and geochronological constraints. Precambrian Res. **173**, 137–145 (2009)

Neuhaus-Url, G., Schweiger, H.G.: The lid forming apparatus in cysts of the green alga *Acetabularia mediterranea*. Protoplasma. **122**, 120–124 (1984)

Pace, N.R.: Mapping the tree of life: progress and prospects. Microbiol Mol Biol Rev. **73**(4), 565–576 (2009)

Pahlow, M., Riebesell, U., Wolf-Gladrow, D.A.: Impact of cell shape and chain formation on nutrient acquisition by marine diatoms. Limnol. Oceanogr. **42**, 1660–1672 (1997)

Pang, K., Tang, Q., Yuan, X.L., Wan, B., Xiao, S.: A biomechanical analysis of the early eukaryotic fossil Valeria and new occurrence of organic-walled microfossils from the paleo-Mesoproterozoic Ruyang Group. Palaeoworld. **24**, 251–262 (2015)

Payne, J.L., Boyer, A.G., Brown, J.H., Finnegan, S., Kowalewski, M., Krause, R.A., Lyons, S.K., McClain, C.R., McShea, D.W., Novack-Gottshall, P.M., Smith, F.A.: Two-phase increase in the maximum size of life over 3.5 billion years reflects biological innovation and environmental opportunity. PNAS. **106**, 24–27 (2009)

Pederson, T.: The nucleus introduced. CSH Perspect. Biol. **3**, a000521 (2011)

Peng, Y., Bao, H., Yuan, X.: New morphological observations for paleoproterozoic acritarchs from the Chuanlinggou formation, North China. Precambrian Res. **168**, 223–232 (2009)

Porter, S.M.: Tiny vampires in ancient seas: evidence for predation via perforation in fossils from the 780–740 million-year-old Chuar group, grand canyon, USA. Proc. R. Soc. B. **283**, 20160221 (2016)

Porter, S.M.: Insights into eukaryogenesis from the fossil record. Interface Focus. **10**, 20190105 (2020)

Porter, S.M., Knoll, A.H.: Testate amoebae in the Neoproterozoic era: evidence from vase-shaped microfossils in the Chuar Group, Grand Canyon. Paleobiology. **26**, 360–385 (2000)
Porter, S.M., Riedman, L.A.: Systematics of organic-walled microfossils from the ca. 780–740 Ma Chuar Group, Grand Canyon. Arizona. J. Paleontol. **90**, 815–853 (2016)
Porter, S.M., Agić, H., Riedman, L.A.: Anoxic ecosystems and early eukaryotes. Emerg. Top. Life Sci. **2**, 299–309 (2018)
Prasad, B., Uniyal, S.N., Asher, R.: Organic-walled microfossils from the Proterozoic Vindhyan Supergroup of Son Valley, Madhya Pradesh, India. Palaeobotanist. **54**, 13–60 (2005)
Quigg, A., Irwin, A.J., Finkel, Z.V.: Evolutionary inheritance of elemental stoichiometry in phytoplankton. Proc. R. Soc. B. **278**, 526–534 (2011)
Reinhard, C.T., Planavsky, N.J., Ward, B.A., Love, G.D., Le Hir, G., Ridgwell, A.: The impact of marine nutrient abundance on early eukaryotic ecosystems. Geobiology. **18**, 139–151 (2020)
Riedman, L.A., Porter, S.: Organic-walled microfossils of the mid-Neoproterozoic Alinya Formation, Officer Basin, Australia. J Paleontol. **90**(5), 854–887 (2016)
Sadoff, H.L.: Encystment and germination in *Azotobacter vinelandii*. Microbiol. Rev. **39**, 516–539 (1975)
Sagan, L.: On the origin of mitosing cells. J. Theor. Biol. **14**, 225–274 (1967)
Samuelsson, J., Butterfield, N. J.: Neoproterozoic fossils from the Franklin Mountains, northwestern Canada: stratigraphic and palaeobiological implications. Precambrian Research. **107**, 235–251 (2001)
Schaap, P., Schilde, C.: Encystation: the most prevalent and underinvestigated differentiation pathway of eukaryotes. Microbiology. **164**, 727–739 (2018)
Schiffbauer, J.D., Wallace, A.F., Hunter Jr., J.L., Kowalewski, M., Bodnar, R.J., Xiao, S.: Thermally-induced structural and chemical alteration of organic-walled microfossils: an experimental approach to understanding fossil preservation in metasediments. Geobiology. **10**, 402–423 (2012)
Schirrmeister, B.E., Sanchez-Baracaldo, P., Wacey, D.: Cyanobacterial evolution during the Precambrian. Int. J. Astrobiol. **15**, 187–204 (2016)
Schopf, J.W.: Atlas of representative Proterozoic microfossils. In: Schopf, J.W., Klein, C. (eds.) The Proterozoic biosphere: a multidisciplinary study, pp. 1055–1117. Cambridge University Press, Cambridge (1992)
Schopf, J.W., Kudryavtsev, A.B., Agresti, D.G., Czaja, A.D., Wdowiak, T.J.: Raman imagery: a new approach to assess the geochemical maturity and biogenicity of permineralized Precambrian fossils. Astrobiology. **5**, 333–371 (2005)
Schulz, H.N., Brinkhoff, T., Ferdelman, T.G., Marine, M.H., Teske, A., Jørgensen, B.B.: Dense populations of a giant sulfur bacterium in Namibian shelf sediments. Science. **284**, 493–495 (1999)
Sergeeva, N.G., Gooday, A.J., Mazlumyan, S.A., Kolesnikova, E.A., Lichtschlag, A., Kosheleva, T.N., Anikeeva, O.V.: Meiobenthos of the oxic/anoxic interface in the southwestern region of the Black Sea: abundance and taxonomic composition. In: Altenbach, A.V., Bernhard, J.M., Seckbach, J. (eds.) Anoxia, pp. 369–401. Springer, Dordrecht (2012)
Sergeev, V. N., Knoll, A. H., Vorob'eva, N. G., Sergeeva, N. D.: Microfossils from the lower Mesoproterozoic Kaltasy Formation, East European Platform. Precambrian Research. **278**, 87–107 (2016)
Smith, M.R.: Cord-forming Palaeozoic fungi in terrestrial assemblages. Bot. J. Linn. Soc. **180**, 452–460 (2016)
Spang, A., Saw, J.H., Jørgensen, S.L., Zaremba-Niedzwiedzka, K., Martijn, J., Lind, A.E., Van Eijk, R., Schleper, C., Guy, L., Ettema, T.J.: Complex archaea that bridge the gap between prokaryotes and eukaryotes. Nature. **521**, 173–179 (2015)
Speijer, D.: What can we infer about the origin of sex in early eukaryotes? Philos. Trans. R. Soc. B. **371**, 20150530 (2016)
Sperling, E.A., Rooney, A.D., Hays, L., Sergeev, V.N., Vorob'Eva, N.G., Sergeeva, N.D., Selby, D., Johnston, D.T., Knoll, A.H.: Redox heterogeneity of subsurface waters in the Mesoproterozoic ocean. Geobiology. **12**, 373–386 (2014)

Stockner, J.G.: Phototrophic picoplankton: an overview from marine and freshwater ecosystems. Limnol. Oceanogr. **33**, 765–775 (1988)

Strullu-Derrien, C., Kenrick, P., Knoll, A.H.: The Rhynie chert. Curr. Biol. **29**, R1218–R1223 (2019)

Stüeken, E.E.: A test of the nitrogen-limitation hypothesis for retarded eukaryote radiation: nitrogen isotopes across a Mesoproterozoic basinal profile. Geochim. Cosmochim. Acta. **120**, 121–139 (2013)

Suda, S., Watanabe, M.M., Inouye, I.: Electron microscopy of sexual reproduction in *Nephroselmis olivacea* (Prasinophyceae, Chlorophyta). Phycol. Res. **52**, 273–283 (2004)

Tang, Q., Pang, K., Xiao, S., Yuan, X., Ou, Z., Wan, B.: Organic-walled microfossils from the early Neoproterozoic Liulaobei Formation in the Huainan region of North China and their biostratigraphic significance. Precambrian Res. **236**, 157–181 (2013)

Tang, Q., Pang, K., Yuan, X., Wan, B., Xiao, S.: Organic-walled microfossils from the Tonian Gouhou formation, Huaibei region, North China Craton, and their biostratigraphic implications. Precambrian Res. **266**, 296–318 (2015)

Tang, Q., Pang, K., Yuan, X., Xiao, S.: A one-billion-year-old multicellular chlorophyte. Nat. Ecol. Evol. **4**, 543–549 (2020)

Torruella, G., De Mendoza, A., Grau-Bove, X., Anto, M., Chaplin, M.A., Del Campo, J., Eme, L., Pérez-Cordón, G., Whipps, C.M., Nichols, K.M., Paley, R.: Phylogenomics reveals convergent evolution of lifestyles in close relatives of animals and fungi. Curr. Biol. **25**, 2404–2410 (2015)

Tyler, S.A., Barghoorn, E.S.: Occurrence of structurally preserved plants in pre-Cambrian rocks of the Canadian shield. Science. **119**, 606–608 (1954)

Wacey, D., Kilburn, M.R., Saunders, M., Cliff, J., Brasier, M.D.: Microfossils of sulphur-metabolizing cells in 3.4-billion-year-old rocks of Western Australia. Nat. Geosci. **4**, 698–702 (2011)

Wang, Y., Wang, Y., Du, W.: The long-ranging macroalga Grypania spiralis from the Ediacaran Doushantuo Formation, Guizhou, South China. Alcheringa: An Australasian Journal of Palaeontology. **40**, 303–312 (2016)

Waterbury, J., Stanier, R.: Two unicellular cyanobacteria which reproduce by budding. Arch. Microbiol. **115**, 249–257 (1977)

Wei, J.H., Yin, X., Welander, P.V.: Sterol synthesis in diverse bacteria. Front. Microbiol. **7**, 990 (2016)

Woltz, C.R., Porter, S.M., Agić, H., Dehler, C.M., Junium, C.K., Riedman, L.A., Hodgskiss, M.S.W., Wörndle, S. and Halverson, G.P.: Total organic carbon and the preservation of organic-walled microfossils in Precambrian shale. Geology. **49**, 556–560 (2021)

Wood, R.: Exploring the drivers of early biomineralization. Emerg. Top. Life Sci. **2**, 201–212 (2018)

Wörndle, S., Crockford, P.W., Kunzmann, M., Bui, T.H., Halverson, G.P.: Linking the Bitter Springs carbon isotope anomaly and early neoproterozoic oxygenation through I/[Ca+ Mg] ratios. Chem. Geol. **524**, 119–135 (2019)

Xiao, S., Knoll, A.H., Kaufman, A.J., Yin, L., Zhang, Y.: Neoproterozoic fossils in Mesoproterozoic rocks? Chemostratigraphic resolution of a biostratigraphic conundrum from the North China platform. Precambrian Res. **84**, 197–220 (1997)

Xiao, S., Muscente, A.D., Chen, L., Zhou, C., Schiffbauer, J.D., Wood, A.D., Polys, N.F., Yuan, X.: The Weng'an biota and the Ediacaran radiation of multicellular eukaryotes. Nat. Sci. Rev. **1**, 498–520 (2014)

Ye, Q., Tong, J., Xiao, S., Zhu, S., An, Z., Tian, L., Hu, J.: The survival of benthic macroscopic phototrophs on a Neoproterozoic snowball earth. Geology. **43**, 507–510 (2015)

Index

A. Neubeck, S. McMahon (eds.), *Prebiotic Chemistry and the Origin of Life*, Advances in Astrobiology and Biogeophysics,
https://doi.org/10.1007/978-3-030-81039-9